# Catalytic Hydrogenation
# in Organic Syntheses

# Catalytic Hydrogenation in Organic Syntheses

**PAUL RYLANDER**

*Engelhard Industries Division*
*Engelhard Minerals and Chemicals Corporation*
*Menlo Park, Edison, New Jersey*

ACADEMIC PRESS
New York   San Francisco   London   1979
*A Subsidiary of Harcourt Brace Jovanovich, Publishers*

ACADEMIC PRESS, INC.
111 Fifth Avenue, New York, New York 10003

*United Kingdom Edition published by*
ACADEMIC PRESS, INC. (LONDON) LTD.
24/28 Oval Road, London NW1 7DX

Library of Congress Cataloging in Publication Data

Rylander, Paul Nels, (Date)
  Catalytic hydrogenation in organic syntheses.

  Includes bibliographies.
  1. Hydrogenation. 2. Catalysis. 3. Chemistry,
Organic–Synthesis. I. Title.
QD281.H8R89      547'.23      79–9711
ISBN 0–12–605355–3

PRINTED IN THE UNITED STATES OF AMERICA

79 80 81 82    9 8 7 6 5 4 3 2 1

# Contents

# Chapter 4
# Hydrogenation of Acids, Esters, Lactones, and Anhydrides

# Chapter 5
# Hydrogenation of Aldehydes

# Chapter 6
# Hydrogenation of Ketones

# Chapter 7
# Hydrogenation of Nitro Compounds

## *Chapter 13*
## Catalytic Dehydrohalogenation

## *Chapter 14*
## Hydrogenolysis of Small Rings

## *Chapter 15*
## Miscellaneous Hydrogenolyses

## Index

# Preface

Catalytic hydrogenation is one of the most useful and versatile tools available to the synthetic organic chemist. The scope of the reaction is very broad; most functional groups can be made to undergo reduction, frequently in high yield, to any of several possible products. Multifunctional molecules can often be reduced selectively at any of several functions. A high degree of stereochemical control is possible and with considerable predictability, and products free of contaminating reagents are obtained easily. Scale up of laboratory experiments to industrial processes presents little difficulty.

The major disadvantage of catalytic hydrogenation, its complexity and mystic quality, is more apparent than real. Admittedly, the choices of catalysts and conditions would be bewilderingly large if one were to work without guides or precedents, but there is no need for this. A major aim of "Catalytic Hydrogenation in Organic Syntheses" is to give the reader easy access to catalytic history, to show what can be done, and how to do it. A variety of working generalities and common sense guides are given as aids in selecting catalytic metal, catalyst support, concentration of metal and catalyst, solvent, and reaction conditions. Mechanisms of hydrogenation are presented at a level that is useful to the synthetic organic chemist.

The present work is organized around the functional group undergoing reduction, a classification that was used in my earlier volume, "Catalytic Hydrogenation over Platinum Metals" (1967), and one that has proved serviceable. The present work is both an updating and an elaboration of the earlier text. Many working generalities and truths known at the time of the earlier publication have been restated here, mainly in the form of summary statements with reliance on references to the original literature for experimental substantiation; however, the bulk of the text in the present book is derived from material published since the earlier manuscript was completed. All manner of hydrogenation catalysts are considered here instead of being limited, as in the previous work, to heterogeneous noble metal catalysts. Numerous references to catalyst preparation are provided for interested readers.

Catalysts are treated here as if they were organic reagents each with characteristic reactions toward various functional groups, a viewpoint familiar to organic chemists. This appears to be, for the purpose of using catalysts in synthesis, by far the best way to organize the vast array of catalytic data. As with any reagent, behavior of a catalyst toward a function is influenced by steric demands, neighboring functions, and reaction environment. Viewed favorably each of these factors can be considered additional means of exerting control over a catalytic hydrogenation.

I wish to express my appreciation to Mrs. Karen Fedrow for her diligent, careful, and conscientious typing of an often difficult manuscript. Special thanks are due the management of Engelhard Industries for their encouragement in this undertaking and for making time available.

Paul N. Rylander

# Chapter 1

# Hydrogenation Catalysts, Reactors, and Reaction Conditions

## I. GENERAL CONSIDERATIONS

At its best, catalytic hydrogenation is scarcely surpassed as a means of achieving controlled transformations of organic compounds. The reaction, which is carried out easily, has a broad scope. Yields are often very high, and the product obtained is free of contaminating reagents. Satisfactory, if not optimal, results can often be obtained under a wide range of conditions. Other hydrogenations require more exacting conditions, but even these can frequently be guessed at or developed from literature data, without extensive work.

A comment seems in order regarding literature. Most synthetic chemists who report hydrogenation experiments are not interested in hydrogenation per se; they merely use it as a means to an end. The successful experiment may have been the only hydrogenation attempted; the successful catalyst may have been the only one in the laboratory. Therefore, optimal conditions or catalysts are rarely given in the literature. What the literature does offer is one way of achieving the goal and as such is an invaluable guide, but one that need not be followed slavishly if common sense and experience dictate otherwise.

## A. Types of Catalysts

Heterogeneous hydrogenation catalysts are of two types, supported and unsupported, with the former group being further divided into those for use in slurry processes and those for use in fixed-bed operations. Catalysts for slurry processes are usually fine powders, whereas fixed-bed catalysts are usually in the form of cylinders, spheres, or granules with a particle size of approximately $\frac{1}{4}$ to $\frac{1}{32}$ in.

Homogeneous catalysts are metal compounds that are soluble in the reaction medium. These compounds are sometimes anchored on a support, which is insoluble in the reaction medium, in an attempt to combine the best features of both heterogeneous and homogeneous catalysis (Brubaker, 1977). A comprehensive review of the use of homogenous hydrogenation catalysts in organic syntheses is in Birch and Williamson (1976).

## B.  Choosing a Catalyst

Hydrogenation catalysts are many and differ widely in activity and selectivity. The catalytic characteristics are determined mainly by the major metal component. For hydrogenation of each functional group, metals can be ordered into a descending hierarchy of activity, or more generally into categories of good, fair, and poor. The division of catalysts into activity groupings is a major feature of this work. Once this division is known, a logical metal can be chosen easily. One simply chooses a metal that is good for what one wants to do and poor for what one does not want to do. The choice is complicated somewhat when molecules containing two or more reducible functions are involved, for it is not the activity toward the function per se that counts, but the product of its activity and its relative adsorbability in competition with other functions for catalyst sites. A mitigating circumstance is that high activity and relatively strong adsorption often go hand in hand. The activity hierarchy is therefore a reliable guide more often than not.

## C.  Choosing a Support

A great many materials have been used as catalyst supports, but a good carbon or alumina will be suitable for the majority of reactions. Exactly what constitutes "good" cannot be stated easily. In essence "good" supports make efficient use of the catalytic metal. "Poor" supports probably could often be used as well if their deficiences were compensated for by higher metal loadings or more catalyst. Catalyst manufacturers use empirically determined "good" supports, for efficient use of metal is at the heart of economical use of catalysts, especially noble metal catalysts. Of course, supports influence the reaction in other ways as well, but the influence is usually small compared to the effect of metal.

## D.  Metal Loadings

Noble metal batch-type catalysts usually contain between 1 and 10% metal; fixed-bed catalysts usually contain much less metal, often 0.1–1.0%.

On a weight of metal basis, activity is linear with metal concentration only over a limited range. As the metal concentration increases, metal becomes piled on metal and an increasingly large percentage of it becomes unavailable for use. In general, on a weight of metal basis, the more dilute the metal the more efficient the catalyst, but the amount of catalyst (metal plus support) needed to maintain a constant weight of metal increases directly as the metal concentration decreases. A frequently used metal loading for batch-type noble metal catalysts is 5%. This number is derived as a compromise between making the most efficient use of metal and the economic need to minimize the amount of catalyst. Supported base metal catalysts usually contain much more metal than do noble metal catalysts.

Selectivity is sometimes also influenced by metal loadings. The change in selectivity is probably a reflection of various mass transport processes, as well as of trace impurities, which function as catalyst modifiers. The lower the amount of metal in a system the more influence an impurity might have. The relationship between metal loading and selectivity is discussed further in Section II.

## E. Solvents

Solvents are often used in catalytic hydrogenation (Rylander, 1978). They serve to increase the ease of handling and the ease of catalyst recovery, to moderate exothermic reactions, to increase rate and selectivity, and to permit hydrogenation of solid material. Sometimes very marked changes in selectivity can be effected by the solvent. Most liquid materials that are stable under hydrogenation conditions and do not inactivate the catalyst can be used as solvents. Acetic acid, methanol, and ethanol are very commonly used. A convenient solvent, when applicable, is the product of the hydrogenation itself. Dioxane may react *exposively* with hydrogen and Raney nickel above 200°C (Mozingo, 1955). Many specific examples of the influence of solvent on hydrogenation are given throughout this volume.

## F. Safety

Virgin noble metal catalysts are generally nonpyrophoric and can be safely held in the hand. Virgin catalysts, such as Raney nickel, which contain dissolved hydrogen ignite when exposed to air, and due care should be taken in handling. All used catalysts containing adsorbed hydrogen may ignite as they dry. A used, filtered catalyst should be wetted and kept out of contact with combustible vapor.

Metal catalysts on finely divided carbon are subject to dust explosions, just as carbon itself or flour is. Due care should accordingly be exercised in the handling of carbon-supported catalysts.

Catalysts, although completely safe themselves, catalyze the oxidation of combustible organic vapors, and care must be taken when catalysts are brought into contact with organic liquids or combustible vapors in the presence of oxygen. Lower alcohols and compounds such as cyclohexene that can dehydrogenate are particularly prone to ignition. Most other solvents are ignited only with difficulty, but it would be foolish to depend on it. The possibility of fire can be diminished greatly by cooling both the catalyst and solvent before mixing, and it can be eliminated completely by removing oxygen from the system.

Commercial carbon-supported noble metal catalysts can be obtained as 50% water-wet, free-flowing powders (manufactured by Engelhard Industries, Newark, New Jersey). The chance of fire, as well as of dust explosion, is greatly diminished when these catalysts are used. In addition, they are often somewhat more active than the corresponding dry catalysts.

### G. Catalyst Reuse

In commercial operations a catalyst should be reused as many times as possible, with or without intervening regenerations, for each reuse lowers the cost of catalyst per pound of product. In laboratory practice, reuse introduces an uncertainty, and the small saving in catalyst cost is rarely worth the risk of failure.

### H. Catalyst Poisons

A catalyst poison is here considered to be anything that causes a partial or complete loss of catalyst activity. It is not easy to enumerate catalyst poisons. They vary from reaction to reaction. Moreover, whereas small amounts of a substance may be beneficial to catalyst functioning, larger amounts may be poisonous. Poisons include heavy metal cations, halides, divalent sulfur compounds, carbon monoxide, amines, phosphines (Maxted, 1951; Baltzly, 1976a, b, c), as well as, in some instances, the substrate itself (Smutny et al., 1960). A common poison, although usually unknown, is some product of the reaction.

The presence of a poison in a system should be suspected when a reaction does not run as well as experience and literature suggest it should. A check on the catalyst itself, reactor cleanliness, hydrogen, and solvent purity can be made simultaneously by hydrogenating some easily reduced substance. A quantitative measure of catalyst poisons can be made by carrying out the same hydrogenation at different catalyst loadings. If the rate increases more rapidly than the increase in amount of catalyst, the presence of a poison is confirmed.

*Agglomeration*

Occasionally catalysts show a marked tendency to agglomerate. Agglomeration always affects the rate adversely and, if severe, may cause the reduction to fail. This cause of catalyst deactivation can be seen readily, and a visual examination of the catalyst should be made in the case of an unsuccessful reduction. Frequently, agglomeration can be overcome by changing the pH of the medium or by changing the solvent, the solvent–substrate ratio, or the catalyst support.

## I.  Catalyst Regeneration and Reclamation

- Lost catalyst activity can sometimes be restored by regeneration. Many procedures are known, but in essence they are all some variation or combination of oxidation, hydrogenation, steaming, heating, or solvent wash. One can never be certain in advance which procedures will work. A reasonable approach to the problem is to guess what might have caused deactivation and then treat the catalyst accordingly. Physical measurements on the catalyst are helpful in determining why the catalyst has lost activity, but meanful measurements require complex and expensive equipment.

Eventually a catalyst can no longer be regenerated sufficiently. Noble metal catalysts are then returned to a refiner (if volume warrants) and destroyed, and the pure metal is recovered. Reclamation of metal is an integral part of the economic use of noble metals. Base metals may or may not be reclaimed.

## J.  Synergism

Two catalysts used together sometimes give better results than either separately. This effect may occur when two catalytic metals make a single catalyst and also when two separate catalysts are used together. Synergism by mixtures of two catalysts has been accounted for by the assumption that the hydrogenation involves two or more discrete stages, or multiple intermediates, some of which may be reduced more easily by one catalyst and some by the other (Rylander and Cohn, 1961). The second catalyst may also function by its superior ability to remove catalyst inhibitors formed in the reaction. Synergism in coprecipitated or cofused mixed-metal catalysts may be accounted for similarly, as well as by formation of alloys, by alterations in electronic properties, and by changes in particle size and surface area.

## K.  Promoters

Small quantities of various substance that have favorable effects on activity, selectivity, or catalyst life may be loosely termed promoters (Innes,

1954). Promotion is not an intrinsic property of the modifier and catalyst; it also depends on the reaction in which the catalyst is used. The effects produced by promoters are sometimes quite remarkable, as various examples cited later illustrate. By and large there is very little theory to guide one in the use of promoters. Success in their use comes most easily from an extension or modification of something already known.

## II. REACTION CONDITIONS AND CATALYST VARIABLES

Temperature, pressure, and agitation can affect both activity and selectivity in catalytic hydrogenation. In general, as each of these variables is increased over the usual range of operating conditions the rate of hydrogenation increases until some limit is reached. However, hydrogenation rates may decrease with increasing temperature due to lower solubility of hydrogen or cease completely at the boiling point of the solvent (Cowan and Eisenbraun, 1976b), and rates may reach maximums at intermediate pressures (Siegel and Garti, 1977). The effect of these reaction variables on selectivity is less straightforward, but it can often be linked to the effect of these variables and others on hydrogen availability at the catalyst surface.

### A. Mass Transport of Hydrogen

In any liquid-phase catalytic hydrogenation, hydrogen moves from the gas phase across a gas–liquid interface and from the liquid phase across a liquid–solid interface to the external surface of the catalyst and hence into the porous structure of the catalyst. This net movement of hydrogen to the catalyst is the result of concentration gradients, which develop when hydrogen is consumed by the catalytic reaction. In many catalytic hydrogenations, the rate is limited in whole or in part by how fast hydrogen is transported to the catalyst surface. The more active the catalyst, the more likely it is that mass transport will be a limiting factor.

Theoretical analysis of mass transport resistances has led to the conclusion that, if the inverse of the rate of hydrogenation is plotted against the inverse of the weight of the catalyst, the intercept will be exactly equal to the rate that would exist if the reaction were controlled by gas–liquid hydrogen transport. To be certain that gas–liquid hydrogen transport is not controlling, the actual experimental value of rate$^{-1}$ should be much larger than the intercept. The slope of the line in this type of plot is related to resistances due to liquid–solid mass transport plus the chemical reaction and pore diffusion resistances. It is not easy to separate these terms, but some measure of their relative contribution can be obtained from the slope of the line as a function of temperature (Roberts, 1976).

## B.   Effect of Hydrogen Availability on Selectivity

Because of limitations imposed on the rate of hydrogenation by hydrogen transport, hydrogen availability at the catalyst surface may vary from a condition in which the rate of reaction is controlled almost entirely by the intrinsic rate of the chemical reaction (a condition in which the catalyst may be said to be "hydrogen rich") to one in which the rate is controlled completely by the rate of hydrogen transport to the active catalyst site (a condition in which the catalyst may be said to be "hydrogen poor"). These distinctions are convenient in considering how various parameters may affect the selectivity of competing hydrogenations. Consider a reaction of the type $A \begin{smallmatrix} \nearrow B \\ \searrow C, \end{smallmatrix}$ where the reaction order with respect to hydrogen is higher for formation of B than for C. Two examples of this situation are saturation of an olefin versus its isomerization and, in general, hydrogenation versus hydrogenolysis. Conditions leading to high hydrogen availability at the catalyst surface ("hydrogen rich") will favor the higher-order reaction relative to the lower-order reaction and vice versa for "hydrogen-poor" catalysts.

From these considerations a guide can be derived that is useful in efforts to maximize selectivity. Assume that the desired product is favored by a "hydrogen-rich" catalyst. This product is then favored by (a) increased agitation; (b) decreased catalyst loadings; (c) decreased concentrations of metal; (d) increased pressure; (e) deactivation of the catalyst by various inhibitors, with the assumption that they will decrease only the rate and hence increase hydrogen availability; (f) use of less active catalyst; (g) use of a solvent in that a solvent may give better hydrogen transport through increasing hydrogen solubility and/or decreased viscosity and surface tension (there may be other complex effects of solvents as well) and (h) decreased temperature in that it lowers the rate of chemical reaction relative to the rate of mass transport.

For maximization of a product whose formation is of lower reaction order, the various parameters should all be changed in the reverse sense.

An easy diagnostic test for the possibility of influencing selectivity by changing hydrogen availability and for determining the direction in which the changes should be made is simply to measure selectivity under two widely different degrees of agitation with all other variables held constant.

### Measurement of Selectivity

Occasionally hydrogenations are reported to be nonselective when hydrogen absorption fails to stop at the stoichiometric amount or when there is no break in the hydrogenation rate curve. These are unreliable criteria of selectivity. For instance, very selective reductions of acetylenes to olefins may continue at an undiminished or even accelerated rate after the acetylene has been consumed. The only sure way of determining selectivity is by examination of the product. Maximal selectivity usually, but not always, occurs at the stoichiometric absorption for the desired reaction.

## III. HYDROGENATION REACTORS

Hydrogenation reactors come in a great variety of designs and sizes. They all serve the purpose of bringing hydrogen, the catalyst, and the substrate into contact in the absence of air. Most hydrogenations are carried out in batch-type reactors, but in some hydrogenations, especially large-scale processes, continuous reactors are used.

### A. Atmospheric Pressure Reactors

Reactors for hydrogenation at atmospheric pressure can be built readily. Many consist of a flask attached to a burette filled with water or mercury (Nickon and Bagli, 1961) and provided with stopcocks that permit evacuation of air from the system before the introduction of hydrogen. Agitation is provided by a stirrer of some sort, by a flow of hydrogen, or by shaking the reaction flask (Fieser and Hershberg, 1938; Noller and Barusch, 1942; Joshel, 1943; Frampton et al., 1951; Morritz et al., 1953; Meschke and Hartung, 1960). Reactors can be provided with sampling devices (Cowan and Eisenbraun, 1976a) and with means of recycling hydrogen (Cowan and Eisenbraun, 1976b).

A number of workers have described microreactors suitable for hydrogenation on a very small scale (Brown et al., 1963; Cheronis and Levin, 1944; Clauson-Kaas and Limborg, 1947; Engelbrecht, 1957; Gould and Drake, 1951; Harrison and Harrison, 1964; Hyde and Scherp, 1930; Miller and DeFord, 1958; Ogg and Cooper, 1949; Pack et al., 1952; Siggia, 1963; Southworth, 1956; Vandenheuvel, 1952; Weygand and Werner, 1937).

### B. Low-Pressure Reactors

Apparatus for carrying out hydrogenation at several atmospheres can be constructed readily (Adams and Voorhees, 1932; Snyder et al., 1957; Rohwedder, 1968). The most commonly used low-pressure apparatus is the commercial Parr hydrogenator (manufactured by Parr Instrument Co.,

Moline, Illinois). The apparatus is recommended for hydrogenations not exceeding 60 psig or 100°C. Various workers have suggested modifications of this useful equipment (Buck and Jenkins, 1929; Russotto, 1964; Pearlman, 1969).

## C. High-Pressure Reactors

High-pressure processing requires specially constructed equipment. Expert advice and equipment of all kinds and sizes can be obtained from manufacturers. Some companies design and build completely packaged units. General descriptions of high-pressure reactors together with various aspects of safety, design, and control have been given by Adkins (1937), Komarewsky et al. (1956), Bowen and Jenkins (1957), Rebenstorf (1961, 1967), Blackburn and Reiff (1967), Bowen (1967), Friedrich (1970, 1971), and Lavagnino and Campbell (1973).

## D. Fixed-Bed Reactors

Fixed-bed reactors are useful in the hydrogenation of large volumes of material. The reactor may be either *trickle bed*, in which a liquid phase and hydrogen flow concurrently downward over a fixed bed of catalyst particles, or *flooded bed*, in which hydrogen and the liquid pass concurrently upward. Trickle-bed reactors have been well reviewed by Satterfield (1975).

## E. Loop Reactors

A loop type of reactor, which may be operated in batch or continuous mode, is manufactured by Buss Ltd, Basel, Switzerland. Catalyst and substrate are circulated continuously through a nozzle, in which most of the reduction occurs. The technique provides excellent mass transfer and temperature control.

## IV. PREPARATION OF CATALYSTS

Good reviews of general techniques of catalyst preparation have been given by Ciapetta and Plank (1954) and Innes (1954). Peterson (1977) has reviewed U. S. patents since 1970 that deal with the preparation of hydrogenation catalysts. Some literature references to the preparation of various catalysts are given in Table I.

**TABLE I**

**Preparation of Hydrogenation Catalysts**

| Catalyst | Reference |
|---|---|
| Platinum oxide | Adams and Shriner (1923), Carothers and Adams (1923), Frampton et al. (1951), Keenan et al. (1954) |
| Platinum black | Willstatter and Waldschmidt-Leitz (1921), Baltzly (1952), Theilacker and Drössler (1954) |
| Platinum-on-carbon | Baltzly (1952, 1976a) |
| Platinum–rhodium oxides | S. Nishimura (1960, 1961), A. Nishimura (1961) |
| Platinum–ruthenium oxides | Bond and Webster (1964) |
| Palladium oxide | Shriner and Adams (1924), Starr and Hixon (1943) |
| Palladium-on-barium carbonate | Mozingo (1955) |
| Palladium-on-strontium carbonate | Johnson et al. (1956) |
| Palladium-on-carbon | Mozingo (1955) |
| Palladium hydroxide-on-carbon | Pearlman (1967) |
| Rhodium-on-carbon | Baltzly (1976a) |
| Nickel boride P1 | Brown (1970) |
| Nickel boride P2 | Brown and Ahuja (1973) |
| Nickel (NiC) | Brunet et al. (1977) |
| Nickel–copper | Takai (1968) |
| Nickel-on-kieselguhr | Covert et al. (1932) |
| Urushibara nickel | Taira (1961), Motoyama (1960), Urushibara (1967) |
| Raney nickel | Pavlic and Adkins (1946), Mozingo (1941) |
| Raney nickel W1 | Covert and Adkins (1932) |
| Raney nickel W2 | Mozingo (1955) |
| Raney nickel W3 | Adkins and Pavlic (1947) |
| Raney nickel W4 | Pavlic and Adkins (1946) |
| Raney nickel W5 | Adkins and Billica (1948) |
| Raney nickel W6 | Billica and Adkins (1955) |
| Raney nickel W7 | Adkins and Billica (1948) |
| Raney nickel W8 | Khan (1952) |
| Raney cobalt | Reeve and Eareckson (1950) |
| Copper chromite | Adkins et al. (1932), Adkins (1937) |

Most workers describing synthetic applications of hydrogenation catalysts do not know or do not report the method or details of catalyst preparation. However, this need not cause undue concern. In all probability, most active catalysts of the type specified will give yields close to those reported.

Hydrogenation catalysts in great variety can be purchased from various suppliers. There are some advantages in using commercial catalysts beyond the labor saved. One is that the catalyst will be active and probably will represent the culmination of efforts to prepare a superior catalyst of its type. Another is that if the reaction reaches commercial development a supply of the needed catalyst is ensured.

# REFERENCES

Adams, R., and Shriner, R. L., *J. Am. Chem. Soc.* **45**, 2171 (1923).

Adams, R., and Voorhees, V., *Org. Synth.*, *Collect. Vol.* **1**, 61 (1932).

Adkins, H., "Reactions of Hydrogen." Univ. of Wisconsin Press, Madison, 1937.

Adkins, H., and Billica, H. R., *J. Am. Chem. Soc.* **70**, 695 (1948).

Adkins, H., and Pavlic, A. A., *J. Am. Chem. Soc.* **69**, 3039 (1947).

Adkins, H., Connor, R., and Folkers, K., *J. Am. Chem. Soc.* **54**, 1138 (1932).

Baltzly, R., *J. Am. Chem. Soc.* **74**, 4586 (1952).

Baltzly, R., *J. Org. Chem.* **41**, 920 (1976a).

Baltzly, R., *J. Org. Chem.* **41**, 928 (1976b).

Baltzly, R., *J. Org. Chem.* **41**, 933 (1976c).

Billica, H. R., and Adkins, H., Org. *Synth.*, *Collect. Vol.* **3**, 176 (1955).

Birch, A. J., and Williamson, D. H., *Org. React.* **24**, 1 (1976).

Blackburn, D. W., and Reiff, H. E., *Ann. N.Y. Acad. Sci.* **145**, 192 (1967).

Bond, G. C., and Webster, D. E., *Proc. Chem. Soc.* p. 398 (1964).

Bowen, J. C., *Ann. N.Y. Acad. Sci.* **145**, 169 (1967).

Bowen, J. C., and Jenkins, R. L., *Ind. Eng. Chem.* **49**, 2019 (1957).

Brown, C. A., *J. Org. Chem.* **35**, 1900 (1970).

Brown, C. A., and Ahuja, V. K., *J. Chem. Soc.*, *Chem. Commun.* p. 553 (1973).

Brown, H. C., Sivasankaran, K., and Brown, C. A., *J. Org. Chem.* **28**, 214 (1963).

Brubaker, C. H., Jr., *in* "Catalysis in Organic Syntheses, 1977" (G. V. Smith, ed.), p. 25. Academic Press, New York, 1977.

Brunet, J. J., Gallois, P., and Caubere, P., *Tetrahedron Lett.* p. 3955 (1977).

Buck, J. S., and Jenkins, S. S., *J. Am. Chem. Soc.* **51**, 2163 (1929).

Carothers, W. H., and Adams, R., *J. Am. Chem. Soc.* **45**, 1071 (1923).

Cheronis, N. D., and Levin, N., *J. Chem. Educ.* **21**, 603 (1944).

Ciapetta, F. G., and Plank, C. J., *Catalysis* **1**, 315 (1954).

Clauson-Kaas, N., and Limborg, F., *Acta Chem. Scand.* **1**, 884 (1947).

Covert, L. W., and Adkins, H., *J. Am. Chem. Soc.* **54**, 4116 (1932).

Covert, L. W., Connor, R., and Adkins, H., *J. Am. Chem. Soc.* **54**, 1651 (1932).

Cowan, K. D., and Eisenbraun, E. J., *Chem. Ind.* (*London*) p. 416 (1976a).

Cowan, K. D., and Eisenbraun, E. J., *Chem. Ind.* (*London*) p. 221 (1976b).

Engelbrecht, R. M., *Anal. Chem.* **29**, 1556 (1957).

Fieser, L. F., and Hershberg, E. B., *J. Am. Chem. Soc.* **60**, 940 (1938).

Frampton, V. L., Edwards, J. D., Jr., and Henze, H. R., *J. Am. Chem. Soc.* **73**, 4432 (1951).

Friedrich, J. P., *Ann. N.Y. Acad. Sci.* **172**, 155 (1970).

Friedrich, J. P., *Chemtech* p. 52 (1971).

Gould, C. W., and Drake, H. J., *Anal. Chem.* **23**, 1157 (1951).

Harrison, I. T., and Harrison, S., *Chem. Ind.* (*London*) p. 834 (1964).

Hyde, J. F., and Scherp, H. W., *J. Am. Chem. Soc.* **52**, 3359 (1930).

Innes, W. B., *Catalysis* **1**, 245 (1954).

Johnson, W. S., Rogier, E. R., Szmuszkovicz, J., Hadler, H. I., Ackerman, J., Bhattacharyya, B. K., Bloom, B. M., Stalmann, L., Clement, R. A., Bannister, B., and Wynberg, H., *J. Am. Chem. Soc.* **78**, 6289 (1956).

Joshel, L. M. *Ind. Eng. Chem.*, *Anal. Ed.* **15**(9), 590 (1943).

Keenan, C. W., Giesemann, B. W., and Smith, H. A., *J. Am. Chem. Soc.* **76**, 229 (1954).

Khan, N. A., *J. Am. Chem. Soc.* **74**, 3018 (1952).

Komarewsky, V. I., Riesz, C. H., and Morritz, F. L., *in* "Technique of Organic Chemistry" (A. Weissberger, ed.), 2nd Ed., Vol. 11, pp. 18–93. Wiley (Interscience), New York, 1956.

Lavagnino, E., and Campbell, J., *Ann. N.Y. Acad. Sci.* **214**, 3 (1973).
Maxted, E. B., *Adv. Catal.* **3**, 129 (1951).
Meschke, R. W., and Hartung, W. H., *J. Org. Chem.* **25**, 137 (1960).
Miller, J. W., and Deford, D. D., *Anal. Chem.* **30**, 295 (1958).
Morritz, F. L., Lieber, E., and Bernstein, R. B., *J. Am. Chem. Soc.* **75**, 3116 (1953).
Motoyama, I., *Bull. Chem. Soc. Jpn.* **33**, 232 (1960).
Mozingo, R., *Org. Synth.* **21**, 15 (1941).
Mozingo, R., *Org. Synth., Collect. Vol.* **3**, 181 (1955).
Nickon, A., and Bagli, F. J., *J. Am. Chem. Soc.* **83**, 1498 (1961).
Nishimura, S., *Bull. Chem. Soc. Jpn.* **34**, 32 (1961).
Nishimura, S., *Bull. Chem. Soc. Jpn.* **33**, 566 (1960).
Nishimura, S., *Bull. Chem. Soc. Jpn.* **34**, 1544 (1961).
Noller, C. R., and Barusch, M. R., *Ind. Eng. Chem., Anal. Ed.* **14**(11), 907 (1942).
Ogg, C. L., and Cooper, F. J., *Anal. Chem.* **21**, 1400 (1949).
Pack, F. C., Planck, R. W., and Dollear, F. G., *J. Am. Oil Chem. Soc.* **29**, 227 (1952).
Pavlic, A. A., and Adkins, H., *J. Am. Chem. Soc.* **68**, 1471 (1946).
Pearlman, W. M., *Tetrahedron Lett.* p. 1663 (1967).
Pearlman, W. M., *Ann. N.Y. Acad. Sci.* **158**, 577 (1969).
Peterson, R. J., "Hydrogenation Catalysts." Noyes Data Corp., Park Ridge, New York, 1977.
Rebenstorf, M. A., *Ind. Eng. Chem.* **53**, 40A (1961).
Rebenstorf, M. A., *Ann. N.Y. Acad. Sci.* **145**, 178 (1967).
Reeve, W., and Eareckson, W. M., III, *J. Am. Chem. Soc.* **54**, 1138 (1932).
Roberts, G. W., *in* "Catalysis in Organic Syntheses, 1976" (P. N. Rylander and H. Greenfield, eds.), p. 1. Academic Press, New York, 1976.
Rohwedder, W. K., *J. Catal.* **10**, 47 (1968).
Russotto, P. T., *Chemist-Analyst* **53**(3), 85 (1964).
Rylander, P. N., *in* "Catalysis in Organic Syntheses, 1978" (W. H. Jones, ed.), Academic Press, New York, 1978.
Rylander, P. N., and Cohn G., *Actes Congr. Int. Catal., 2nd, Paris, 1960* **1**, 977 (1961).
Satterfield, C. N., *AIChE J.* **21**(2), 209 (1975).
Shriner, R. L., and Adams, R., J. Am. Chem. Soc. **46**, 1683 (1924).
Siegel, S., and Garti, N., *in* "Catalysis in Organic Syntheses, 1977" (G. V. Smith, ed.), p. 9. Academic Press, New York, 1977.
Siggia, S., "Quantitative Organic Analysis," 3rd Ed., pp. 318–341. Wiley, New York, 1963.
Smutny, E. J., Caserio, M. C., and Roberts, J. D., *J. Am. Chem. Soc.* **82**, 1793 (1960).
Snyder, J. R., Hagerty, P. F., and Molstad, M. C., *Ind. Eng. Chem.* **49**, 689 (1957).
Southworth, B. C., *Anal. Chem.* **28**, 1611 (1956).
Starr, D., and Hixon, R. M., *Org. Synth., Collect. Vol.* **2**, 566 (1943).
Taira, S., *Bull. Chem. Soc. Jpn.* **34**, 1294 (1961).
Takai, E., *Sci. Pap. Inst. Phys. Chem. Res.* **62**, 24 (1968).
Theilacker, W., and Drössler, H. G., *Chem. Ber.* **87**, 1676 (1954).
Urushibara, Y., *Ann. N.Y. Acad. Sci.* **145**, 52 (1967).
Vandenheuvel, F. A., *Anal. Chem.* **24**, 847 (1952).
Weygand, C., and Werner, A., *J. Prakt. Chem.* **149**, 330 (1937).
Willstatter, R., and Waldschmidt-Leitz, E., *Chem. Ber.* **54**, 121 (1921).

*Chapter* 2

# Hydrogenation of Acetylenes

The acetylenic group is one of great synthetic utility because of the ability of acetylenes to form carbon–carbon bonds with retention of the triple bond, which subsequently can be made to undergo a broad range of transformations. One of the most useful of these reactions is conversion to a a *cis*-olefin through selective hydrogenation. The reaction is of particular value because it can be carried out selectively in the presence of a variety of other reducible functional groups. Acetylene itself or butadiyne (Inhoffen *et al.*, 1950) can serve as a "linch-pin" (Raphael, 1955) to join two molecules. These reactions have proved to be of value in the synthesis of complex natural polyenes (Inhoffen and Raspé, 1955; Isler *et al.*, 1956a,b; Akhtar and Weedon, 1959; Rüegg *et al.*, 1961), as well as a variety of other compounds.

## I. GENERAL CONSIDERATIONS

Alkanes are formed on complete hydrogenation of acetylenes. Formation of alkanes may or may not involve an intermediate olefin detached from the catalyst surface. If the olefin intermediate is bypassed, the reaction route is described by

$$RC{\equiv}CR \begin{cases} \longrightarrow \text{alkane} \\ \longrightarrow \text{olefin} \end{cases}$$

whereas a route via an intermediate olefin is described by

$$RC{\equiv}CR \longrightarrow \text{olefin} \longrightarrow \text{alkane}$$

The first reaction gives a constant ratio of alkane to olefin as the reaction progresses, with alkane present even in the earliest stages of the reaction. The second reaction produces increasing ratios of alkane to olefin, as the

olefin formed competes for catalyst sites with the acetylene. The two pathways are thus easily distinguished experimentally. Selective formation of
an olefin requires that both pathways be selective. Selectivity in the first
route has been termed mechanistic selectivity and, in the second route,
thermodynamic selectivity, since it is related to the relative free energies of
adsorption of the acetylene and olefin (Bond, 1962). Ratios of thermodynamic and mechanistic selectivities vary characteristically with the metal
(Mann and Khulbe, 1968, 1969, 1970; Brown, 1970), and either may be the
more important (Bond and Rank, 1965).

High thermodynamic selectivity demands that the acetylene be much more
strongly adsorbed than the corresponding olefin and that the rate of displacement of the olefin by the acetylene be high relative to the rate of olefin
hydrogenation. As the acetylene concentration becomes smaller, its effectiveness in displacing olefin will diminish, and thermodynamic selectivity
in the last stages will fall. Thermodynamic selectivity is also affected by
depletion of the acetylenes in the vicinity of the catalyst due to intraparticle
or interparticle diffusion resistance (Matsumoto et al., 1972). Reaction
parameters such as temperature and flow rate or agitation tend to become
more influential at the transition into the range of external mass transfer
(Chuchvalec et al., 1971).

Over palladium both double-bond migration and cis–trans isomerization
tend to be faster than olefin hydrogenation. Consequently, hydrogenations
proceeding beyond 1 mole of hydrogen absorption will result in more loss
of initial cis-olefin than the stoichiometry of the overrun would indicate.
Saturation relative to isomerization is favored by high hydrogen availability
at the catalyst surface and retarded by low hydrogen availability.

In view of the above considerations it is not surprising to find selectivity
sometimes markedly influence by small variations in the amount of hydrogen
absorbed, by numerous additives that change the rate of hydrogenation
relative to the rate of mass transport and the rate of acetylene hydrogenation
relative to that of the olefin, and by reaction variables such as amount of
catalyst (Csuros et al., 1951; Gensler and Thomas, 1951; Mondon, 1952;
Schinz, 1955; Dobson et al., 1961; Sokol'skii, 1964), agitation (Ege et al.,
1961), temperature (Takei and Ono, 1942), and catalyst support (Dobson
et al., 1961).

Selectivity in hydrogenation of vinylacetylenes (or conjugated diynes) to
dienes is more difficult to achieve than selectivity in hydrogenation of an
isolated triple bond to a double bond (Blomquist and Marvel, 1933; Celmer
and Solomons, 1953; Marvell and Tashiro, 1965; Garanti et al., 1968;
Gunstone and Jie, 1970; Morris et al., 1972). The problem may be increased
if divalent sulfur is present in the molecule (Vasil'ev et al., 1971). Nonetheless,
selective hydrogenation of vinylacetylenes is practical. The steps in the

hierarchy of adsorbabilities (diyne > enyne > diene > ene) are sufficiently steep so that it is possible on an industrial scale to use palladium for selective hydrogenation of traces of vinylacetylene in massive amounts of butadiene and for selective hydrogenation of traces of butadiene in massive amounts of butene. In the synthetic realm, many selective hydrogenations of vinyl-acetylenes have been achieved. It is easier to obtain high selectivity if the acetylene is in the terminal position (Bal'yan and Borovikova, 1959a) or if the olefin is trisubstituted.

## A.  Effect of Catalyst Loading and Agitation

A number of investigators have noted that catalyst loading may influence selectivity. An important, but overlooked, aspect of the effect of catalyst loading is the ability of catalysts to simply act as a sop for poisons, inhibitors, and catalyst modifiers. This effect could be of major influence, for selectivity in acetylene hydrogenations is demonstrably sensitive to small amounts of extraneous material. Beyond this effect, changing the amount of catalyst would not be expected to have much influence on selectivity, unless the transition between reaction-controlled and mass-transfer-controlled rates of reduction were crossed. Changing agitation can likewise effect this transition, and its influence might be expected to be much the same as the effect brought about by changing the catalyst loading.

## B.  Effect of Temperature

It does not seem possible at present to predict the effect of temperature on reaction selectivity. Various steps have different energies of activation, and the ratio of adsorption constants for the acetylene and olefin may change with temperature (Sporka et al., 1972). It has been recommended that selective reductions of acetylenes to the cis-olefin be carried out at sub-ambient temperatures (Gutmann and Lindlar, 1969), and this advice is often heeded (Voerman et al., 1974; Roelofs et al., 1975; Descoins and Samain, 1976). Excellent selectivity was obtained by Henrick (1977) over Lindlar catalyst inhibited by synthetic quinoline at $-10°$ to $-30°C$ in pentane, hexane, or hexane–tetrahydrofuran. trans-Olefin was kept to less than 0.5%. Of theoretical interest is Henrick's observation that the trans isomer was formed throughout the reaction and not simply when the acetylene was nearly exhausted. Henrick (1977) found that subambient temperatures were necessary if this excellent selectivity were to be consistently achieved. On the other hand, hydrogenation of phenylacetylenecarboxylic acid over pal-ladium-on-barium sulfate gave 75% cis-cinnamic acid at $-18°C$ and 100% cis isomer at 100°C (Takei and Ono, 1942).

## C.  Effect of Solvent

The role of solvent is obscure. Rates generally increase with the dielectric constant of the solvent. The extent of solvent sensitivity depends on the substrate as well as the catalyst (Sokol'skii et al., 1972). The rate of hydrogenation of 1-hexyne over platinum black in the aprotic solvents hexane, cyclohexane, decane, and hexadecane increased with increasing dipole moment and decreasing chain length for both moles of hydrogen absorbed. In protic solvents, the rate of hydrogenation of the double bond increased with increasing dielectric constant of the solvent, but the rate of hydrogenation of the triple bond was virtually unchanged (Sokol'skaya et al., 1972). The rates of hydrogenation of both bonds decreased with increasing solvent density, viscosity, and surface tension.

The effect of solvent on selectivity is infrequently reported. Many solvents have been used successfully. Some workers prefer nonpolar solvents such as pentane or hexane over alcohols (Lindlar and Dubuis, 1966; Gutmann and Lindlar, 1969; Henrick, 1977).

## II.  CATALYSTS

Many workers have compared various metals for the hydrogenation of acetylenes, especially lower acetylenes (Bond et al., 1958, 1962; Freidlin and Kaup, 1963). It always emerges from these studies that palladium is by far the most selective metal for the conversion of acetylenes to olefins, a conclusion confirmed in many synthetic applications. Bond and Wells (1966) established the following order of decreasing selectivity for hydrogenation of acetylene to ethylene: $Pd \gg Pt \sim Rh > Ir$. The same order holds for selective hydrogenation of 1-pentyne and 2-pentyne, as well as the diolefinic compounds allene, 1,3-butadiene, and 1,3-pentadiene. In hydrogenation of 2-pentyne (Bond and Wells, 1966), selectivity to form the olefin decreased in the series $Pd > Pt > Ru > Rh > Ir$, whereas the stereoselectivities to the cis-olefin, which lie between 91 and 98%, fall in the sequence $Pd > Rh > Pt > Ru \simeq Ir$. Selectivity of unsupported metals for conversion of methylacetylene to propylene falls in the order Pd (0.98) > Pt (0.92) > Rh (0.87) > Ni (0.76) > Ru (0.44) > Ir (0.29) (Yoshida and Hirota, 1975).

In practice most hydrogenations of acetylenes are carried out over palladium, frequently in conjugation with various modifiers that increase the selectivity. In special circumstances, such as the hydrogenation of the acetylenic bond in di($\Delta^2$-cyclopentyl)acetylene, platinum may prove to be more selective than palladium (Plate and Stanko, 1960). Palladium probably fails here because of its strong tendency to adsorb trienes.

The catalyst support as well as the metal may influence selectivity (Rieche *et al.*, 1961; Dobson *et al.*, 1961). Palladium-on-barium sulfate (Cram and Allinger, 1956; Sheads and Beroza, 1973; Su *et al.*, 1973; Carlson *et al.*, 1974; Sonnet, 1974; Hendry *et al.*, 1975) or palladium-on-calcium carbonate (Inhoffen *et al.*, 1950; Milas *et al.*, 1950; Raphael and Sondheimer, 1950) is widely used, but good results have also been achieved with palladium-on-carbon or palladium-on-alumina. Russian workers (Batkov *et al.*, 1972), after examining the hydrogenation of diacetylene glycols, concluded that the support controlled selectivity by determining the spacing of catalyst atoms. Optimal catalysts varied, depending on the goals, with both metal and support, as indicated below. Some isomerized product, $HOC(CH_3)_2CH_2CH=CHCH_2C(CH_3)_2OH$, was obtained as an intermediate with all catalysts.

$$HOC(CH_3)_2(CH_2)_4C(CH_3)_2OH \qquad\qquad HOC(CH_3)_2CH=CHCH=CHC(CH_3)_2OH$$

$$\text{Rh-on-Al}_2\text{O}_3$$

$$\text{Pd-on-BaSO}_4 \qquad\qquad\qquad \text{RaNi}$$

$$HOC(CH_3)_2C\equiv CC\equiv CC(CH_3)_2OH$$

$$\text{Pd-on-C} \qquad\qquad\qquad \text{Pd-on-C}$$

$$(CH_3)_2CHCH=CH(CH_2)_2C(CH_3)_2OH \qquad\qquad (CH_3)_2CH(CH_2)_4C(CH_3)_2OH$$

Highly stereospecific hydrogenations of acetylenes to *cis*-olefins can be achieved with a nickel (P2) catalyst, prepared by sodium borohydride reduction of nickel acetate in ethanol, and ethylenediamine as a promoter. Ratios of cis to trans isomers as high as 200 : 1 are obtained (Brown and Ahuja, 1973). Presumably ethylenediamine should be effective with other catalysts as well. This catalyst was used in the preparation of (*Z*)-6-heneicosen-11-one (**5**), a sex attractant of the Douglas fir tussock moth, through selective hydrogenation of 6-heneicosyn-11-ol (**2**) and subsequent oxidation of (*Z*)-9-heneicosen-11-ol (**3**), which was obtained in better than 97% purity; (*E*)-6-heneicosen-11-ol (**1**) is obtained from the same acetylene by reduction with sodium in liquid ammonia (Smith *et al.*, 1975).

The same sex pheromone (**5**) was obtained by selective hydrogenation of 6-heneicosyn-11-one (**4**) carried out over 5% palladium-on-barium sulfate in methanol containing small amounts of pyridine (Kocienski and Cerni-gliaro, 1976). Many other sex pheromones have been prepared by similar hydrogenations of acetylenic compounds; Lindlar catalyst and palladium-on-barium sulfate usually are used in these syntheses (Henrick, 1977). Nickel (P2) inhibited by ethylenediamine was used in the synthesis of (Z)-7-(Z)-11-hexadecadien-1-yl acetate (Mori et al., 1975).

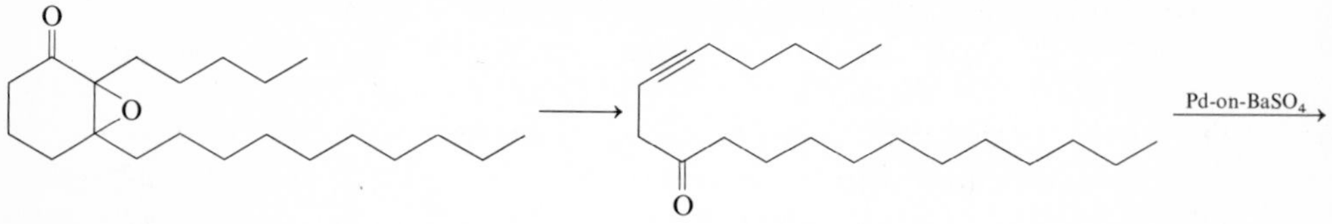

4

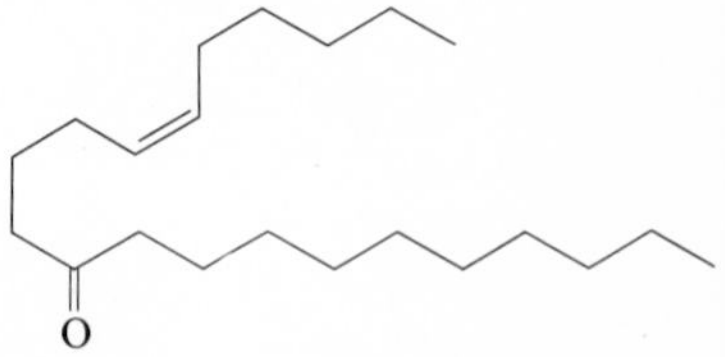

5

Hydrogenations over nickel may also come to a halt at the olefin stage under conditions in which palladium-catalyzed reductions do not. Hydrogenation of 5-cyclononynone over 5% palladium-on-carbon in ethyl acetate consumed 2 moles of hydrogen to afford cyclononanone, whereas over nickel boride (P2) (Brown and Brown, 1963) the reaction virtually ceased after the consumption of 1 mole of hydrogen to afford cis-5-cyclononenone (Lange and Hall, 1974).

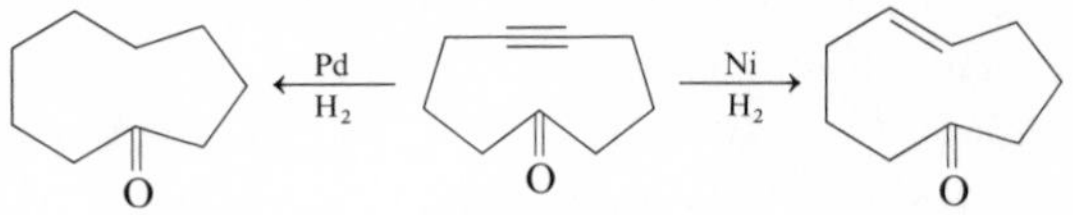

Various other nickel catalysts have been used by other investigators, but the use of nickel is relatively infrequent and rarely notable (Marvell and Li, 1973). Nickel catalysts may give different results than palladium catalysts due to differences in isomerizing ability. This is illustrated by the hydrogena-tion of compounds substituted by a 1,3-bis(trimethylsilyl)acetylene. Reduc-tion of the acetylenes 7 over Raney nickel (W2) in ethyl acetate affords 8, whereas over palladium-on-barium sulfate the isomerized products 6 are formed. However, when R is hydroxyl or formyl, reduction over palladium-on-barium sulfate tends to give the nonisomerized olefin (Atsumi and Kuwajima, 1977).

$$\underset{\overset{|}{SiMe_3}}{RCH_2C}=CHCH_2SiMe_3 \xleftarrow[\text{hexane}]{\text{Pd-on-BaSO}_4} \underset{\overset{|}{SiMe_3}}{RCH_2CHC}\equiv CSiMe_3 \xrightarrow[\text{EtOAc}]{\text{RaNi}} \underset{\overset{|}{SiMe_3}}{RCH_2CHCH}=CHSiMe_3$$

**6**             **7**             **8**

$R = CH_3CO$
$R = HC(OCH_3)_2$

## A.  Metal Additives

Many metal additives have been found that alter the characteristics of catalysts in the hydrogenation of acetylenes. The additives may be incorporated into the catalysts as it is made or added subsequently to preformed catalysts at the time or use. Among metals that have been used as modifiers are zinc (Naidin *et al.*, 1970; Chukhadzhyan *et al.*, 1970), copper (Hort, 1960; Bizhanova *et al.*, 1969), zirconium (Yagudeev and Azerbaev, 1972), lead (Lindlar and Dubuis, 1966), gold (Rylander and Himelstein, 1964), iron (Reppe *et al.*, 1955), tin (French Patent 1,224,182), ruthenium (Sokol'skii *et al.*, 1970), and mercury (Maxwell, 1970). Additives sometimes increase the rate sharply, but they may do so at the expense of selectivity (Bizhanova *et al.*, 1969). The best known promoted catalyst is the Lindlar catalyst (Lindlar and Dubuis, 1966), palladium–lead-on-calcium carbonate, which has been used successfully by many investigators for selective hydrogenation of acetylenic compounds (Jacobson *et al.*, 1970; Meyer *et al.*, 1973; Tamaki *et al.*, 1973; Su and Mahany, 1974; Voerman *et al.*, 1974; Tumlinson *et al.*, 1975). Sometimes the substrate itself contains some adventitious inhibitor, which together with the inhibiting effect of lead renders the catalyst too slow for use. In this circumstance, satisfactory results often can be obtained by filtering the mixture and adding fresh catalyst (Henrick, 1977).

The function of metal promoters is often obscure. For instance, ruthenium–palladium catalysts have been shown to be much more active than palladium alone in the hydrogenation of acetylenic carbinols (Sokol'skii *et al.*, 1969, 1976). Yet simple mixtures of palladium-on-carbon and ruthenium-on-carbon were shown to be strongly synergistic in similar reactions, even under conditions in which the ruthenium catalyst itself was inactive. The salutary effect of ruthenium in this reduction was attributed to its ability to reduce inhibiting aldehydes, which can be formed in the hydrogenation of certain acetylenic carbinols (Rylander and Cohn, 1961). So many modifiers of widely diverse types give improved performance in the hydrogenation of acetylenes that one is inclined to the view that modifiers may function in part by increasing the availability of hydrogen at the catalyst surface, a consequence of the generally lower rate of hydrogenation. Many more complex explanations of the function of modifiers are possible; the area cannot be summarized succinctly.

## B.   Other Additives

Acetylenes are frequently hydrogenated in the presence of additives such as amines, sulfur compounds, or alkali to improve selectivity. Some investigators believe that additives such as chlorobenzenes and phenyl thiocyanates (Bal'yan, 1951) or sodium or potassium hydroxide (Tedeschi, 1962; Tedeschi et al., 1967) act through binding with the substrate rather than act on the catalyst. A variety of amines such as quinoline, pyridine, piperidine, aniline, and diethylamine have been used effectively as catalyst modifiers, but these amines are not necessarily equivalent in action (Fukuda and Kusama, 1958; Bal'yan and Borovikova, 1959a,b). Amines have also been used as solvents in selective hydrogenations of acetylenes (Bowers et al., 1958; Chase and Galender, 1959). One drop of quinoline may be sufficient to cause hydrogenation of an acetylene to come to an abrupt stop after 1 mole of hydrogen is absorbed (Berkowitz, 1972).

Selective hydrogenation of acetylenes may be very sensitive to the presence of extraneous materials, some of which may enter the system inadvertently (Gensler and Schlein, 1955; Newman, 1964). Adventitious modification of the catalyst in this way undoubtedly accounts for some of the conflicting reports in the literature.


## III.   ACETYLENIC CARBINOLS AND GLYCOLS

Acetylenic carbinols and glycols, which are formed readily by condensation of an acetylene and a carbonyl compound, appear frequently in synthetic work. Compounds of this type have been studied extensively by Russian workers, mainly from the standpoint of hydrogenation mechanisms. In general, the reduction can be controlled without much difficulty to give either the semihydrogenated or fully hydrogenated derivative in excellent yield. The glycols are more sensitive to catalyst and conditions than are carbinols, for glycols undergo hydrogenolysis more readily, as well as cyclization to furans in the presence of acids. Side reactions can account for as much as 75% of the total product, even under mild conditions, if the system is inappropriate (Tedeschi et al., 1967). Many catalysts have been used in these reductions, but palladium is generally the preferred metal for either partial or complete hydrogenation.

Various bases such as quinoline (Isler et al., 1947) or butylamine (Sokol'skii et al., 1971) are often used alone or in conjugation with a metallic modified catalyst such as palladium–lead-on-calcium carbonate to improve yields. A variety of metal additives have been used. Improved yields of semihydrogenated products are obtained by addition of zinc or cadmium cations to palladium catalyst (Zakarina et al., 1972; Naidin et al., 1970, 1972). Inhibition of palladium-on-calcium carbonate by ammonia and zinc acetate permits yields of 98% unsaturated carbinol from hydrogenation of dimethylethynylcarbinol in aqueous solution (Chukhadzhyan et al., 1970). The reaction is a key step in an industrial synthesis of isoprene (Heath, 1973).

$$C_2H_2 + CH_3\underset{\underset{O}{\parallel}}{C}CH_3 \longrightarrow CH_3\underset{\underset{CH_3}{|}}{\overset{\overset{OH}{|}}{C}}C{\equiv}CH \xrightarrow[H_2]{Pd} CH_3\underset{\underset{CH_3}{|}}{\overset{\overset{OH}{|}}{C}}CH{=}CH_2 \xrightarrow{-H_2O} CH_2{=}\underset{\underset{CH_3}{|}}{C}CH{=}CH_2$$

Fair selectivities can be obtained with smaller carbinols even without modification of the catalyst, but as the carbon chain increases selectivities fall. Excellent selectivities have been obtained in the hydrogenation of a variety of acetylenic carbinols and glycols by the use of palladium and potassium hydroxide (Tedeschi et al., 1967). The technique is successful even in compounds in which other modifiers failed to give satisfactory results. Reductions using palladium-on-barium carbonate stop spontaneously after the first mole of hydrogen is absorbed, but reductions over palladium-on-carbon have to be stopped. With the latter catalyst, excellent yields of either the partially or fully hydrogenated derivatives can be obtained. In a typical example, 1 mole 3,6-dimethyl-4-octyne-3,6-diol is reduced over 2.0 gm 5% palladium-on-carbon in 200 ml heptane containing 0.05 gm potassium hydroxide at 27–45°C and 30–60 psig (Tedeschi et al., 1967).

$$\underset{\underset{OH}{|}}{\overset{\overset{CH_3}{|}}{C_2H_5C}}C{\equiv}C\underset{\underset{OH}{|}}{\overset{\overset{CH_3}{|}}{CC_2H_5}} \longrightarrow \underset{\underset{OH}{|}}{\overset{\overset{CH_3}{|}}{C_2H_5C}}CH{=}CH\underset{\underset{OH}{|}}{\overset{\overset{CH_3}{|}}{CC_2H_5}} \longrightarrow \underset{\underset{OH}{|}}{\overset{\overset{CH_3}{|}}{C_2H_5C}}CH_2CH_2\underset{\underset{OH}{|}}{\overset{\overset{CH_3}{|}}{CC_2H_5}}$$

98%            99%

Reductions of glycols and carbinols in general seem to be less stereospecific than those of nonsubstituted acetylenes. Cis isomers are usually the major product of the reduction, although their percentage may vary widely. Since the conversion of cis-olefin to the trans isomer occurs readily once the acetylene is depleted (Dobson et al., 1961), it might be expected that the cis–trans ratio will depend markedly on when the reduction is stopped. The presence of large amounts of alkali seems to promote mainly trans isomers in certain reductions, such as butynediol and acetylenedicarboxylic acids (McQuillin and Ord, 1959).

Hydrogenation of the ketal propargyl alcohol **9** over platinum oxide in ethyl acetate containing triethylamine affords 99% of the ketal allylic alcohol **10** in a 3 : 1 mixture of cis to trans isomers. An extensive investigation of other catalysts and conditions failed to improve this ratio. The authors suggested that the hydroxyl was bound tightly to the catalyst surface, facilitating subsequent isomerization of the initially formed *cis*-olefin (Klioze and Darmory, 1975).

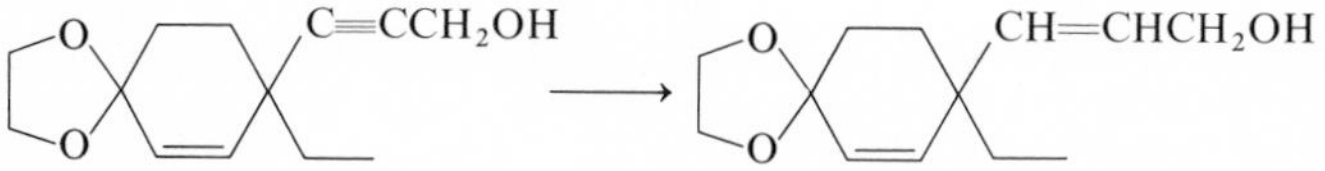

**9**                                              **10**

A synthesis of unsaturated aldehydes involves semihydrogenation of an acetylenic silane over palladium-on-carbon in ethyl acetate containing pyridine. It is illustrated here by the preparation of cinnamaldehyde. The procedure is adaptable to saturated aldehydes and ketones as well (Stork and Colvin, 1971).

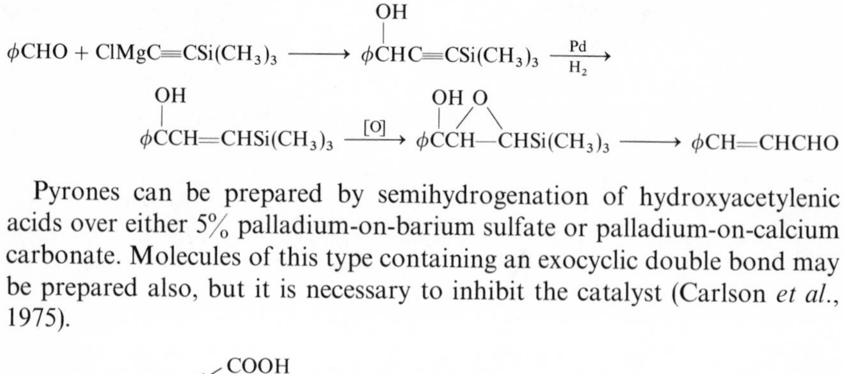

Pyrones can be prepared by semihydrogenation of hydroxyacetylenic acids over either 5% palladium-on-barium sulfate or palladium-on-calcium carbonate. Molecules of this type containing an exocyclic double bond may be prepared also, but it is necessary to inhibit the catalyst (Carlson *et al.*, 1975).

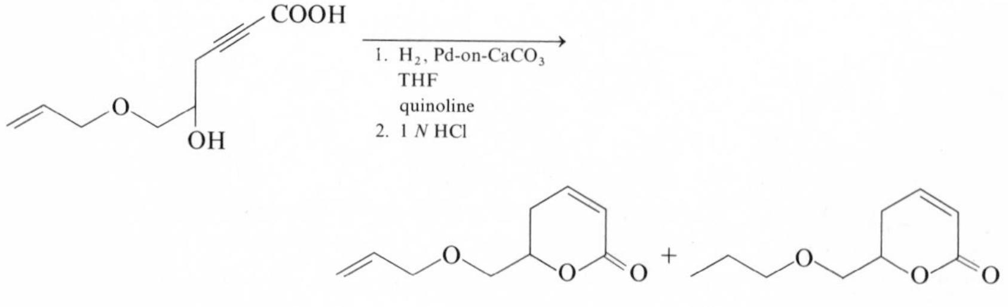

80%                                              20%

## IV.  HYDROGENOLYSIS

Hydrogenation of acetylenes carrying adjacent oxygen or nitrogen functions may be accompanied by loss of that function to various degrees. Allylic functions are particularly susceptible to hydrogenolysis, and it is probably generally true that hydrogenolysis during reduction of acetylenes occurs sometime after formation of the olefin and before complete saturation (Tedeschi, 1962). Hydrogenolysis can occur after saturation in certain sensitive compounds (Nogaideli and Vardosanidze, 1963) or in reductions carried out at high temperatures (Tedeschi *et al.*, 1967). Substituents influence the extent of hydrogenolysis, but it is difficult to formulate an encompassing generality (Tedeschi, 1962).

Acetylenic glycols are much more susceptible to hydrogenolysis than are acetylenic carbinols; hydrogenolysis in the latter is usually a minor side reaction. Compounds containing epoxides adjacent to the acetylenic function are very sensitive to hydrogenolysis, as might be expected (Ghera *et al.*, 1962). Platinum would probably be generally more useful than palladium in the reduction of these compounds if hydrogenolysis were to be avoided. Some acetylenic amines undergo hydrogenolysis readily; the extent of hydrogenolysis may be very sensitive to minor structural variations (Hennion and Perrino, 1961). Hydrogenolysis of acetylenic amines has been prevented by the use of platinum oxide and the hydrochloride salt instead of the free amine (Easton *et al.*, 1961), and, even with the free amine, platinum oxide produces much less hydrogenolysis than does palladium-on-carbon (Morrow *et al.*, 1965). Conversion of **11** to the hydrochloride salt or to a quaternary salt prevented hydrogenolysis of the acetoxy group **12** (Marszak and Marszak-Fleury, 1949, 1950).

$$\underset{\textbf{11}}{H_3C\overset{\overset{\textstyle O}{\|}}{C}OCH_2C{\equiv}CCH_2NR_2} \overset{Pd}{\longrightarrow} \underset{\textbf{12}}{H_3CCH{=}CHCH_2NR_2 + CH_3COOH}$$

Ethynylcyclohexanol undergoes extensive hydrogenolysis during complete hydrogenation over palladium catalysts (Tedeschi and Clark, 1962) but, if the reaction is carried out at subambient temperature ($-25°C$), there is no detectable loss of hydroxyl (Brown, 1969). Nickel (P2) is said to saturate propargylic and allylic hydroxy compounds without hydrogenolysis (Brown and Ahuja, 1973). Acetylenes of the structure **13** readily undergo hydrogenolysis to give the olefinic alcohol **14** (Sabirov *et al.*, 1972).

$$\underset{\textbf{13}}{ROCR^1R^2C{\equiv}CR^2R^4COCH_2N(C_2H_5)_2} \longrightarrow \underset{\textbf{14}}{ROCR^1R^2HC{=}CHR^2R^4COH + CH_3N(C_2H_5)_2}$$

## V. ACETYLENIC ALDEHYDES AND KETONES

Reduction of $\alpha,\beta$-ynones apparently proceeds without undue difficulty. The products may be readily isomerized, however, to *trans*-enones (Theus *et al.*, 1955; Theus and Schinz, 1956; Kugatova-Shemyakina and Vidugirene, 1964) as well as further reduced. Both problems may be helped by the use of an inhibited catalyst. Selective reduction of acetylenes containing carbonyl functions not conjugated with the triple bond seems to present no difficulty at all (Hershberg *et al.*, 1951; Sandoval *et al.*, 1955; Venkataramani *et al.*, 1966).

Successful reduction of enynones to dienones depends in large measure on their structure. Selectivity in compounds with the acetylenic bond attached directly to the carbonyl group is low (Heilbron *et al.*, 1949a,b,c; Surber *et al.*, 1956; Marvell *et al.*, 1972). For instance, attempted selective hydrogenation of 1-phenyl-3-(cyclohexen-1-yl)-2-propynone under a variety of conditions gave a mixture of cyclized products (Marvell *et al.*, 1972).

If the double bond in conjugated enynones is tetrasubstituted, selective reduction to the dienone is achieved much more readily (Eschenmoser *et al.*, 1953; Schiess and Chia, 1970; Schiess *et al.*, 1970). Selectivity is also much better for enynones if the acetylenic bond is terminal in the conjugated system (Schiess and Chia, 1970; Schiess *et al.*, 1970) or, if not terminal, at least not adjacent to the carbonyl (Inhoffen and von der Bey, 1953; Mildner and Weedon, 1953; Marvell *et al.*, 1972).

## VI. POLYACETYLENES

Diacetylenes have been reduced stepwise to an enyne, a diene, an olefin, and a hydrocarbon. The preference is for hydrogenation of a terminal triple bond rather than an internal one and of an internal acetylene rather than an external olefin (Dobson *et al.*, 1961; Sokol'skaya and Anisimova, 1973). This is possible because of the operation of a hierarchy of adsorbabilities, a phenomenon to which the term *molecular queuing* has been affixed (Crombie and Jenkins, 1969). The selectivity at each stage depends in large measure on whether the diacetylene is conjugated; high selectivity is more difficult to achieve in a conjugated system (Dale *et al.*, 1963). Palladium-catalyzed reductions, especially if inhibited, are apt to stop spontaneously at an olefin stage (McCrae, 1964). Platinum-catalyzed reductions are more likely to proceed to complete saturation (Zal'kind and Zhuravleva, 1948; Zal'kind and Iremadze, 1948).

Burgstahler and Widiger (1973) converted the acetylenic triol **16** to the dicisoid dienetriol **17** in 95% yield over 5% palladium-on-barium sulfate inhibited by quinoline. The partial reduction was unsatisfactory over a Lindlar catalyst (Lindlar and Dubuis, 1966). The saturated triol **15** was obtained in 93% yield by hydrogenation of **16** over 5% palladium-on-carbon. Hydrogenolysis of the activated allylic hydroxyls, a possible trouble spot, was unimportant.

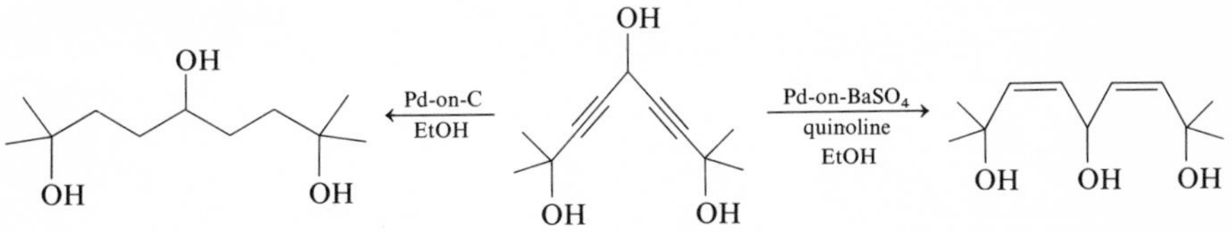

A synthesis of termite trail-following pheromone *cis*-3,*cis*-6,*trans*-8-dodecatrien-1-ol (**19**) involves hydrogenation under ambient conditions of 8-dodecene-3,6-diyn-1-ol (**18**) over Lindlar catalyst in methanol containing a small amount of quinoline. Isomeric *cis*-3,*trans*-6,*cis*-8-dodecatrien-1-ol was prepared by a similar selective hydrogenation of *trans*-6-dodecene-3,8-diyn-1-ol (Tai *et al.*, 1969). The presence of potassium hydroxide in a hydrocarbon medium is said to improve selectivity in hydrogenation of $\beta$-, $\gamma$-, $\delta$-acetylenes (Sokol'skaya *et al.*, 1973).

$$C_3H_7CH{=}CHC{\equiv}CCH_2C{\equiv}CCH_2CH_2OH \longrightarrow C_3H_7CH{=}CHCH{=}CHCH_2CH{=}CHCH_2CH_2OH$$

**18**                                                                                       **19**

All-cis long-chain unsaturated acids are prepared readily by selective hydrogenation of long-chain polyacetylenic compounds. For instance, hydrogenation of 5,8,11,14-eicosatetraynoic acid over Lindlar catalyst affords arachidonic acid (Fryer *et al.*, 1975). Quinoline-inhibited palladium-on-barium sulfate has been used similarly (Scheffer and Wostradowski, 1972).

An improved synthesis of {18}annulene (**21**) involves selective hydrogenation of 1,3,7,9,13,15-hexadehydro{18}annulene (**20**). The best results were achieved by the use of palladium-on-barium sulfate in benzene inhibited by a drop of quinoline. The reduction was apparently nonselective over either 10% palladium-on-carbon or Lindlar catalyst; the product was hydrogenated much more rapidly than the starting material (Figeys and Gelbcke, 1970). The same product can be obtained by hydrogenation of tridehydro{18}annulene (**22**) over 10% palladium-on-carbon in benzene (Sondheimer *et al.*, 1962). Reductions over Lindlar catalyst were erratic, and it has been suggested that 10% palladium-on-carbon was more effective due to its greater isomerizing power; three cis double bonds in **22** have to be isomerized to trans to obtain {18}annulene (Marvell and Li, 1973).

20                                    21                                    22

An unusual reaction takes place in the hydrogenation of diacetylenes held in enforced propinquity (Staab and Ipaktschi, 1971):

Some carbon–carbon bond formation may also occur intermolecularly (Hall and Steuck, 1971):

Major                                    Minor

# REFERENCES

Akhtar, M., and Weedon, B. C. I., *J. Chem. Soc.* p. 4058 (1959).

Atsumi, K., and Kuwajima, I., *Tetrahedron Lett.* p. 2001 (1977).

Bal'yan, K. V., *Zh. Obshch. Khim.* **21**, 720 (1951); *J. Gen. Chem. USSR* **21**, 793 (1951).

Bal'yan, K. V., and Borovikova, N. A., *Zh. Obshch. Khim.* **29**, 2882 (1959a).

Bal'yan, K. V., and Borovikova, N. A., *Zh. Obshch. Khim.* **29**, 2553 (1959b).

Batkov, A. A., Kirilyus, I. V., Akhmetova, T. A., and Azerbaev, I. N., *Tr. Khim.-Metall. Inst.*, *Akad. Nauk Kaz. SSR* **18**, 126 (1972); *Chem. Abstr.* **79**, 31405w (1973).

Berkowitz, W. F., *J. Org. Chem.* **37**, 341 (1972).

Bizhanova, N. B., Erzhanova, M. S., and Sokol'skii, D. V., *Izv. Akad. Nzuk Kaz. SSR, Ser. Khim.* **19**(6), 15 (1969).

Blomquist, A. T., and Marvel, C. S., *J. Am. Chem. Soc.* **55**, 1655 (1933).

Bond, G. C., "Catalysis by Metals." Academic Press, New York, 1962.

Bond, G. C., and Rank, J. S., *Proc. Int. Congr. Catal., 3rd, Amsterdam, 1964*, **11**, 1225 (1965).

Bond, G. C., and Wells, P. B., *J. Catal.* **5**, 419 (1966).

Bond, G. C., Dowden, D. A., and Mackenzie, N., *Trans. Faraday Soc.* **54**, 1537 (1958).

Bond, G. C., Webb, G., Wells, P. B., and Winterbottom, J. B., *J. Catal.* **1**, 74 (1962).

Bowers, A., Ringold, H. J., and Denot, E., *J. Am. Chem. Soc.* **80**, 6115 (1958).

Brown, C. A., *J. Am. Chem. Soc.* **91**, 5901 (1969).

Brown, C. A., *Chem. Commun.* p. 139 (1970).

Brown, C. A., and Ahuja, V. K., *J. Chem. Soc., Chem. Commun.* p. 553 (1973).

Brown, H. C., and Brown, C. A., *J. Am. Chem. Soc.* **85**, 1005 (1963).

Burgstahler, A. W., and Widiger, G. N., *J. Org. Chem.* **38**, 3652 (1973).

Carlson, D. A., Doolittle, R. E., Beroza, M., Rogoff, W. M., and Gretz, G. H., *J. Agric. Food Chem.* **22**, 194 (1974).

Carlson, R. M., Oyler, A. R., and Peterson, J. R., *J. Org. Chem.* **40**, 1610 (1975).

Celmer, W. D., and Solomons, I. A., *J. Am. Chem. Soc.* **75**, 3430 (1953).

Chase, G. O., and Galender, J., U.S. Patent 2,883,431, Apr. 21, 1959.

Chuckvalec, P., Havlicek, M., and Horak, J., *Chem. Prum.* **21**(10), 488 (1971); *Chem. Abs.* **76**, 7039x (1972).

Chukhadzhyan, G. A., Movsisyan, G. V., Daglyan, L. G., Margaryan, D. A., and Pososyan, Y. M., *Arm. Khim. Zh.* **23**(9), 784 (1970); *Chem. Abstr.* **74**, 41596v (1971).

Cram, D. J., and Allinger, N. L., *J. Am. Chem. Soc.* **78**, 2518 (1956).

Crombie, L., and Jenkins, R. A., *Chem. Commun.* p. 394 (1969).

Csuros, Z., Geczy, I., and Polgar, J., *Acta Chim. Acad. Sci. Hung.* **1**, 417 (1951).

Dale, J., Hubert, A. J., and King, G. S. D., *J. Chem. Soc.* p. 73 (1963).

Descoins, C. E., and Samain, D., *Tetrahedron Lett.* p. 745 (1976).

Dobson, N. A., Eglinton, G., Krishnamurti, M., Raphael, R. A., and Willis, R. G., *Tetrahedron* **16**, 16 (1961).

Easton, N. R., Dillard, R. D., Doran, W. J., Livezey, M., and Morrison, D. E., *J. Org. Chem.* **26**, 3772 (1961).

Ege, S. N., Wolovsky, R., and Gensler, W. J., *J. Am. Chem. Soc.* **83**, 3080 (1961).

Eschenmoser, A., Schreiber, J., and Julia, S. A., *Helv. Chim. Acta* **36**, 483 (1953).

Figeys, H. P., and Gelbcke, M., *Tetrahedron Lett.* p. 5139 (1970).

Freidlin, L. K., and Kaup, Y., *Dokl. Akad. Nauk SSSR* **152**(6), 1383 (1963).

Fryer, R. I., Gilman, N. W., and Holland, B. C., *J. Org. Chem.* **40**, 348 (1975).

Fukuda, T., and Kusama, T., *Bull. Chem. Soc. Jpn.* **31**, 339 (1958).

Garanti, L., Marchesini, A., U. M. Pagnoni, U. M., and Trave, R., *Gazz. Chim. Ital.* **106**, 187 (1968).

Gensler, W. J., and Schlein, H. N., *J. Am. Chem. Soc.* **77**, 4846 (1955).

Gensler, W. J., and Thomas, G. R., *J. Am. Chem. Soc.* **73**, 4601 (1951).
Ghera, E., Gibson, M., and Sondheimer, F., *J. Am. Chem. Soc.* **84**, 2953 (1962).
Gunstone, G. D., and Jie, M. L. K., *Chem. Phys. Lipids* **4**, 1 (1970).
Gutmann, H., and Lindlar, H., *in* "Chemistry of Acetylenes" (H. G. Viehe, ed.), p. 355. Dekker, New York, 1969.
Hall, H. K., Jr., and Steuck, M. J., *Chem. Ind. (London)* p. 545 (1971).
Heath, A., *Chem. Eng.* Oct., p. 48 (1973).
Heilbron, I., Jones, E. R. H., Ricardson, R. W., and Sondheimer, F., *J. Chem. Soc.* p. 737 (1949a).
Heilbron, I., Jones, E. R. H., Toogood, J. B., and Weedon, B. C. L., *J. Chem. Soc.* p. 1827 (1949b).
Heilbron, I., Jones, E. R. H., Toogood, J. B., and Weedon, B. C. L., *J. Chem. Soc.* p. 2028 (1949c).
Hendry, L. B., Korzeniowski, S. H., Hindenlang, D. M., Kosarych, Z., Mumma, R. O., and Jugovich, J., *J. Chem. Ecol.* **1**, 317 (1975).
Hennion, G. F., and Perrino, A. C., *J. Org. Chem.* **26**, 1073 (1961).
Henrick, C. A., *Tetrahedron* **33**, 1845 (1977).
Hershberg, E. B., Oliveto, E. P., Gerold, C., and Johnson, L., *J. Am. Chem. Soc.* **73**, 5073 (1951).
Hort, E. V., U.S. Patent 2,953,604, Sept. 20, 1960.
Inhoffen, H. H., and Raspé, G., Liebigs *Ann. Chem.* **592**, 214 (1955).
Inhoffen, H. H., and von der Bey, G., *Justus Liebigs Ann. Chem.* **583**, 100 (1953).
Inhoffen, H. H., Bohlmann, F., Bartram, K., Rummert, G., and Pommer, H., *Justus Liebigs Ann. Chem.* **570**, 54 (1950).
Isler, O., Huber, W., Ronco, A., and Kofler, M., *Helv. Chim. Acta* **30**, 1911 (1947).
Isler, O., Lindlar, H., Montavon, M., Rüegg, R., Saucy, G., and Zeller, P., *Helv. Chim. Acta* **39**, 2041 (1956a).
Isler, O., Gutmann, H., Lindlar, H., Montavon, M., Rüegg, R., and Zeller, P., *Helv. Chim. Acta* **39**, 449 (1956b).
Jacobson, M., Redfern, R. E., Jones, W. A., and Aldridge, M. H., *Science* **170**, 542 (1970).
Klioze, S. S., and Darmory, F. P., *J. Org. Chem.* **40**, 1588 (1975).
Kocienski, P. J., and Cernigliaro, G. J., *J. Org. Chem.* **41**, 2927 (1976).
Kugatova-Shemyakina, G. P., and Vidugirene, V. I., *J. Gen. Chem. USSR* **34**, 1742 (1964).
Lange, G. L., and Hall, T. W., *J. Org. Chem.* **39**, 3819 (1974).
Lindlar, H., and Dubuis, R., *Org. Synth.* **46**, 89 (1966).
McCrae, W., *Tetrahedron* **20**, 1773 (1964).
McQuillin, F. J., and Ord, W. O., *J. Chem. Soc.* p. 2902 (1959).
Mann, R. S., and Khulbe, K. C., *J. Catal.* **10**, 401 (1968).
Mann, R. S., and Khulbe, K. C., *J. Catal.* **13**, 25 (1969).
Mann, R. S., and Khulbe, K. C., *J. Catal.* **17**, 46 (1970).
Marszak, J., and Marszak-Fleury, A., *C. R. Acad. Sci., Ser. C* **228**, 1501 (1949).
Marszak, J., and Marszak-Fleury, A., *Bull. Soc. Chim. Fr.* **17**, 1305 (1950).
Marvell, E. N., and Li, T. H., *Synthesis* No. 8, p. 457 (1973).
Marvell, E. N., and Tashiro, J., *J. Org. Chem.* **30**, 3991 (1965).
Marvell, E. N., Gosink, T., Churchley, P., and Li, T. H., *J. Org. Chem.* **37**, 2989 (1972).
Matsumoto, S., Fukui, H., Imanaka, T., and Teranishi, S., *Nippon Kagaku Kaishi* No. 8, p. 1527 (1972); *Chem. Abs.* **77**, 128572v (1972).
Maxwell, C. E., U.S. Patent 3,522,192, July 28, 1970.
Meyer, A. I., Nabeya, A., Adickes, H. W., Politzer, I. R., Malone, G. R., Kovelesky, A. C., Nolen, R. L., and Portnoy, R. C., *J. Org. Chem.* **38**, 36 (1973).
Milas, N. A., Davis, P., Belié, I., and Fles, D. A., *J. Am. Chem. Soc.* **72**, 4844 (1950).
Mildner, P., and Weedon, B. C. L., *J. Chem. Soc.* p. 3294 (1953).

Mondon, A., *Justus Liebigs Ann. Chem.* **577**, 181 (1952).

Mori, K., Tominaga, M., and Matsui, M., *Tetrahedron* **31**, 1846 (1975).

Morris, S. G., Herb, S. F., Magidman, P., and Luddy, F. E., *J. Am. Oil Chem. Soc.* **49**, 92 (1972).

Morrow, D. F., Butler, M. E., and Huang, E. C. Y., *J. Org. Chem.* **30**, 579 (1965).

Naidin, V. A., Zakumbaeva, G. D., and Sokol'skii, D. V., *Kinet. Katal.* **11**(4), 1072 (1970); *Chem. Abstr.* **73**, 113329g (1970).

Naidin, V. A., Zakumbaeva, G. D., and Sokol'skii, D. V., *Katal. Reakts. Zhidk. Faze* p. 417 (1972); *Chem. Abstr.* **79**, 115042p (1973).

Newman, H., *J. Org. Chem.* **29**, 1461 (1964).

Nogaideli, A. I., and Vardosanidze, T. N., *Zh. Obshch. Khim.* **33**, 379 (1963).

Plate, A. F., and Stanko, V. I., *Izv. Akad. Nauk SSSR, Otdel Khim. Nauk* p. 1481 (1960).

Raphael, R. A., "Acetylene in Organic Synthesis," p. 56. Academic Press, New York, 1955.

Raphael, R. A., and Sondheimer, F., *J. Chem. Soc.* p. 2100 (1950).

Reppe, W., *et al.*, *Justus Liebigs Ann. Chem.* **596**, 25 (1955).

Rieche, A., Grimm, A., and Albrecht, H., *Brennstoff.-Chem.* **42**, 177 (1961).

Roelofs, W. L., Kochansky, J. P., Cardé, R. T., Henrick, C. A., Labovitz, J. N., and Corbin, V. L., *Life Sci.* **17**, 699 (1975).

Rüegg, R., Schwieter, U., Ryser, G., Schudel, P., and Isler, O., *Helv. Chim. Acta* **44**, 994 (1961).

Rylander, P. N., and Cohn, G., *Actes Congr. Int. Catal., 2nd, Paris, 1960* **1**, 977 (1961).

Rylander, P. N., and Himelstein, N., unpublished observations, Engelhard Ind. Res. Lab. Newark, New Jersey, 1964.

Sabirov, S. S., Oripov, S., and Nikitin, V. I., *Dokl. Vses. Konf. Khim. Atsetilena, 4th* **1**, 416 (1972); *Chem. Abstr.* **79**, 17959y (1973).

Sandoval, A., Thomas, G. H., Djerassi, C., Rosenkranz, G., and Sondheimer, F., *J. Am. Chem. Soc.* **77**, 148 (1955).

Scheffer, J. R., and Wostradowski, R. A., *J. Org. Chem.* **37**, 4317 (1972).

Schiess, P., and Chia, H. L., *Helv. Chim. Acta* **53**, 485 (1970).

Schiess, P., Seeger, R., and Suter, C., *Helv. Chim. Acta* **53**, 1713 (1970).

Schinz, H., *Kosmet.-Parfum-Drogen Rundsch.* p. 1 (1955).

Sheads, R. E., and Beroza, M., *J. Agric. Food Chem.* **21**, 751 (1973).

Sheridan, J., *J. Chem. Soc.* p. 470 (1944).

Smith, R. G., Daves, G. D., Jr., and Daterman, G. E., *J. Org. Chem.* **40**, 1593 (1975).

Sokol'skaya, A. M., and Anisimova, N. V., *Vestn. Akad. Nauk Kaz. SSR* **29**(6), 37 (1973); *Chem. Abstr.* **79**, 77959f (1973).

Sokol'skaya, A. M., Bosyakova, E. N., Ryabinina, S. A., and Sokol'skaya, D. V., *Zh. Fiz. Khim.* **46**(11), 2939 (1972); *Chem. Abstr.* **78**, 57586k (1973).

Sokol'skaya, A. M., Fasman, A. B., Shoshenkova, V. A., Lokhmatova, V. F., and Sokol'skii, D. V., *Otkrytiya, Izobret., Prom. Obraztsy, Tovarnye Znaki* **50**(27), 59 (1973); *Chem. Abstr.* **79**, 125799e (1973).

Sokol'skii, D. V., "Hydrogenations in Solutions." Davey, New York, 1964.

Sokol'skii, D. V., Dzhardamalieva, K. K., and Dukhovnaya, T. M., *Russ. J. Phys. Chem.* **43**, 275 (1969).

Sokol'skii, D. V., Kuzora, T. V., and Dzhardamalieva, K. K., *Izv. Akad. Nauk Kaz. SSR, Ser. Khim.* **20**(6), 72 (1970).

Sokol'skii, D. V., Pak, A. M., and Ten, E. I., *Izv. Akad. Nauk Kaz. SSR, Ser. Khim.* **21**(2), 71 (1971); *Chem. Abstr.* **75**, 34877e (1971).

Sokol'skii, D. V., Pak, A. M., and Rozmanova, L. D., *Dokl. Vses. Konf. Khim. Atsetilena, 4th* **3**, 278 (1972).

Sokol'skii, D. V., Dzhardamalieva, K. K., Dukhovnaya, T. M., and Dinasylova, S. D., *Izv. Vyssh. Uchebn. Zaved., Khim. Khim. Tekhnol.* **19**(7), 1063 (1976).

Sondheimer, F., Wolovsky, R., and Amiel, Y., *J. Am. Chem. Soc.* **84**, 274 (1962).

Sonnet, P. E., *J. Org. Chem.* **39**, 3793 (1974).

Sporka, K., Hanika, J., Ruzicka, V., and Vostry, B., *Collect. Czech. Chem. Commun.* **37**(1), 52 (1972); *Chem. Abstr.* **76**, 112299e (1972).

Staab, H. A., and Ipaktschi, J., *Chem. Ber.* **104**, 1170 (1971).

Stork, G., and Colvin, E., *J. Am. Chem. Soc.* **93**, 2080 (1971).

Su, H. C. F., and Manhany, P. G., *J. Econ. Entomol.* **67**, 319 (1974).

Su, H. C. F., Manhany, P. G., and Brady, U. E., *J. Econ. Entomol.* **66**, 845 (1973).

Surber, W., Theus, V., Colombi, L., and Schinz, H., *Helv. Chim. Acta* **39**, 1299 (1956).

Tai, A., Matsumura, F., and Coppel, H. C., *J. Org. Chem.* **34**, 2180 (1969).

Takei, S., and Ono, M., *Nippon Nogei Kagaku Kaishi* **18**, 119, 625 (1942).

Tamaki, Y., Noguchi, H., and Yushima, T., *Appl. Entomol. Zool.* **8**, 200 (1973).

Tedeschi, R. J., *J. Org. Chem.* **27**, 2398 (1962).

Tedeschi, R. J., and Clark, G., Jr., *J. Org. Chem.* **27**, 4232 (1962).

Tedeschi, R. J., McMahon, H. C., and Pawlak, M. S., *Ann. N.Y. Acad. Sci.* **145**, 91 (1967).

Theus, V., and Schinz, H., *Helv. Chim. Acta* **39**, 1290 (1956).

Theus, V., Surber, W., Colombi, L., and Schinz, H., *Helv. Chim. Acta* **38**, 239 (1955).

Tumlinson, J. H., Hendricks, D. E., Mitchell, E. R., Doolittle, R. E., and Brennan, M. M., *J. Chem. Ecol.* **1**, 203 (1975).

Vasil'ev, G. S., Mikos, E. P., Petrov, V. N., Polkovnikov, B. D., Shekhtman, R. I., and Prilezhaeva, E. I., *Izv. Akad. Nauk SSSR, Ser. Khim.* No. 5, p. 1079 (1971); *Chem. Abstr.* **75**, 109777h (1971).

Venkataramani, P. S., John, J. P., Ramakrishnan, V. T., and Swaminothan, S., *Tetrahedron* **22**, 2021 (1966).

Voerman, S., Minks, A. K., and Houx, N. W. H., *Environ. Entomol.* **3**, 701 (1974).

Yagudeev, T. A., and Azerbaev, T. G., *Tr. Inst. Khim. Nefti Prir. Solei, Akad. Nauk Kaz. SSR* No. 4, p. 159 (1972).

Yoshida, N., and Hirota, K., *Bull. Chem. Soc. Jpn.* **48**, 184 (1975).

Zakarina, N. A., Curina, D. K., Zakumbaeva, G. D., and Sokol'skii, D. V., *Dokl. Vses. Konf. Khim. Atsetilena, 4th* **3**, 246 (1972); *Chem. Abstr.* **79**, 115040m (1973).

Zal'kind, Y. S., and Iremadze, I., *Zh. Obshch. Khim.* **18**, 1554 (1948).

Zal'kind, Y. S., and Zhuravleva, L., *Zh. Obshch. Khim.* **18**, 984 (1948).

# Chapter 3

# Hydrogenation of Olefins

Catalytic hydrogenation of olefins presents no problem per se. Olefins, except highly hindered ones, are reduced very easily with a variety of catalysts, solvents, and operating conditions. Problems connected with olefin hydrogenation involve some aspect of regio- or stereoselectivity. Catalysts and operating environment often have an important bearing on the outcome of reduction, and emphasis is placed here on means of controlling the course of an olefin hydrogenation through these agencies.

## I. MECHANISM OF HYDROGENATION

The mechanism of olefin saturation is complex, and despite extensive studies many of the finer details are still elusive. It appears, in fact, that the search for unifying principles leads to ever more diversity. The scheme proposed long ago by Horiuti and Polanyi (1934) has undergone many modifications and refinements, but it contains generally accepted elements, which account for two aspects of olefin hydrogenation that are of interest to the synthetic organic chemist, i.e., double-bond migration and cis–trans isomerization.

Hydrogen is dissociatively adsorbed on two catalyst sites,

$$2* + H_2 \rightleftharpoons \underset{\underset{*}{|}}{H} + \underset{\underset{*}{|}}{H}$$

and the olefin is diadsorbed on two adjacent sites,

$$-CH_2-CH{=}CH-CH_2- + 2* \rightleftharpoons -CH_2-\underset{\underset{*}{|}}{CH}-\underset{\underset{*}{|}}{CH}-CH_2-$$

Hydrogen adds to a carbon, leaving a monoadsorbed radical, which is often referred to as a "half-hydrogenated species."

31

$$-CH_2-CH-CH-CH_2- + H \rightleftharpoons -CH_2-CH-CH_2-CH_2-$$

An important point is that these reactions are reversible. If the reverse steps are identical to the forward steps, the original olefin will be regenerated. However, if the monoadsorbed species first undergoes a configurational change, cis–trans isomerization may occur, returning the double bond to its original position, but perhaps isomerized. Reversal of the monoadsorbed species to form a different diadsorbed species may also occur, resulting in double-bond migration.

$$-CH_2-CH-CH_2-CH_2- + 2* \rightleftharpoons -CH-CH-CH_2-CH_2- + H$$

$$-CH-CH-CH_2-CH_2 \rightleftharpoons -CH=CH-CH_2-CH_2- + 2*$$

Cis–trans isomerization may or may not occur as a consequence of double-bond migration, depending on the configuration of the carbon chain at the time of migration.

Catalysts show marked differences in tendencies to promote cis–trans isomerization and double-bond migration during hydrogenation. Catalyst differences can be readily explained in terms of adsorption–desorption equilibria, since any step may be rate or product controlling. Catalysts such as platinum oxide that permit relatively little olefin desorption tend to produce very little isomerization, whereas catalysts such as palladium that bind adsorbed intermediates less strongly permit extensive reversal and consequently produce extensive isomerization (Siegel, 1966). In these differences also lie the explanation for the frequent observation that platinum produces a kinetically controlled product, whereas palladium tends to produce a thermodynamically controlled product (Jardine and McQuillin, 1968).

Hydrogen is consumed in saturation but not in isomerization. It is to be expected, therefore, that an increase in hydrogen pressure, and consequently an increase in hydrogen surface concentration, should increase the rate of saturation relative to the rate of isomerization. This is usually found to be the case; isomerization relative to hydrogenation is favored by low hydrogen pressures and impeded by higher pressures, a point of considerable practical consequence. The effect of pressure on reaction products has proved to be a powerful tool for probing mechanistic detail (Siegel and Ohrt, 1972).

The Horiuti–Polanyi mechanism is used heuristically. Other mechanisms can accommodate the same facts (Bond and Wells, 1964; Siegel, 1966; Burwell, 1969; Smith and Desai, 1973; Thomson and Webb, 1976). Most mechanisms have in common a stepwise addition of hydrogen and a series of reversible elemental steps. Some workers have evidence for addition of hydrogen to the double bond on the side of the molecule opposite the catalyst (Farina *et al.*, 1976), whereas others (Smith and Menon, 1969; Pecque and Maurel, 1970) have evidence for a trans addition in which one hydrogen atom adds to one side of the double bond and one atom to the other.

## II.  CATALYSTS

Olefins usually are reduced very easily, and many heterogeneous and homogeneous catalysts have been used. The intrinsic rate of hydrogenation over palladium, platinum, rhodium, or nickel is often so high that unless precautions are taken the rate will be determined largely by mass transport (Albright, 1970). Supported palladium or various nickel catalysts are often used.

Catalysts differ greatly, however, and in compounds containing more than one function the yield may depend to a large extent on the catalyst. Catalysts differ in their ability to promote double-bond migration and cis–trans isomerization, in their thermodynamic and mechanistic selectivities in diene hydrogenation, in their tendencies toward 1,2 3,4, or 1,4 addition, and in their activity toward other functions present in the molecule. Hydrogenation of *cis*-penta-1, 3-diene over cobalt affords 90% yields of *trans*-pent-2-ene, whereas over copper 70% pent-1-ene is obtained (Wells and Wilson, 1970).

Platinum, as noted later, is particularly useful when double-bond migration is to be avoided. It is useful when reductions fail to go or to go to completion, but the comparison often made is of supported palladium versus platinum oxide, with the amount of metal used much higher when the latter catalyst is used. The comparison is hardly a fair one (Larrabee and Craig, 1951; Cope and Campbell, 1952; Woodward *et al.*, 1952; Kidwai and Devasia, 1962; Tallent, 1964; Cheung *et al.*, 1968; Kupchan *et al.*, 1969; Bahner *et al.*, 1970). In some cases prereduced rhodium–platinum (3:1) oxide is much more effective than platinum alone (Murai *et al.*, 1973). The support may also have an influence on the depth of reduction. Hydrogenation of the acetate of $\Delta^{16}$-22-ketocholesterol over 10% palladium-on-carbon in ethyl acetate saturated both double bonds, whereas over 10% palladium-on-calcium carbonate the 16–17 double bond was selectively reduced (Chaudhuri *et al.*, 1969).

Iridium, which is seldom used, has given highly stereospecific reductions of steroids, probably as a consequence of its low isomerization activity (Phillipps, 1963; Gregory *et al.*, 1966). Both rhodium (Newman and Addor, 1955; Tarbell *et al.*, 1961; Roy and Wheeler, 1963; Bonner *et al.*, 1964;

Ham and Coker, 1964) and ruthenium (Cope *et al.*, 1957; Eisenbraun *et al.*, 1960) have proved to be valuable in the hydrogenation of substituted olefins when hydrogenolysis is to be avoided.

Homogeneous catalysts augment heterogeneous catalysts and enlarge the capabilities for reaction control (Lyons *et al.*, 1970; Harmon *et al.*, 1973; Fahey, 1973; Rylander, 1973; Froborg *et al.*, 1975). Homogeneous catalysts are most useful in those areas in which heterogeneous catalysts are in some respect deficient. These areas include reactions in which some aspect of selectivity is involved, reactions that poison heterogeneous catalysts, selective labeling, avoidance of disproportionation, and, of exceptional interest, asymmetric hydrogenation.

Tricarbonyl chromium complexes are useful for 1,4 addition of hydrogen to 1,3-dienes to afford monoenes selectively. With 1,4-dienes, conjugation to the 1,3-diene precedes addition. The complexes are highly stereoselective for trans, trans–conjugated dienes, which are much more rapidly reduced than are *cis, cis-* or *cis,trans*-dienes (Frankel and Little, 1969; Frankel and Butterfield, 1969; Frankel *et al.*, 1969, 1970; Frankel, 1970b). A synthesis of citronellol involves 1,4 addition to the conjugated system of the triene **1** in which tricarbonyl chromium–methyl benzoate is the catalyst. Hydration of the resulting diene **2** by hydroboration affords citronellol (**3**) (Hidai *et al.*, 1975).

Many other homogeneous catalysts have been used similarly for selective hydrogenation of dienes (Murahashi *et al.*, 1975).

## III.  EFFECT OF OLEFIN STRUCTURE

A useful, but not infallible, generality is that, in a molecule containing more than one double bond, the least hindered bond will be reduced preferentially (Newhall, 1958; Paquette and Rosen, 1968; Blomquist and Himics, 1968). Haptophilic effects, among other things, may help to render this generality invalid (DePuy and Story, 1960). With steric hindrance about the bonds approximately equal, the more strained bond will be reduced preferentially (Cope *et al.*, 1952; Huffman, 1959; Chapman *et al.*, 1963; Cristol *et al.*, 1965). A generality formulated from the study of terpenes is that exocyclic double bonds are reduced more easily than those in the ring (Smith *et al.*, 1949).

The relative reactivity toward hydrogenation of double bonds has been altered through first complexing the more reactive bond. Nicholas (1975) obtained selective reductions opposite to those usually achieved by first complexing the more reactive double bond with $C_5H_5Fe(CO)_2^+$ (denoted by $F_p^+$) and then hydrogenating the less reactive bond. The free olefin is conveniently regenerated by treatment with sodium iodide in acetone. For instance,

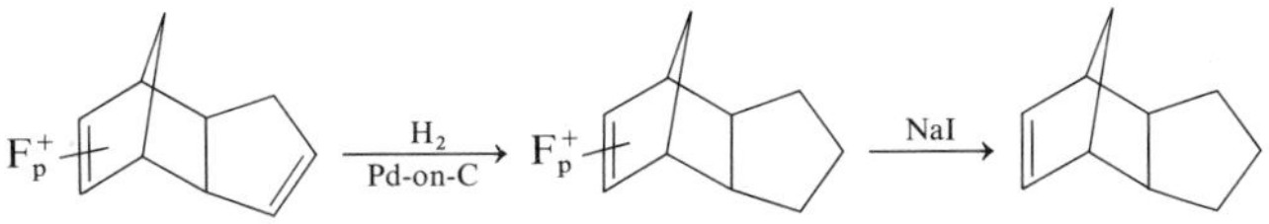

whereas normally the isomeric product is formed:

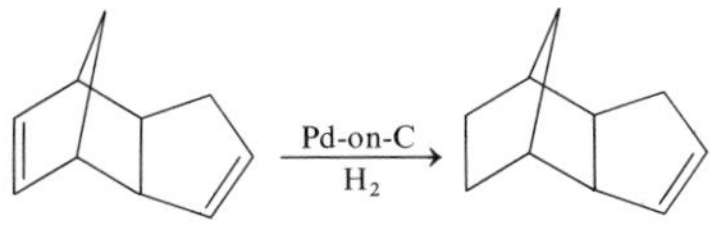

## Allenes and Cumulenes

The selectivity of reduction of an allene to a monoolefin varies appreciably with the metal. For allene itself, selectivity declined in the series palladium > rhodium $\simeq$ platinum > ruthenium > osmium $\gg$ iridium (Bond and Wells, 1964). Allenes with terminal double bonds are selectively reduced in the terminal position (Eglinton *et al.*, 1954). Interesting use was made of this fact in the synthesis of the antibiotic phosphonomycin (**5**) through a highly selective and stereospecific hydrogenation of di-*tert*-butyl propadienylphosphonate (**4**) (Glamkowski *et al.*, 1970). Similar reductions have employed 5% palladium-on-calcium carbonate in alcoholic solution (Petrov *et al.*, 1968).

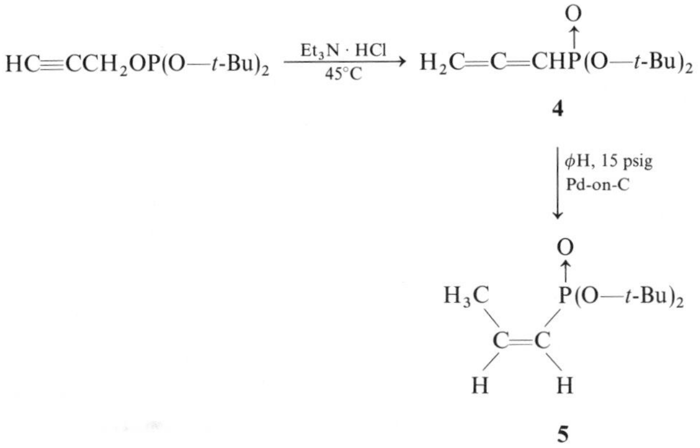

Tetraenes add hydrogen to the central bond. The *cis*-cumulene **6** was smoothly reduced to the *cis,cis*-1,3-butadiene **7** over palladium-on-calcium carbonate in chloroform. A small amount of the trans,trans isomer was also formed (Westmijze *et al.*, 1975). Similar results were obtained in the reduction of the corresponding tetraphenyl derivative (Kuhn and Fischer, 1960) over palladium–lead-on-calcium carbonate (Lindlar, 1952).

Hydrogenation of the allene **8** over Lindlar catalyst gave mainly the cis,trans compound **10**, whereas over palladium-on-kieselguhr a 3:2 mixture of cis,cis and cis,trans isomers formed. The authors suggested that the basic Lindlar catalyst first isomerized the allene into the *trans*-acetylene **9**, which then was reduced to the cis,trans compound **10** (Fischer and Fischer, 1969).

Stereospecific and stereoselective hydrogenations of allenes have been achieved by the use of chlorotris(triphenylphosphine) rhodium in benzene under ambient conditions (Bhagwat and Devaprabhakara, 1972).

## IV. DOUBLE-BOND MIGRATION

Double-bond migrations during the hydrogenation of olefins are probably very common but, unless tracers are used (Cookson *et al.*, 1962) or special products result or asymmetry is lost (Huntsman *et al.*, 1963; Smith *et al.*, 1973) or the new bond is resistant to hydrogenation, no evidence for the migration remains on completion of the reduction. The extent of double-bond migration and the amount of cis–trans isomerization are of major importance in determining the properties of margarines and shortenings produced by partial hydrogenation of natural oils (Bailey, 1951; Rylander, 1970; Frankel, 1970a). Isomerization also has a major influence on many other hydrogenations and is frequently the key to success or failure (Rylander, 1971). In conjugated systems, double-bond migration may occur as a result of 1,4 or 1,6 addition of hydrogen (Garland *et al.*, 1976).

The extent of isomerization highly depends on the substrate, and certain conditions must be met for double-bond migrations to occur (Bream *et al.*, 1957). In order to be removed the allylic hydrogen must be sterically accessible to the catalyst and on the same side of the molecules as the entering hydrogen (Fischer *et al.*, 1968). Carboxyl groups conjugated to double bonds seem to reduce greatly the extent of double-bond migration (Smith and Roth, 1965).

## A.   Catalysts

Catalysts differ widely in the extent to which they affect double-bond migration. The usual measurement of this tendency, and the point of interest, is not the rate of this reaction per se, but its rate relative to the rate of hydrogenation. Table I, compiled from studies of low molecular weight olefins, shows how widely metals differ in their tendency to promote migration and how this relative tendency varies with olefin structure. The authors (Gostunskaya *et al.*, 1967) noted that the order for double-bond migration (palladium > nickel ≫ rhodium ~ ruthenium > osmium ~ iridium ~ platinum) is the same as the order for the ability of the metals to dissolve

TABLE I

**Effect of Metal on Selectivity Coefficient**

| Metal | Selectivity coefficient[a] | | | |
|---|---|---|---|---|
| | 1-Hexene | 2-Methyl-1-pentene | 3-Methyl-1-pentene | 2,3-Dimethyl-1-butene |
| Ni | 1.45 | 0.69 | 0.25 | 0.13 |
| Ru | 0.41 | 0.63 | 0.06 | 0.11 |
| Rh | 0.49 | 1.1 | 0.06 | 0.13 |
| Pd | 1.7 | 12.7 | 1.1 | 5.3 |
| Os | 0.19 | 0.09 | — | — |
| Ir | 0.1 | 0.1 | — | — |
| Pt | 0.12 | 0.13 | — | — |

[a] Selectivity coefficient of isomerization is equal to percent isomerization divided by percent hydrogenation at 40°C ethanol solvent.

hydrogen. Essentially the same ordering of these metals was deduced from a large number of experiments on small olefins, and the observation was made that ordering is unchanged by support or solvent (Bond and Rank, 1965; Bond and Wells, 1964). Nickel boride catalysts formed by reduction of nickel salts with sodium borohydride in aqueous solution (Brown, 1970) or in ethanol (Brown and Ahuja, 1973), designated P1 and P2, respectively, give considerably less double-bond migration than does Raney nickel.

Rhodium is unique among metals in that the rate of isomerization relative to the rate of hydrogenation is affected markedly by temperature. At low temperatures the isomerization is low and rhodium resembles platinum, whereas at temperatures above 80°C isomerization is high and rhodium resembles palladium (Bond and Wells, 1964).

## B.  Conditions

The relative rates of isomerization and hydrogenation may be altered by the reaction environment as well as by the catalyst. Isomerization is favored by "hydrogen-poor" catalysts (catalysts operating under conditions in which the rate of hydrogen consumption is limited by the rate of mass transport of hydrogen to the catalyst) and retarded by "hydrogen-rich" catalysts (catalysts operating under conditions in which the rate does not depend on mass transport of hydrogen to the catalyst) (Smith, 1966). Isomerization is therefore impeded by conditions that increase hydrogen availability at the catalyst surface (i.e., increased pressure, increased agitation, lower catalyst loadings, lower metal loadings, and various additives such as alkalis, amines, and heavy metals (which decrease the rate of hydrogen consumption), and perhaps also by solvents, which lower the surface tension and permit a more rapid transport of hydrogen from the gas phase to the liquid (Yao and Emmett, 1959). Conversely, isomerization is favored by low pressure, poor agitation, high catalyst loadings, high metal loadings, and catalysts of high intrinsic activity. It has been suggested that the effect of additives in inhibiting isomerization is due to the blocking of catalyst sites with enhanced activity for migration (Huntsman et al., 1963). In competitive hydrogenation of diolefins and monoolefins, isomerization of the latter is prevented by the use of carbon monoxide in the hydrogen gas (Xomatsu et al., 1972).

## V.   EFFECTS OF DOUBLE-BOND MIGRATION

Double-bond migration can influence the results of a catalytic hydrogenation in a variety of ways. Selected examples are given as follows.

## A.  Migration to an Inaccessible Position

The migration of double bonds in complex molecules often leads to compounds in which the isomerized olefin can be reduced only with difficulty or not at all. Examples are the hydrogenation of pulchellin (Herz *et al.*, 1963), of mexicanol (Connolly *et al.*, 1967), of estafiatone (Sanchez–Viesca and Romo, 1963), of flexuosin A over platinum oxide in acetic acid (Herz *et al.*, 1964), of azepine esters over platinum in cyclohexane (Anderson and Johnson, 1964), of multiflorenol over platinum oxide in acetic acid (Sengupta and Khastgir, 1963), and of quinine (Cheung *et al.*, 1968). When isomerization is to be avoided, palladium should also be avoided, and a metal of low isomerizing power, such as platinum oxide, should be used instead, but as noted in the above examples the driving force toward isomerization is so great in some compounds that even platinum fails (Bankar and Kulkarni, 1973). Ruthenium dioxide at elevated pressure gave good results in cases in which platinum oxide caused isomerization (deVivar *et al.*, 1966). In some cases homogeneous catalysts proved to be satisfactory when heterogeneous catalysts produced excessive isomerization, as, for example, in the hydrogenation of coronopilin (Rüesch and Mabry, 1969), psilostachyine and confertiflorin (Biellman and Jung, 1968), and seychellene (Piers *et al.*, 1969).

## B.  Oxidation of Alcohols

The hydroxyl function of olefinic alcohols may become oxidized during hydrogenation as a result of double-bond migration. Reduction of cyclohexen-2-ol over palladium affords a mixture of about 67% cyclohexanol and 33% cyclohexanone; over platinum only cyclohexanol is formed (Rylander and Himelstein, 1964). The same sort of change can be effected even when the alcohol is far removed from the double bond, as in the reduction of 5-cyclodecen-1-ol. However, in this case, the olefin is in spatial proximity to the alcohol, making a hydrogen transfer reaction possible as well (Cope *et al.*, 1955). When saturation relative to migration is inhibited sufficiently, isomerization acquires synthetic utility (Herz *et al.*, 1968; Barnes and MacMillan, 1967). Allylic migrations of sterically impeded olefinic alcohols resulting in a saturated carbonyl compound are well documented (McQuillin, 1963).

## C.  Catalyst Inhibition

One cause of catalyst inhibition is the accumulation of strongly adsorbed products, which are formed as a result of the hydrogenation. An example is the inhibition of hydrogenation of benzyl alcohol to cyclohexylcarbinol through the formation of by-product cyclohexanecarboxaldehyde. The aldehyde arises through isomerization of intermediate 1-cyclohexenylcarbinol. Unexpectedly, platinum was better than rhodium in this reduction. The intermediate unsaturated carbinol was isomerized less over platinum than over rhodium, and the aldehyde, once formed, was hydrogenated about five times more rapidly over platinum (Nishimura and Hama, 1966).*

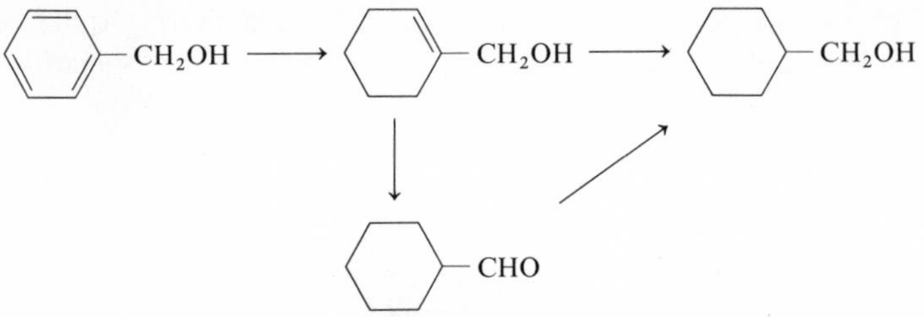

In this sequence of events can be seen a reason for the frequent occurrence of synergism. When two catalysts are used together, the chances of removing an inhibiting species increases.

## D.  Stereochemistry

Many factors affect the stereochemistry of olefin hydrogenation. One of them is the extent of isomerization before saturation. For instance, hydrogenation of 1,2-dimethylcyclohexene over platinum affords mainly *cis*-1,2-dimethylcyclohexane, whereas over palladium-on-alumina in acetic acid 73% of the *trans*-cycloparaffin forms (Siegel and Smith, 1960a,b). Similarly, hydrogenation of the isopropenyl function in the sesquiterpene alcohol paradisiol (11) over palladium-on-carbon or platinum-on-carbon gave different dihydro compounds depending on the metal used. Inversion occurred over palladium through migration of the double bond before saturation (Sulser *et al.*, 1971). Isomerization of axial isopropenyl groups during hydrogenation is well documented (Bream *et al.*, 1957).

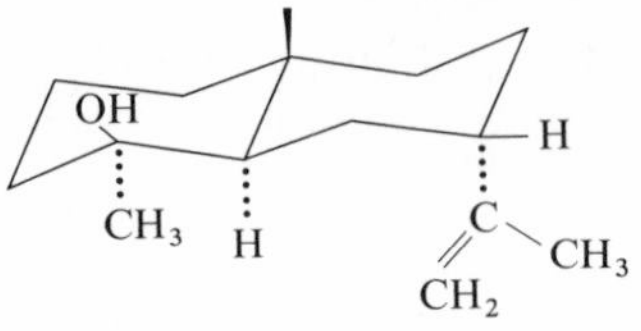

11

---

* Some investigators eschew the use of methanol and ethanol as solvents when using rhodium to eliminate the possibility of inhibition by aldehydes produced through oxidative dehydrogenation.

It is not always easy to decide in advance the extent to which isomerization will occur, its direction, or how much it will affect the stereochemistry. Nonetheless, many examples in the literature attest to the marked influence of catalysts on stereochemistry, an influence that often parallels the isomerizing power of the metal. In the hydrogenation of various disubstituted cycloolefinic compounds the order of increasing isomerization activity and increasing trans content of the product was Ir < Os < Pt, Ru, Rh ≪ Pd (Nishimura et al., 1973b). With Raney nickel, the isomer ratio varied widely with the age of the catalyst; newer catalysts tended to produce more isomerization (Tyman and Wilkins, 1973).

## E. Products

Isomerization of double bonds before saturation may permit a variety of reductive reactions that are not possible in the original compound (Canonica et al., 1969). For instance, the oxygen atoms of hept-3-yne-1,7-diol are not activated toward hydrogenolysis, yet reduction over palladium gives a mixture of saturated diol and n-heptanol. The latter compound is accounted for best by the assumption of a double-bond migration in the intermediate olefin to an allylic position, followed by hydrogenolysis and saturation (Crombie and Jacklin, 1957).

A striking example of the effect of isomerization is found in the hydrogenation of car-3-ene, in which two different products are formed in essentially quantitative yields depending on the catalyst used. Platinum, which has a low activity for isomerization, affords carane, whereas over palladium the product is 1,1,4-trimethylcycloheptane (Cocker et al., 1966).

Ring opening of conjugated cyclopropyl systems, as if by 1,4 addition, occurs frequently over palladium. For example, reduction of (−)-thujopsene over 5% palladium-on-carbon at 30°C affords dihydrothujopsene in about 95% yield (Hochstetler, 1972).

Similarly, 1,4 addition followed by isomerization accounts for ring opening of 12 and loss of an acetoxy group (13) not originally activated. About 40% of 14 is formed as well (Zelnik et al., 1977).

12                                   13                                   14

## F.   Disproportionation

Disproportionation of olefins can be considered an intermolecular double-bond migration. It may occur readily in incipient aromatic systems with or without hydrogen present. Because of this, hydrogenation of certain olefins may be incomplete due to the formation of aromatic systems resistant to reduction (House and Rasmusson, 1963). An order of effectiveness for disproportionation of dialkylcyclohexenes is $Pd \gg Pt > Rh$ (Hussey et al., 1968). Palladium also proved to be much more effective than rhodium, ruthenium, or platinum in disproportionation of dihydrophthalic acids (Cerefice and Fields, 1974). Palladium appears to be the catalyst of choice when disproportionation is to be achieved and, conversely, palladium should be avoided when disproportionation is apt to be troublesome.


## VI.   CIS-TRANS ISOMERIZATION

Geometric isomerization occurs readily over noble metal catalysts and especially so if hydrogen is present. Isomerization may occur without a shift in the position of the double bond, or it may occur as a consequence of migration. One might reasonably expect a parallel relationship in catalyst activity for isomerization and migration, and indeed the same ordering of metals, $Pt < Ir < Ru < Rh < Pd$, has been found for both (Zajcew, 1960a; Riesz and Weber, 1964). Cis–trans isomerization activity, like double-bond migration activity, has been shown to be related to process variables in a way that can be correlated with hydrogen availability at the catalyst surface (Rylander, 1970); the greater the deficiency of hydrogen at the catalyst surface, the more hydrogenation will be retarded relative to isomerization (Zajcew, 1960b). Refluxing alcohol provides a convenient, limited source of hydrogen for isomerization.

In molecules not capable of double-bond migration, the ordering of catalysts for geometric isomerization may differ from that given above. For *cis*-stilbene the order of increasing isomerization activity in the presence of hydrogen is Ru < Pt ≪ Pd ≪ Rh (Bellinzona and Bettinetti, 1960). Activity ordering may depend on whether hydrogen is present. Isomerization is achieved easily by heating the substrate with or without solvent in the presence of a small amount of catalyst (Puterbaugh and Newman, 1959; Williams *et al.*, 1961). Nickel–boron catalysts (Brown and Brown, 1963) have very low activity for cis–trans isomerization (Bell *et al.*, 1967).

Smith and Menon (1969), in a study of the isomerization of *cis*-cyclododecene to the trans isomer during hydrogenation, discovered a remarkable effect of carbon tetrachloride solvent on the rate ratio of these reactions. Over palladium-on-carbon saturation is strongly inhibited relative to isomerization, whereas over platinum-on-carbon the reverse is true, and saturation is hurried relative to isomerization. It is not known if this phenomenon is general.

## VII.  STEREOCHEMISTRY

Correctly predicting the stereochemical outcome of olefin hydrogenation is not an easy task. The usual assumptions in making a prediction are that the olefin will adsorb on the catalyst is such a way as to minimize steric interactions between the substrate and catalyst and that the product will arise by cis addition of hydrogen to the side of the molecule adsorbed on the catalyst. This simple scheme often adequately accounts for the stereochemistry of the major product (Siegel, 1966). There are, however, various complications. It is not always easy to decide which side of the molecule offers the least hindrance to approach of the catalyst and, as Siegel and Cozort (1975) pointed out, torsional strains and intramolecular nonbonding interactions (Marshall *et al.*, 1970) distort the geometry of the adsorbed species. Furthermore, adsorption is only one step in the reaction sequence (Horiuti and Polanyi, 1934; Burwell, 1969; Mitsui *et al.*, 1972) and need not be product controlling. Isomerization of the olefin frequently occurs during hydrogenation, and the major product is the resultant of a number of competing reactions. Experimentally, stereochemistry has been found to vary with the purity of the olefin, with the reaction conditions (Siegel *et al.*, 1975), and, sometimes to a marked extent, with the catalyst (Hussey *et al.*, 1968; Mutsui *et al.*, 1973b).

Some of the above points are illustrated by the results of hydrogenation of 2-methylcyclopentylidenecyclopentane (Table II). Classic theory would suggest that this molecule should be adsorbed on the catalyst in such a way as to afford mainly the *cis*-paraffin. However, Siegel and Cozort (1975) anticipated that repulsive interactions between the 2,2′ and 5,5′ ring positions

TABLE II[a]

| Catalyst | % Cis |
|----------|-------|
| 5% Pt-on-C | 13 |
| 5% Pd-on-C | 13 |
| 5% Rh-on-C | 45 |
| 5% Ru-on-Al$_2$O$_3$ | 45 (at 45% completion) |

[a] Data of Siegel and Cozort (1975). Used with permission.

would be greater in the transition state leading to the cis isomer than in the transition state leading to the trans isomer. This expection was fulfilled; over platinum oxide, a catalyst ordinarily exhibiting high stereospecificity, the product contained about 78% trans isomer. The catalyst also has a pronounced effect on this stereochemistry (Table II).

Siegel and Cozort (1975) emphasize that stereochemistry also depends on the purity of the olefin. Olefins that have stood for several days give substantially different results than do freshly prepared olefins, presumably because of the formation of peroxides. They also attribute various anomalies and contradictions in the literature to the use of impure olefins.

## A. Haptophilicity

Haptophilicity refers to the propensity of a functional group to bind to the catalyst surface during hydrogenation in such a way as to enforce addition of hydrogen from its own side of the molecule. This has also been referred to as an "anchor effect." An interesting generality has been made (Thompson et al., 1976) concerning the influence of solvent on the stereochemistry of reduction of compounds displaying haptophilicity (Thompson and Naipawer, 1973): To the extent that haptophilic effects operate, the extremes of stereospecificity are likely to be found with extremes of solvent dielectric constant. The effect is assumed to operate through competition of solvent with the haptophilic group for catalyst sites. The effect is illustrated in Table III; palladium and platinum parallel each other, with the latter always giving more cis isomer than palladium.

Haptophilicity of the hydroxymethyl group can be enhanced by conversion to an anion and further controlled by the choice of cation (Thompson et al., 1976). The technique is applicable to both heterogeneous and homo-

**TABLE III**[a]

|                | | Relative amounts in product | |
| Solvent | Dielectric constant | Cis | Trans |
| --- | --- | --- | --- |
| Hexane | 1.9 | 39 | 61 |
| Dioxane | 2.2 | 74 | 26 |
| $Bu_2O$ | 3.1 | 73.5 | 26.5 |
| Diglyme | 7.2 | 81.5 | 18.5 |
| Tetrahydrofuran | 7.6 | 82 | 18 |
| EtOH | 24.6 | 94 | 6 |
| Dimethylformamide | 36.7 | 94 | 6 |

[a] Data of Thompson *et al.* (1976). Used with permission.

geneous catalysts. Thompson and McPherson (1974) markedly increased the activity of $(\phi_3 P)_3 RhCl$ toward hydrogenation of **15** by coordination of the catalyst with an anionic species. The cis isomer **16** was formed exclusively in contrast to mixtures, which result from heterogeneous catalysis.

An interesting, but not yet successful, attempt has been made to relate haptophilicity with steric and electronic properties of a variety of substituents

(Thompson, 1971, 1973). The problem is complicated, for other effects may be operating as well. Haptophilicity depends on the catalyst as well as the functional group. The point is illustrated by hydrogenation of the allylic alcohol, 2-butylidenecyclopentanol, over various catalysts (Table IV). Inspection of molecular models suggests that either side of the double bond can be presented in an equally planar conformation to the catalyst, yet one-sided addition of hydrogen occurs overwhelmingly, affording mainly *trans*-2-butylcyclopentanol. The result suggests that the substrate is adsorbed by interaction of the oxygen electron pairs as well as $\pi$ electrons of the double bond with the catalyst. Isomerization of the allylic alcohol to 2-butylcyclopentanone is an important side reaction, especially over palladium (Sehgal *et al.*, 1975).

Ring size has an important influence on the stereochemistry, affecting both the isomer ratio and the ordering of catalysts. The problems of predicting stereochemistry have been discussed in some detail (Sehgal *et al.*, 1975). Here suffice it to note that large variations in isomer ratios can be achieved through changes in catalyst.

Haptophilic effects have been recorded by a number of investigators (Minckler *et al.*, 1956; Haynes and Timmons, 1958; Halsall *et al.*, 1959; Dart and Henbest, 1960; Dauben *et al.*, 1961; Howard, 1963; Nishimura and Mori, 1963; Mitsui *et al.*, 1966; Minyard *et al.*, 1968; Watanabe *et al.*, 1968; Micheli *et al.*, 1969; Beak and Chaffin, 1970; McMurry, 1970; Mori *et al.*, 1971; Stolow and Sachdev, 1971; Powell *et al.*, 1974; Warawa and Campbell, 1974). Haptophilicity of hydroxyl functions is often of importance in determining the stereochemistry of steroid hydrogenation (Augustine, 1972).

TABLE IV[a]

| Catalyst | % of alcohol | | % Ketone |
|---|---|---|---|
| | Trans | Cis | |
| Raney nickel $W_3$ | 91 | 7 | 2 |
| $Ni_2B$ P1 | 73 | 27 | — |
| Ru-on-C | 69 | 31 | 2 |
| Pt-on-C | 55 | 45 | 5 |
| Rh-on-C | 64 | 36 | 12 |
| Pd-on-C | 66 | 34 | 60 |

[a] Data of Sehgal *et al.* (1975). Used with permission.

## B.   Exo Addition

Addition of hydrogen to certain bridged polycyclic systems such as bicyclo[2.2.1]hept-2-enes (Alder et al., 1936; Alder and Roth, 1954; Cristol and LaLonde, 1959), dicyclopentadienes (Brown et al., 1972), benzonorbornadienes (Martin and Koster, 1968), or α-pinene (Eigenmann and Arnold, 1959) follows an exo addition rule to give endo substituents. The stereochemistry can be influenced markedly by reaction temperature, as illustrated in Table V by selected data on reduction of (+)-α-pinene over palladium-on-carbon in propionic acid (Cocker et al., 1966). More cis isomer forms over platinum than over palladium. The temperature effect has been interpreted by Mitsui et al. (1973a) in terms of increasing temperature causing a decrease in hydrogen availability at the catalyst with its consequent influence on the reversibility of the elemental steps.

An interesting effect of substituents has been reported by Baird and Surridge (1972). Exclusive exo–cis addition of hydrogen occurs with 7-substituted norbornenes when the substituent is small. However, with syn-7-tert-butylnorbornene hydrogenation occurs 80% endo. When the tert-butyl group is anti, there is still 25% endo addition even though no direct steric or electronic interaction is evident. Endo attack is thought to be encouraged by the development of repulsive interactions between the anti-7-tert-butyl group and the exo–cis 5,6-hydrogens in the transition state. Reduction in these systems is sensitive to the oxidation state of the catalyst (Baird et al., 1969).

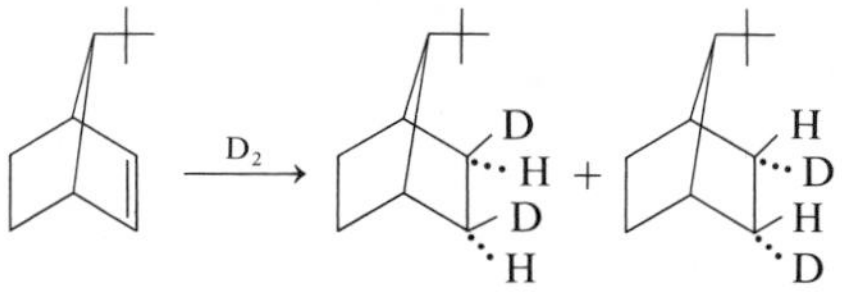

20%          80%

**TABLE V**

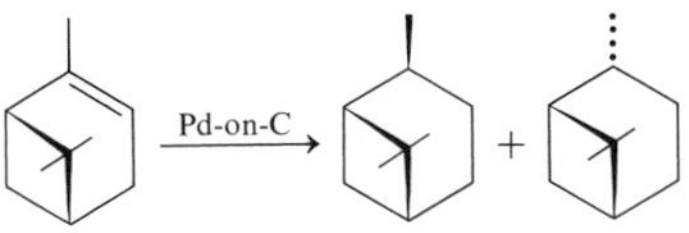

| Temperature (°C) | % in product | |
| --- | --- | --- |
| | cis | trans |
| 0 | 80 | 20 |
| 20 | 73.5 | 26.5 |
| 89 | 55.5 | 44.5 |
| 138 | 48.5 | 51.5 |

## C.  Asymmetric Hydrogenation

Asymmetric hydrogenation over chiral heterogeneous catalysts has long been known, but despite much effort only partial success has been obtained in developing a practical synthesis. Homogeneous asymmetric hydrogenation, on the other hand, has blossomed in a few years from a laboratory curiosity into commercial reality. Some leading references to this fast-growing area are James (1973), Knowles *et al.* (1972, 1974), and Morrison *et al.* (1975, 1976). Two examples will suffice for illustration. Both examples stress conformational rigidity as a key to high optical yields. The first ligand is an example of a general group carrying a chiral center at the phosphorus atom, and the second is an example of ligands carrying chirality in the carbon moiety. Ligands have been made carrying chiral centers at both phosphorus and carbon (Fisher and Mosher, 1977). Marked solvent effects have been reported with certain ligands (Pracejus and Pracejus, 1977).

Enantiomeric excesses of 96% were achieved in asymmetric hydrogenation of α-acylamidoacrylic acids using a rhodium complex and chiral 1,2-bis-(o-anisylphenylphosphino)ethane as a ligand. The reduction can be carried out with either the free acid or the carboxylate anion, but with this ligand better results are obtained with the anion. The high optical yields are not sensitive to either temperature or pressure, an insensitivity attributed to formation of a rigid, five-membered ring formed between the ligand and rhodium. The methoxy group is thought to contribute to the high selectivity by hydrogen bonding with the amide substrate (Knowles *et al.*, 1975).

The chiral ligand *d-trans*-1,2-bis(diphenylphosphinoxy)cyclopentane **(17)** is especially useful for rhodium-catalyzed asymmetric hydrogenation of unsaturated substrates bearing no functional group, e.g., α-ethylstyrene, which cause a secondary interaction between the substrate and catalyst (Hayashi *et al.*, 1977). The ligand has greater conformational rigidity than the corresponding chiral cyclohexane derivative (Tanaka and Ogata, 1975).

17

Ligands that carry only an axial element of chirality, such as $2,2^1$-bis-(diphenylphosphinomethyl)-$1,1^1$-binaphthyl **(18)** have also been used (Tamao *et al.*, 1977).

**18**

## VIII.   VINYL FUNCTIONS

Vinyl functions can undergo hydrogenolysis as well as hydrogenation, and one might expect the product composition to be sensitive to substrate structure, solvent, and catalyst. Nishimura *et al.* (1971) examined in detail the hydrogenation of ethyl 4-methyl-1-cyclohexenyl ether over all six platinum group metals. In ethanol the product was a mixture of 4-methylcyclohexanone diethyl ketal, 4-methylcyclohexanone, ethyl 4-methylcyclohexyl ether, 4-methylcyclohexanol, and methylcyclohexane in widely varying amounts depending on metal and solvent. The amount of hydrogenolysis to give methylcyclohexane increased in the order Pd $\simeq$ Ru $\ll$ Os $<$ Rh $<$ Ir $\ll$ Pt; hydrogenolysis tended to decrease with solvent in the order ethanol $>$ *tert*-butyl alcohol $>$ isopropyl ether. Ketal formation relative to hydrogenation increased in the order Os $<$ Ru $<$ Ir $\ll$ Rh $<$ Pd $<$ Pt. The relationship between catalyst, solvent, and product composition is complex. Yields of ethyl 4-methylcyclohexyl ether exceeding 96% were obtained by hydrogenation over palladium in ethanol, ruthenium or rhodium in *tert*-butanol, and ruthenium in isopropyl ether. The 31% yield of 4-methylcyclohexanol obtained by hydrogenation over rhodium in anhydrous ethanol is noteworthy. Cis–trans ratios of the ether and alcohol vary widely with the catalyst, ranging from 8.9 with palladium to 3.0 with osmium.

   The stereochemistry of hydrogenation of isomeric ethyl 2-methyl-1-cyclo-hexenyl ether **(19)** is determined more rigidly. Hydrogenation of this compound with platinum group metal catalysts yields the saturated cis isomer **21** with high stereospecificity. The small amount of trans isomer formed is thought to arise from isomerization to ethyl 6-methyl-1-cyclohexenyl ether, which can afford either the cis or trans saturated isomer. Formation of this particular unsaturated isomer is attributed to the directive effect of the ethoxy group in the formation of the half-hydrogenated state **20** (Nishimura and Kano, 1972).

**19**                              **20**                              **21**

   The above data support a generality made earlier (Rylander, 1967), which is that platinum favors hydrogenolysis and palladium favors hydrogenation (Inhoffen *et al.*, 1950; Nesmeyanov *et al.*, 1955; Jacobson *et al.*, 1957; Fajkos, 1958; Smyrniotis *et al.*, 1960), as do ruthenium (Martin *et al.*, 1966) and rhodium (Cookson *et al.*, 1962). These trends in catalyst performance are, of course, only relative; the substrate may be of such a structure that the reduction will follow only one course regardless of catalyst. Ketonic hydroxymethylene compounds may require conversion to a benzoate in order to effect hydrogenolysis (Astill and Boekelheide, 1955; Harnik, 1963).
   Polar solvents or acids (Knox *et al.*, 1961) favor hydrogenolysis of vinyl functions; nonpolar solvents favor hydrogenation. Reduction of **23** over palladium-on-carbon in ethyl acetate gave **24** in 65% yield, whereas in methanol **22** was the major product, regardless of catalyst (Rosenthal and Shudo, 1972).

**22**                              **23**                              **24**

   Vinylogous urethanes, $\beta$-amino-$\alpha,\beta$-unsaturated esters, are hydrogenated with some difficulty in that they are sensitive to hydrogenolysis. Rhodium-on-alumina at 85°C and 500 psig has been used with success (Liska, 1964). Excellent yields of saturated products have also been obtained over 5% palladium-on-carbon at 85°C and 1000–1500 psig in methanol. The reaction is sensitive to temperature; at room temperature there is little reduction, whereas above 120°C hydrogenolysis is extensive (Augustine et al., 1968).


## IX.   UNSATURATED CARBONYL COMPOUNDS

Reduction of unsaturated aliphatic carbonyl compounds to the saturated carbonyl is achieved readily under a variety of conditions, unless the olefinic function is highly hindered. Selective reduction is obtained automatically when palladium is used, for the reduction virtually stops after the absorption of 1 equivalent of hydrogen, except under vigorous conditions (Carson et al., 1970). Nickel can also be an effective catalyst, but it has a greater tendency to reduce the carbonyl function (Ames and Davey, 1956). For example, citronellal was reduced preferentially at the carbon–carbon double bond over palladium-on-barium sulfate, but with nickel-on-kieselguhr the carbonyl was reduced first. Similar differences between metals are described in Chapter 6.

An interesting use of diketones as inhibitors for carbonyl reduction was discovered during a synthesis of 3-substituted 2,4-pentanediones. The latter compounds are conveniently prepared by Knoevengal condensation of acetylacetone with aldehydes followed by hydrogenation.

$$
\begin{array}{ccc}
CH_3 & CH_3 & CH_3 \\
| & | & | \\
C{=}O & C{=}O & C{=}O \\
| & | & | \\
RCHO + CH_2 \xrightarrow{\ \ \text{NH}\ \ } RCH{=}C & \xrightarrow[\substack{70^{\circ}C \\ 100\ \text{atm}}]{\substack{RaNi \\ H_2}} & RCH_2CH \\
| & | & | \\
C{=}O & C{=}O & C{=}O \\
| & | & | \\
CH_3 & CH_3 & CH_3
\end{array}
$$

At higher temperatures, such as 120°C, some deacylation occurs, affording monoketones. These monoketones resist hydrogenation to the alcohol when in the presence of diketones. Addition of $\alpha,\beta$-diketones to hydrogenations of unsaturated ketones can, in fact, improve the selectivity. Only 2-heptanone was formed in the reduction of 4-hepten-2-one in the presence of 3-butyl-2,4-pentanedione; the diketone was recovered unchanged (Uehara et al., 1973).

$$
\underset{\overset{\|}{O}}{CH_3CCH_2CH{=}CHCH_2CH_3} \longrightarrow \underset{\overset{\|}{O}}{CH_3CCH_2CH_2CH_2CH_2CH_3}
$$

Unsaturated ketones can be formed and reduced in a single step to saturated ketones by the use of a bifunctional catalyst (Watanabe *et al.*, 1974; Onoue *et al.*, 1977). Yields of methylisobutyl ketone as high as 95% at 20–40% conversion of acetone were achieved by the use of a palladium-on-zirconium phosphate catalyst in continuous processing at 110°C and pressures greater than 10 atm. The reaction is believed to proceed by the following consecutive steps:

$$2CH_3COCH_3 \rightleftharpoons (CH_3)_2C\!\!=\!\!CHCOCH_3 + H_2O$$

$$(CH_3)_2C\!\!=\!\!CHCOCH_3 + H_2 \longrightarrow (CH_3)_2CHCH_2COCH_3$$

## A.  Stereochemistry

The various factors controlling the sterochemistry of hydrogenation of α,β-unsaturated ketones are interrelated in a complex fashion. Reaction parameters can sometimes be varied to produce marked changes in the stereochemistry of hydrogenation. In other cases, some structural feature of the molecule may so dominate the course of reduction that reaction variables have little effect on the outcome.

Augustine (1976) made an attempt to rationalize the effect of various reaction parameters according to the way in which they affect the relative proportions of 1,2 and 1,4 addition and the reversibility of half-hydrogenated states. The stereochemistry of β-octalone hydrogenation in neutral media is cleanly related to the dielectric constant of the solvent, if the solvents are divided into aprotic and protic groups. When this is done the percentage of *cis-β*-decalone falls steadily with decreasing dielectric constant in aprotic solvent and increases with dielectric constant in protic solvent (Augustine, 1967). The extremes of each of these series are shown in Table VI, illustrating that large changes in stereochemistry can be achieved by use of the appropriate solvent. Similar results are obtained in hydrogenation of cholestenone and testosterone (McQuillin *et al.*, 1963).

These results were accounted for by the assumption that 1,4 addition predominates in polar aprotic solvents, whereas in nonpolar aprotic solvents hydrogenation occurs mainly by 1,2 addition. In protic media, smaller polar solvents promote 1,2 addition, whereas more bulky solvents, which interact less with the carbonyl function, tend to give 1,4 addition.

Acids and bases influence stereochemistry, as many investigators have noted (Augustine, 1958, 1972; Brewster, 1954; Gabbard and Segaloff, 1962; Howe and McQuillin, 1958; Huckel *et al.*, 1958; Lowenthal, 1959; Nishimura *et al.*, 1966; Scott *et al.*, 1972; Slomp *et al.*, 1955; Wilds *et al.*,

**TABLE VI**

| Solvent | Dielectric constant | % cis-β-Decalone |
|---------|--------------------|-----------------|
| Dimethylformamide | 38.0 | 79 |
| n-Hexane | 1.89 | 48 |
| Methanol | 33.6 | 41 |
| tert-Butanol | 10.9 | 91 |

1950). Hydrogenation in basic media is thought to occur through an enolate anion, which adsorbs irreversibly on the catalyst. Hydrogenation proceeds by hydride ion transfer from the catalyst, followed by protonation of the β-carbon. In acid, protonation occurs first, followed by adsorption and hydride ion transfer (Augustine, 1976; this reference gives details of this concept and the views of other investigators). Not all acids are equal. In the hydrogenation of 3-oxo-4-ene steroids, hydrobromic acid gives much greater yields of 4,5-β-dihydro-ketones than does added hydrochloric or sulfuric acid. Also, hydrogenolysis products are completely suppressed in the presence of hydrobromic acid (Nishimura and Shimahara, 1966).

Some success has been achieved in relating stereochemistry to catalyst activity, catalyst concentration, pressure, and agitation and to the effects that these variables have on hydrogen availability at the catalyst surface. Under conditions of low hydrogen availability, a condition that favors reversible adsorption, the product stereochemistry is determined by the relative stability of the cis- and trans-adsorbed species, whereas under conditions of high hydrogen availability the product stereochemistry is determined by the mode of initial adsorption (Augustine, 1967, 1976). Extensive comparisons of catalysts have not been made, but it would appear that platinum, more than palladium, gives products determined by the initial adsorption.

Substrate structure itself has an important influence on the stereochemical outcome of reduction. Often the products are as reasonably expected, but occasionally the results run counter to reason and to precedent. Two examples of this type are given below, with rationalizations for the phenomenon.

Hydrogenation of the β,γ-unsaturated keto ester **26** over either 10% palladium-on-carbon or platinum oxide resulted in a mixture of keto esters in which the major component was cis B/C-ring-fused **25** (Ireland *et al.*, 1969). Formation of the cis isomer contradicts the experience of other investigators (Ghatak *et al.*, 1960; Stork and Schulenberg, 1962; Ogiso and Pelletier, 1967), who obtained trans B/C ring fusion on hydrogenation of similar compounds. Together these results suggest that the outcome of saturation of the 10(10a) double bond in these tricyclic molecules depends on the character of ring C and its substituents. For example, the hydroxy ester in contrast to the keto ester **26** gives the trans B/C isomer **27**. The latter differences were rationalized on the grounds that the preferred conformation for reaction of the keto compound is quasi-boat, whereas the the hydroxy compound is reduced in a quasi-chair conformation (Ireland *et al.*, 1969).

Hydrogenation of the hydrindanone **28** gives largely the cis product

whereas compounds with substitution at C-4 (**30**) tend to give more trans isomer (**31**), with the amount of trans isomer correlating approximately with the size of the substitutent.

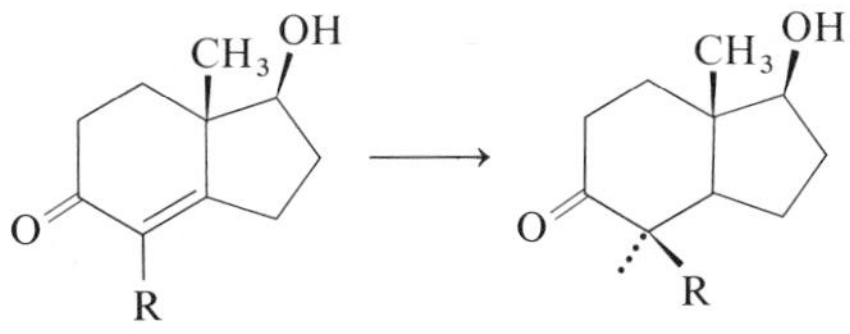

30                                  31

The tendency of the C-4 substitutent to promote trans isomer formation is puzzling, for the least stable ring junction is formed, and the side chain is required to become axial. The phenomenon has been rationalized in terms of the preferred configuration of the adsorbed olefin, which minimizes steric repulsions between the side chain and 3-α-hydrogen atom (McKenzie, 1974). Hydrogenation of the keto acid 32 or keto ester 33 also gives mainly a trans ring junction (34). Intramolecular hydrogen bonding is thought to diminish 1,4 addition, which should lead to preferential cis hydrogenation and increase 1,2 addition as well as favour a half-chair conformation in the hydrogen-bonded structure. Trans ring formation is also favored by the bulky β-oriented C-1 substituent (Hajos and Parrish, 1973).

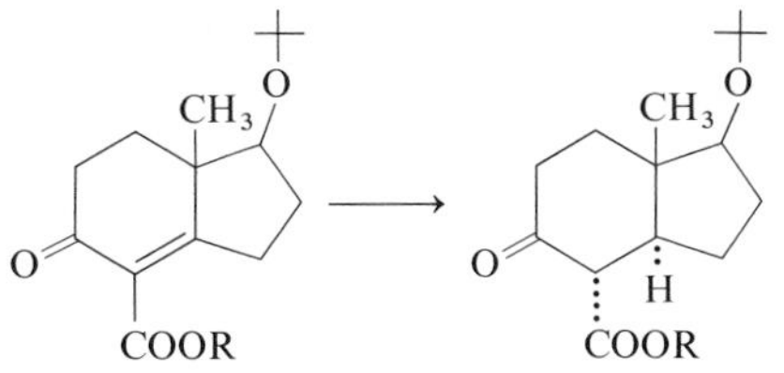

32, R = H
33, R = $CH_3$

34
(93–98%)

## B.   Multiple Unsaturation

Selectivity problems are increased in carbonyl compounds containing multiple unsaturation. Selectivity is sensitive to the catalyst, modifiers, and reaction conditions. For instance, hydrogenation of 3,5-heptadien-2-one or 6-methyl-3,5-heptadien-2-one over nickel-on-alumina or nickel-on-zinc oxide occurs mainly by 5,6 addition, but if the catalyst is modified by the addition of lead (Borunova et al., 1972) or cadmium (Freidlin et al., 1972) mainly 3,4 addition occurs. Silver was ineffective in changing the direction of addition. It is uncertain at present as to how many types of catalysts this technique can be extended. Unpromoted palladium-on-calcium carbonate behaves similarly to unpromoted nickel. Some participation of the carbonyl is suggested by the reduction of the more hindered double bond; very little hydrogenation of this bond occurs in reductions of the corresponding dienol (Miropol'skaya *et al.*, 1962).

$$CH_3CHCH_2CH{=}CHCCH_3 \xleftarrow{\ Ni\ } CH_3C{=}CHCH{=}CHCCH_3 \xrightarrow{\ Ni,\ Pb,\ Cd\ } CH_3C{=}CHCH_2CH_2CCH_3$$

$$R = H, CH_3$$

Selective reduction of the $\gamma,\delta$ double bond in **36** or **37** proceeds quantitatively in methanol containing triethylamine over palladium-on-carbon inhibited by quinoline and sulfur, affording **38**. Lindlar catalyst can also be used, but the results are less satisfactory. Complete quantitative reduction to the saturated aldehyde **35** is obtained over palladium-on-carbon (Traas *et al.*, 1976).

**35**                    **36**, R = H                    **38**
                         **37**, R = CH₃

Selective reduction of conjugated olefins in $\alpha,\beta$-unsaturated carbonyl compounds that also contain an isolated double bond can be achieved in excellent yield by the use of triethylsilane in the presence of catalytic amounts of tris(triphenylphosphine) rhodium chloride. Dihydroionone (**40**) was obtained from $\alpha$-ionone (**39**) in 96% yield by this technique. Similarly, citral was reduced selectively to citronellal in 97% yield (Ojima *et al.*, 1972).

**39**                                                    **40**

Selective hydrogenation of the tetrasubstituted double bond in estra-4,9(10)-dien-17$\beta$-ol-3-one (**41**) to form 10$\beta$-estr-4-en-17$\beta$-ol-3-one (**42**) was unsatifactory over most catalysts; significant quantities of mixed tetrahydro and aromatized steroid formed as well (Debono *et al.*, 1969). However, good results were obtained with either palladium-on-strontium carbonate or palladium-on-barium sulfate in benzene. As a general observation, these supports together with calcium carbonate are often effective in improving selectivity. Frequently, selective reductions of dienones are profoundly affected by solvent (Woodward *et al.*, 1952; Shepherd *et al.*,

1955; Garrett *et al.*, 1956; Nickon and Bagli, 1961; Crandall and Paulson, 1968).

       **41**                              **42**

Homogeneous catalysts such as chlorotris(triphenylphosphine) rhodium (Birch and Walker, 1966; Djerassi and Gutzwiller, 1966) and dichlorotris-(triphenylphosphine) ruthenium (Nishimura and Tsuneda, 1969) have proved to be very effective in the selective hydrogenation of 3-oxo-1,4-diene steroids. The factors influencing selectivity have been examined in detail using dichlorotris(triarylphosphine) ruthenium and 1,4-androstadiene-3,17-dione (**43**). Selectivity to 4-androstene-3,17-dione is favored by elevated pressures (Nishimura and Tsuneda, 1969), low temperatures (Nishimura *et al.*, 1975a), and the presence of optimal amounts of amines, such as triethylamine (Nishimura *et al.*, 1973a). In the presence of amines, the activity of the ruthenium complex is enhanced by electron-releasing *p*-methoxy and *p*-methyl groups and retarded by electron-withdrawing *p*-fluoro groups. Over ruthenium complexes, unlike rhodium, further hydrogenation of the initial product is very slow.

       **43**                             **44**

   (500 mg)         (89% yield after recrystallization from acetone–hexane)

The activity of dichlorotris [tris(*p*-methoxyphenyl)phosphine] ruthenium depends on its method of preparation. An improved method was developed by Nishimura *et al.* (1975b):

$$RuCl_3 + \tfrac{9}{2}Ar_3P + \tfrac{1}{2}H_2O \xrightarrow[\Delta]{C_2H_5OH} RuCl_2(Ar_3P)_3 + \tfrac{1}{2}Ar_3PO + Ar_3P \cdot HCl$$

Selective hydrogenation of unsaturated steroids similar to that just described has been achieved heterogeneously by the use of 10% palladium-on-carbon and benzyl alcohol as a donor (Vitali *et al.*, 1972) or prereduced ruthenium oxide (Tiffany and Rebenstorf, 1970).

Hydrogenation of α-santonin (45) over tris(triphenylphosphine) rhodium chloride in benzene–ethanol selectively affords 1,2-dihydro-α-santonin (46) (Greene *et al.*, 1974). Both double bonds are saturated over 5% palladium-on-carbon in acetic acid. The cis isomer, santonin C, with an axial oxygen undergoes carbon–oxygen bond hydrogenolysis over palladium, producing a saturated keto acid. It has been suggested that hydrogenolysis is characteristic of an allylic axial oxygen grouping (Dauben *et al.*, 1960).

45                                            46

# REFERENCES

Albright, L. F., *J. Am. Oil Chem. Soc.* **47**, 490 (1970).
Alder, K., and Roth, W., *Chem. Ber.* **87**, 161 (1954).
Alder, K., Stein, G., Schneider, S., Liebmann, M., Rolland, E., and Schultze, G., *Justus Liebigs Ann. Chem.* **525**, 183 (1936).
Ames, G. R., and Davey, W., *J. Chem. Soc.* p. 3001 (1956).
Anderson, M., and Johnson, A. W., *Proc. Chem. Soc.* p. 263 (1964).
Astill, B. D., and Boekelheide, V., *J. Am. Chem. Soc.* **77**, 4079 (1955).
Augustine, R. L., *J. Org. Chem.* **23**, 1853 (1958).
Augustine, R. L., *Ann. N.Y. Acad. Sci.* **145**, 19 (1967).
Augustine, R. L., *in* "Organic Reactions in Steroid Chemistry" (J. Fried and J. Edwards, eds.), p. 111. Van Nostrand-Reinhold, New York, 1972.
Augustine, R. L., *Adv. Catal.* **25**, 63 (1976).
Augustine, R.L., Bellina, R. F., and Gustavsen, A. J., *J. Org. Chem.* **33**, 1287 (1968).
Bahner, C. T., Brotherton, D., and Harmon, T., *J. Med. Chem.* **13**, 570 (1970).
Bailey, A. E., "Industrial Oil and Fat Products," 2nd Ed. Wiley (Interscience), New York, 1951.
Baird, W. C., Jr., and Surridge, J. H., *J. Org. Chem.* **37**, 1182 (1972).
Baird, W. C., Jr., Franzus, B., and Surridge, J. H., *J. Org. Chem.* **34**, 2944 (1969).
Bankar, N. S., and Kulkarni, G. H., *Chem. Ind. (London)* p. 481 (1973).
Barnes, M. F., and MacMillan, J., *J. Chem. Soc.* p. 361 (1967).
Beak, P., and Chaffin, T. L., *J. Org. Chem.* **35**, 2275 (1970).
Bell, J. M., Garrett, R., Jones, V. A., and Kubler, D.G., *J. Org. Chem.* **32**, 1307 (1967).
Bellinzona, G., and Bettinetti, F., *Gazz. Chim. Ital.* **90**, 426 (1960).
Bhagwat, M. M., and Devaprabhakara, D., *Tetrahedron Lett.* p. 1391 (1972).
Biellman, J. F., and Jung, M. J., *J. Am. Chem. Soc.* **90**, 1673 (1968).
Birch, A. J., and Walker, K. A. M., *J. Chem Soc. C* p. 1894 (1966).
Blomquist, A. T., and Himics, J., *J. Org. Chem.* **33**, 1156 (1968).
Bond, G. C., and Rank, J. S., *Proc. Int. Congr. Catal., 3rd, Amsterdam, 1964* **2**, 1225 (1965).
Bond, G. C., and Wells, P. B., *Adv. Catal.* **15**, 91 (1964).

Bonner, W. A., Burke, N. I., Fleck, W. E., Hill, R. K., Joule, J. A., Sjöberg, B., and Zalkow, J. H., *Tetrahedron* **20**, 1419 (1964).

Borunova, N. V., Freidlin, L. K., Gvinter, L. I., Atabekov, T., Zamureenko, V. A., and Kustanovich, I. M., *Izv. Akad. Nauk SSSR, Ser. Khim.* (6) p. 1299 (1972); *Chem. Abstr.* **77**, 87461 (1972).

Bream, J. B., Eaton, D. C., and Henbest, H. B., *J. Chem. Soc.* p. 1974 (1957).

Brewster, J. H., *J. Am. Chem. Soc.* **76**, 6361 (1954).

Brown, C. A., *J. Org. Chem.* **35**, 1900 (1970).

Brown, C. A., and Ahuja, V. K., *J. Org. Chem.* **38**, 2226 (1973).

Brown, H. C., and Brown, C. A., *J. Am. Chem. Soc.* **85**, 1003, 1005 (1963).

Brown, H. C., Rothberg, I., and Jagt, D. L. V., *J. Org. Chem.* **37**, 4098 (1972).

Burwell, R. L., Jr., *Acc. Chem. Res.* **2**, 289 (1969).

Canonica, L., Corbella, A., Gariboldi, P., Jommi, G., and Krepinsky, J., *Tetrahedron* **25**, 3903 (1969).

Carson, M. S., Cocker, W., Evans, S. M., and Shannon, P. V. R., *J. Chem. Soc. C* p. 1447 (1970).

Cerefice, S. A., and Fields, E. K., *J. Org. Chem.* **39**, 971 (1974).

Chapman, O. L., Smith, H. G., and Barks, P. A., *J. Am. Chem. Soc.* **85**, 3171 (1963).

Chaudhuri, N. K., Nickolson, R., Williams, J. G., and Gut, M., *J. Org. Chem.* **34**, 3767 (1969).

Cheung, A. P., Benitez, A., and Lim, P., *J. Org. Chem.* **33**, 3005 (1968).

Cocker, W., Shannon, P. V. R., and Staniland, P. A., *J. Chem. Soc. C* p. 41 (1966).

Connolly, J. D., Handa, K. L., McCrindle, R., and Overton, K. H., *Tetrahedron Lett.* p. 3449 (1967).

Cookson, R. C., Hamon, D. P. G., and Parker, R. E., *J. Chem. Soc.* p. 5014 (1962).

Cope, A. C., and Campbell, H. C., *J. Am. Chem. Soc.* **74**, 179 (1952).

Cope, A. C., Haven, A. C., Jr., Ramp, F. L., and Trumbull, E. R., *J. Am. Chem. Soc.* **74**, 4867 (1952).

Cope, A. C., Cotter, R. J., and Roller, G. G., *J. Am. Chem. Soc.* **77**, 3594 (1955).

Cope, A. C., Liss, T. A., and Wood, G. W., *J. Am. Chem. Soc.* **79**, 6287 (1957).

Crandall, J. K., and Paulson, D. R., *J. Org. Chem.* **33**, 3291 (1968).

Cristol, S. J., and LaLonde, R. T., *J. Am. Chem. Soc.* **81**, 1655 (1959).

Cristol, S. J., Russell, T. W., and Davies, D. I., *J. Org. Chem.* **30**, 207 (1965).

Crombie, L., and Jacklin, A. G., *J. Chem. Soc.* p. 1622 (1957).

Dart, M. C., and Henbest, H. B., *J. Chem. Soc.* p. 3563 (1960).

Dauben, W. G., Schwarz, J. S. P., Hayes, W. H., and Hance, P. D., *J. Am. Chem. Soc.* **82**, 2239 (1960).

Dauben, W. G., McFarland, J. W., and Rogan, J. B., *J. Org. Chem.* **26**, 297 (1961).

Debono, M., Farkas, E., Molloy, R. M., and Owen, J. M., *J. Org. Chem.* **34**, 1447 (1969).

DePuy, C. H., and Story, P. R., *J. Am. Chem. Soc.* **82**, 627 (1960).

deVivar, A. R., Bratoeff, E. A., and Rios, T., *J. Org. Chem.* **31**, 673 (1966).

Djerassi, C., and Gutzwiller, J., *J. Am. Chem. Soc.* **88**, 4537 (1966).

Eglinton, G., Jones, E. R. H., Mansfield, G. H., and Whitting, M. C., *J. Chem. Soc.* p. 3197 (1954).

Eigenmann, G. W., and Arnold, R. T., *J. Am. Chem. Soc.* **81**, 3440 (1959).

Eisenbraun, E. J., George, T., Riniker, B., and Djerassi, C. J., *J. Am. Chem. Soc.* **82**, 3648 (1960).

Fahey, D. R., *J. Org. Chem.* **38**, 80 (1973).

Fajkos, J., *Chem. Listy* **52**, 1320 (1958).

Farina, M., Morandi, C., Mantica, E., and Botta, D., *J. Chem. Soc., Chem. Commun.* p. 816 (1976).

Fischer, H., and Fischer, H., *Tetrahedron* **25**, 955 (1969).

60	3. Hydrogenation of Olefins

Fischer, N. H., Mabry, T. J., and Kagan, H. B., *Tetrahedron* **24**, 4091 (1968).
Fisher, C., and Mosher, H. S., *Tetrahedron Lett.* p. 2487 (1977).
Frankel, E. N., *in* "Topics in Lipid Chemistry" (F. D. Gunstone, ed.), p. 6. Logos Press, London, 1970a.
Frankel, E. N., *J. Am. Oil Chem. Soc.* **47**, 11 (1970b).
Frankel, E. N., and Butterfield, R. O., *J. Org. Chem.* **34**, 3930 (1969).
Frankel, E. N., and Little, F. L., *J. Am. Oil Chem. Soc.* **46**, 256 (1969).
Frankel, E. N., Selke, E., and Glass, C. A., *J. Org. Chem.* **34**, 3936 (1969).
Frankel, E. N., Thomas, F. L., and Cowan, J. C., *J. Am. Oil Chem. Soc.* **47**, 497 (1970).
Freidlin, L. K., Gvinter, L. I., Borunova, N. V., Dymova, S. F., and Kustanovich, I. I., *Katal. Reakts. Zhidk. Faze* p. 309 (1972); *Chem. Abstr.* **79**, 115066z (1973).
Froborg, J., Magnusson, G., and Thorén, S., *J. Org. Chem.* **40**, 1595 (1975).
Gabbard, R. B., and Segaloff, A., *J. Org. Chem.* **27**, 655 (1962).
Garland, R. B., Palmer, J. R., and Pappo R., *J. Org. Chem.* **41**, 531 (1976).
Garrett, E. R., Donia, R. A., Johnson, B. A., and Scholten, L., *J. Am. Chem. Soc.* **78**, 3340 (1956).
Ghatak, V. R., Datta, D. K., and Ray, S. C., *J. Am. Chem. Soc.* **82**, 1728 (1960).
Glamkowski, E. J., Gal, G., Purick, R., Davidson, A. J., and Sletzinger, M., *J. Org. Chem.* **35**, 3510 (1970).
Gostunskaya, I. V., Petrova, V. S., Leonava, A. I., Mironava, V. A., Abubaker, M., and Kazanskii, B. A., *Neftekhimiya* **7**(1), 3 (1967).
Greene, A. E., Muller, J.-C., and Ourisson, G., *J. Org. Chem.* **39**, 186 (1974).
Gregory, G. I., Hunt, J. S., May, P. J., Nice, F. A., and Phillipps, G. H., *J. Chem. Soc. C* p. 2201 (1966).
Hajos, Z. G., and Parrish, D. R., *J. Org. Chem.* **38**, 3239 (1973).
Halsall, T. G., Rodewald, W. J., and Willis, D., *J. Chem. Soc.* p. 2798 (1959).
Ham, G. E., and Coker, W. P., *J. Org. Chem.* **29**, 194 (1964).
Harmon, R. E., Gupta, S. K., and Brown, D. J., *Chem. Rev.* **73**, 21 (1973).
Harnik, M., US. Patent 3,107,256, Oct. 15, 1963.
Hayashi, T., Tanaka, M., and Ogata, I., *Tetrahedron Lett.* p. 295 (1977).
Haynes, N. B., and Timmons, C. J., *Proc. Chem. Soc.* p. 345 (1958).
Herz, W., Ueda, K., and Inayama, G., *Tetrahedron* **19**, 483 (1963).
Herz, W., Kishida, Y., and Lakshmikantham, M. V., *Tetrahedron* **20**, 979 (1964).
Herz, W., Subramaniam, P. S., and Geissman, T. A., *J. Org. Chem.* **33**, 3743 (1968).
Hidai, M., Ishiwatari, H., Yagi, H., Tanaka, E., Onozawa, K., and Uchida, Y., *J. Chem. Soc., Chem. Commun.* p. 170 (1975).
Hochstetler, A. R., *J. Org. Chem.* **37**, 1883 (1972).
Horiuti, I., and Polanyi, M., *Trans. Faraday Soc.* **30**, 1164 (1934).
House, H. O., and Rasmusson, G. H., *J. Org. Chem.* **28**, 27 (1963).
Howard, T. J., *Chem. Ind.* (*London*) p. 1899 (1963).
Howe, R., and McQuillin, F. J., *J. Chem. Soc.* p. 1194 (1958).
Huckel, W., Maier, M., Jordan, E., and Seeger, W., *Justus Liebigs Ann. Chem.* **616**, 46 (1958).
Huffman, J. W., *J. Org. Chem.* **24**, 447 (1959).
Huntsman, W. D., Madison, N. L., and Schlesinger, S. I., *J. Catal.* **2**, 498 (1963).
Hussey, A. S., Schenach, T. A., and Baker, R. H., *J. Org. Chem.* **33**, 3258 (1968).
Inhoffen, H. H., Stoeck, G., Kölling, G., and Stoech, U., *Justus Liebigs Ann. Chem.* **568**, 52 (1950).
Ireland, R. E., Evans, D. A., Glover, D., Rubottom, G. M., and Young, H., *J. Org. Chem.* **34**, 3717 (1969).
Jacobson, H. I., Griffin, M. J., Preis, S., and Jensen, E. V., *J. Am. Chem. Soc.* **79**, 2608 (1957).

James, B. R., "Homogeneous Hydrogenation." Wiley, New York, 1973.
Jardine, I., and McQuillin, F. J., *Tetrahedron Lett.* p. 5189 (1968).
Kidwai, A. R., and Devasia, G. M., *J. Org. Chem.* **27**, 4527 (1962).
Knowles, W. S., Sabacky, M. J., and Vineyard, B. D., *Chemtech.* p. 590 (1972).
Knowles, W. S., Sabacky, M. J., and Vineyard, B. D., *Adv. Chem. Ser.* No. 132 p. 274 (1974).
Knowles, W. S., Sabacky, M. J., Vineyard, B. D., and Weinkauff, D. J., *J. Am. Chem. Soc.* **97**, 2568 (1975).
Knox, L. H., Villotti, R., Kinel, F. A., and Ringold, H. J., *J. Org. Chem.* **26**, 501 (1961).
Kuhn, R., and Fischer, H., *Chem. Ber.* **93**, 2285 (1960).
Kupchan, S. M., Aynehchi, Y., Cassady, J. M., Schnoes, H. K., and Burlingame, A. L., *J. Org. Chem.* **34**, 3867 (1969).
Larrabee, C. E., and Craig, L. E., *J. Am. Chem. Soc.* **73**, 5471 (1951).
Lindlar, H., *Helv. Chim. Acta* **35**, 446 (1952).
Liska, K. J., *J. Pharm. Sci.* **53**, 1427 (1964).
Lowenthal, H. J. E., *Tetrahedron* **6**, 269 (1959).
Lyons, J. E., Rennick, L. E., and Burmeister, J. L., *Ind. Eng, Chem., Prod. Res. Dev.* **9** (1), 2 (1970).
McKenzie, T. C., *J. Org. Chem.* **39**, 629 (1974).
McMurry, J. E., *Tetrahedron Lett.* p. 3731 (1970).
McQuillin, F. J., *Tech. Org. Chem.* **9**, 498 (1963).
McQuillin, F. J., Ord, W. O., and Simpson, P. L., *J. Chem. Soc.* p. 5996 (1963).
Marshall, J. A., Andersen, N. H., and Johnson, P. C., *J. Org. Chem.* **35**, 186 (1970).
Martin, J. C., Barton, K. R., Gott, P. G., and Meen, R. H., *J. Org. Chem.* **31**, 943 (1966).
Martin, M. M., and Koster, R. A., *J. Org. Chem.* **33**, 3428 (1968).
Micheli, R. A., Gardner, J. N., Dubuis, R., and Buchshacher, P., *J. Org. Chem.* **34**, 1457 (1969).
Minckler, L. S., Hussey, A. S., and Baker, R. H., *J. Am. Chem. Soc.* **78**, 1009 (1956).
Minyard, J. P., Thompson, A. C., and Hedin, P. A., *J. Org. Chem.* **33**, 909 (1968).
Miropol'skaya, M. A., Fedotova, N. I., Veinberg, A. Y., Yanotovskii, M. T., and Samokhvalov, G. I., *Zh. Obshch. Khim.* **32**, 2214 (1962).
Mitsui, S., Senda, Y., and Satio, H., *Bull. Chem. Soc. Jpn.* **39**, 694 (1966).
Mitsui, S., Saito, H., Sekiguchi, S., Kumagui, Y., and Senda, Y., *Tetrahedron* **28**, 4751 (1972).
Mitsui, S., Saito, Y., Yamashita, Y., Kaminaga, M., and Senda, Y., *Tetrahedron* **29**, 1531 (1973a).
Mitsui, S., Gohke, K., Saito, H., Nanbu, A., and Senda, Y., *Tetrahedron* **29**, 1523 (1973b).
Mori, K., Abe, K., Washida, M., Nishimura, S., and Shiota, M., *J. Org. Chem.* **36**, 231 (1971).
Morrison, J. D., Masler, W. F., and Neuberg, M. K., *Adv. Catal.* **25**, 81 (1975).
Morrison, J. D., Masler, W. F., and Hathaway, S., *in* "Catalysis in Organic Syntheses, 1976" (P. N. Rylander and H., Greenfield, eds.), p. 203. Academic Press, New York, 1976.
Murahashi, S.-I., Yano, T., and Hino, K.-I., *Tetrahedron Lett.* p. 4235 (1975).
Murai, A., Arita, K., and Masamune, T., *Bull. Chem. Soc. Jpn.* **46**, 3536 (1973).
Nesmeyanov, A. N., Kochetkov, N. K., Rybinskaya, M. I., and Uglova, E. V., *Bull. Acad. Sci. USSR, Div. Chem. Sci.* p. 579 (1955).
Newhall, W. F., *J. Org. Chem.* **23**, 1274 (1958).
Newman, M. S., and Addor, R. W., *J. Am. Chem. Soc.* **77**, 3789 (1955).
Nicholas, K. M., *J. Am. Chem. Soc.* **97**, 3254 (1975).
Nickon, A., and Bagli, J. F., *J. Am. Chem. Soc.* **83**, 1498 (1961).
Nishimura, S., and Hama, M., *Bull. Chem. Soc. Jpn.* **39**, 2467 (1966).
Nishimura, S., and Kano, Y., *Chem. Lett.* No. 7, p. 568 (1972).
Nishimura, S., and Mori, K., *Bull. Chem. Soc. Jpn.* **36**, 318 (1963).
Nishimura, S., and Shimahara, M., *Chem. Ind. (London)* p. 1796 (1966).

Nishimura, S., and Tsuneda, K., *Bull. Chem. Soc. Jpn.* **42**, 852 (1969).
Nishimura, S., Shimahara, M., and Shiota, M., *J. Org. Chem.* **31**, 2394 (1966).
Nishimura, S., and Tsuneda, K., *Bull. Chem. Soc. Jpn.* **42**, 852 (1969).
Nishimura, S., Shimahara, M., and Shiota, M., *J. Org. Chem.* **31**, 2394 (1966).
Nishimura, S., Katagiri, M., Watanabe, T., and Uramoto, M., *Bull. Chem. Soc. Jpn.* **44**, 166 (1971).
Nishimura, S., Ichino, T., Akimoto, A., and Tsuneda, K., *Bull. Chem. Soc. Jpn.* **46**, 279 (1973a).
Nishimura, S., Sakamoto, H., and Ozawa, T., *Chem. Lett.* p. 855 (1973b).
Nishimura, S., Ichino, T., Akimoto, A., Tsuneda, K., and Mori, H., *Bull. Chem. Soc. Jpn.* **48**. 2852 (1975a).
Nishimura, S., Yumoto, O., Tsuneda, K., and Mori, H., *Bull. Chem. Soc. Jpn.* **48**, 2603 (1975b).
Ogiso, A., and Pelletier, S. W., *Chem. Commun.* p. 94 (1967).
Ojima, I., Kogure, T., and Nagai, Y., *Tetrahedron Lett.* p. 5035 (1972).
Onoue, Y., Mizutani, Y., Akiyama, S., Izumi, Y., and Watanabe, Y., *Chemtech.* Jan., p. 36 (1977).
Paquette, L. A., and Rosen, M., *J. Org. Chem.* **33**, 3027 (1968).
Pecque, M., and Maurel, R., *J. Catal.* **19**, 360 (1970).
Petrov, A. A., Ionin, B. I., and Ignatyev, V. M., *Tetrahedron Lett.* p. 15 (1968).
Phillipps, G. H., US. Patent 3,115,508, Dec. 24, 1963.
Piers, E., Britton, R. W., and deWaal, W., *Chem. Commun.* p. 1069 (1969).
Powell, R. G., Madrigal, R. V., Smith, C. R., Jr., and Mikolajczak, K. L., *J. Org. Chem.* **39**, 676 (1974).
Pracejus, G., and Pracejus, H., *Tetrahedron Lett.* p. 3497 (1977).
Puterbaugh, W. H., and Newman, M. S., *J. Am. Chem. Soc.* **81**, 1611 (1959).
Riesz, C. H., and Weber, H. S., *J. Am. Oil Chem. Soc.* **41**, 400 (1964).
Rosenthal, A., and Shudo, K., *J. Org. Chem.* **37**, 4391 (1972).
Roy, S. K., and Wheeler, D. M. S., *J. Chem. Soc.* p. 2155 (1963).
Rüesch, H., and Mabry, T. J., *Tetrahedron* **25**, 805 (1969).
Rylander, P. N., "Catalytic Hydrogenation over Platinum Metals," p. 448. Academic Press, New York, 1967.
Rylander, P. N., *J. Am. Oil Chem. Soc.* **47**, 482 (1970).
Rylander, P. N., *Adv. Chem. Ser.* **98**, 150 (1971).
Rylander, P. N., "Organic Syntheses with Noble Metal Catalysts," p. 60. Academic Press, New York, 1973.
Rylander, P. N., and Himelstein, N., *Engelhard Ind., Tech. Bull.* **5**, 43 (1964).
Sanchez-Viesca, F., and Romo, J., *Tetrahedron* **19**, 1285 (1963).
Scott, J. W., Widmer, E., Meier, W., Labler, L., Müller, P., and Fürst, A., *J. Org. Chem.* **37**, 3183 (1972).
Sehgal, R. K., Koenigsberger, R. U., and Howard, T. J., *J. Org. Chem.* **40**, 3073 (1975).
Sengupta, P., and Khastgir, H. N., *Tetrahedron* **19**, 123 (1963).
Shepherd, D. A., Donia, R. A., Campbell, J. A., Johnson, B. A., Holysz, R. P., Slomp, G., Jr., Stafford, J. E., Pederson, R. L., and Ott, A. C., *J. Am. Chem. Soc.* **77**, 1212 (1955).
Siegel, S., *Adv. Catal.* **16**, 123 (1966).
Siegel, S., and Cozort, J. R., *J. Org. Chem.* **40**, 3594 (1975).
Siegel, S., and Ohrt, D. W., *Tetrahedron Lett.* p. 5155 (1972).
Siegel, S., and Smith, G. V., *J. Am. Chem. Soc.* **82**, 6082 (1960a).
Siegel, S., and Smith, G. V., *J. Am. Chem. Soc.* **82**, 6087 (1960b).
Siegel, S., Foreman, M., and Johnson, D., *J. Org. Chem.* **40**, 3589 (1975).
Slomp, G., Jr., Shealy, Y. F., Johnson, J. L., Donia, R. A., Johnson, B. A., Holysz, R. P., Pederson, R. L., Jenson, A. D., and Ott, A. C., *J. Am. Chem. Soc.* **77**, 1216 (1955).

Smith, G. V., *J. Catal.* **5**, 152 (1966).

Smith, G. V., and Desai, D. S., *Ann. N.Y. Acad. Sci.* **214**, 20 (1973).

Smith, G. V., and Menon, M. C., *Ann. N.Y. Acad. Sci.* **158**, 501 (1969).

Smith, G. V., and Roth, J. A., *Proc. Int. Congr. Catal., 3rd, Amsterdam, 1964* **1**, 379 (1965).

Smith, G. V., Roth, J. A., Desai, D. S., and Kosco, J. L., *J. Catal.* **30**, 79 (1973).

Smith, H. A., Fuzek, J. F., and Meriwether, H. T., *J. Am. Chem. Soc.* **71**, 3765 (1949).

Smyrniotis, P. Z., Miles, H. T., and Stadman, E. R., *J. Am. Chem. Soc.* **82**, 1417 (1960).

Stolow, R. D., and Sachdev. K., *J. Org. Chem.* **36**, 960 (1971).

Stork, G., and Schulenberg, J. W., *J. Am. Chem. Soc.* **84**, 284 (1962).

Sulser, H., Scherer, J. R., and Stevens, K. L., *J. Org. Chem.* **36**, 2422 (1971).

Tallent, W. H., *Tetrahedron* **20**, 1781 (1964).

Tamao, K., Yamamoto, H., Matsumoto, H., Miyake, N., Hayashi, T., and Kumada, M., *Tetrahedron Lett.* p. 1389 (1977).

Tanaka, M., and Ogata, I., *J. Chem. Soc., Chem. Commun.* p. 735 (1975).

Tarbell, D. S., Carman, R. M., Chapman, D. D., Cremer, S. E., Cross, A. D., Huffman, K. R., Kunstmann, M., McCorkindale, N. J., McNally, J. G., Jr., Rosowsky, A., Varino, F. H. L., and West, R. L., *J. Am. Chem. Soc.* **83**, 3096 (1961).

Thompson, H. W., *J. Org. Chem.* **36**, 2577 (1971).

Thompson, H. W., *Ann. N. Y. Acad. Sci.* **214**, 195 (1973).

Thompson, H. W., and McPherson, E., *J. Am. Chem. Soc.* **96**, 6232 (1974).

Thompson, H. W., and Naipawer, R. E., *J. Am. Chem. Soc.* **95**, 6379 (1973).

Thompson, H. W., McPherson, E., and Lences, B. L., *J. Org. Chem.* **41**, 2903 (1976).

Thompson, S. J., and Webb, G., *J. Chem. Soc., Chem. Commun.* p. 526 (1976).

Tiffany, B. D., and Rebenstorf, M. A., *Ann. N.Y. Acad. Sci.* **172**, 253 (1970).

Traas, P. C., Boelens, H., and Takken, H. J., *Synth. Commun.* **6**, 489 (1976).

Tyman, J. H. P., and Wilkins, S. W., *Tetrahedron Lett.* p. 1773 (1973).

Uehara, K., Ito, M., and Tanaka, M., *Bull. Chem. Soc. Jpn.* **46**, 1566 (1973).

Vitali, R., Caccia, G., and Gardi, R., *J. Org. Chem.* **37**, 3745 (1972).

Warawa, E. J., and Campbell, J. R., *J. Org. Chem.* **39**, 3511 (1974).

Watanabe, Y., Mizuhara, Y., and Shiota, M., *J. Org. Chem.* **33**, 468 (1968).

Watanabe, Y., Matsumura, Y., Izumi, Y., and Mizutani, Y., *Bull. Chem. Soc. Jpn.* **47**, 2992 (1974).

Wells, P. B., and Wilson, G. R., *J. Chem. Soc. A* p. 2442 (1970).

Westmijze, H., Meijer, J., and Vermeer, P., *Tetrahedron Lett.* p. 2923 (1975).

Wilds, A. L., Johnson, J. A., Jr., and Sutton, R. E., *J. Am. Chem. Soc.* **72**, 5524 (1950).

Williams, J. L. R., Webster, S. K., and Van Allen, J. A., *J. Org. Chem.* **26**, 4893 (1961).

Woodward, R. B., Sondheimer, F., Taub, D., Heusler, K., and McLamore, W. M., *J. Am. Chem. Soc.* **74**, 4223 (1952).

Xomatsu, Y., Furukawa, Y., and Yokomizo, T., U.S., Patent 3,674,886, July 4, 1972.

Yao, H.-C., and Emmett, P. H., *J. Am. Chem. Soc.* **81**, 4125 (1959).

Zacjew, M., *J. Am. Oil Chem. Soc.* **37**, 473 (1960a).

Zacjew, M., *J. Am. Oil Chem. Soc.* **37**, 11 (1960b).

Zelnik, R., Lavie, D., Levy, E. C., Wang, A. H. J., and Paul, I. C., *Tetrahedron* **33**, 1457 (1977).

# Chapter 4

# Hydrogenation of Acids, Esters, Lactones, and Anhydrides

## I. HYDROGENATION OF ACIDS

Successful reductions of acids afford alcohols in excellent yields, but both elevated temperatures and high pressures are required (Shuikin *et al.*, 1961). Acetic acid is reduced rapidly over rhodium and, to a lesser extent, over other noble metals under ambient conditions. Both strong acids, such as perchloric acid, and tertiary amines seem to be catalysts for the reduction. Conversion is very small but nonetheless sufficient to introduce appreciable error in quantitative adsorption measurements when carboxylic acids are used as solvents (Chanley and Mezzetti, 1964).

### Catalysts for Acid Hydrogenation

Ruthenium dioxide or ruthenium-on-carbon have given good results in the reduction of mono- and dicarboxylic acids to the corresponding alcohols and glycols. Pressures ranging from 5000 to 10,000 psig and temperatures from 130° to 225°C were used (Ford, 1952; Carnahan *et al.*, 1955; Schreyer, 1958). The presence of water is likely to be beneficial in these reductions, first because of its well-known catalytic effect in ruthenium-catalyzed reductions generally and, second, because it minimizes ester formation and the mixture of products that may be derived from its reduction.

Rhenium oxides are also useful in the reduction of carboxylic acids at high temperatures and pressures (200°C, 250 atm). Aromatic carboxylic acids can be reduced to alcohols without ring saturation (Broadbent and Bartley, 1963; Broadbent and Selin, 1963; Broadbent *et al.*, 1959). Reaction parameters affecting acetic acid hydrogenation have been discussed by Broadbent *et al.* (1970), and various rhenium catalysts have been reviewed by Broadbent (1967).

Copper chromite (Guyer *et al.*, 1947) and barium-promoted copper chromite (Guyer *et al.*, 1955) have been used successfully for acid hydrogenation, but at very high temperatures (300°C).

## II. HYDROGENOLYSIS OF ESTERS

Esters undergo hydrogenolysis more readily than do acids, and it is possible to reduce a half-acid ester to a hydroxy acid in excellent yield (Sayles and Degering, 1949). The products of ester hydrogenolysis may be alcohols, acids, esters (Chanley and Mezzetti, 1964; Edward and Ferland, 1964), or hydrocarbons (Liska and Salerni, 1960; Cavallito and Haskell, 1944). The composition of the product mix seems to be determined largely by the substrate itself.

### A. Hydrogenolysis to Acids

Hydrogenolysis of an ester to an acid is to be expected when the R—O bond has been weakened by some structural feature such as R = benzyl, vinyl or allyl (Shriner and Anderson, 1939; Sakuragi, 1958). The extent of hydrogenolysis in vinyl and allyl derivatives depends on the rate ratio of competing hydrogenation and hydrogenolysis (Astill and Boekelheide, 1955). Esterification of an enol and subsequent hydrogenolysis provide a technique for removing carbonyl oxygen that otherwise is reduced to an alcohol (Harnik, 1963).

Certain bromo esters and bromolactones (Denton *et al.*, 1964; Levine and Wall, 1959) undergo facile dehalogenation and hydrogenolysis to form an acid, perhaps via an intermediate olefin derived by elimination in an organometallic derivative (Campbell and Kemball, 1963).

Tertiary esters undergo hydrogenolysis readily in the presence of strong acids (Peterson and Casey, 1964), probably through an actual or incipient alkene. Elevated temperatures also favor elimination and hydrogenolysis of esters (Shuikin *et al.*, 1961).

### B. Hydrogenolysis to Alcohols

Cleavage of esters to afford alcohols is the most common course of ester hydrogenation. The reduction has been carried out with noble metal catalysts (Rapala *et al.*, 1957), but they are usually not well suited for this purpose.

The most frequently used catalyst for ester hydrogenation is copper chromite, which is often promoted with some substance, such as barium, to maintain stability (Lazier and Arnold, 1953; Riener, 1949; Folkers and Adkins, 1932; Adkins *et al.*, 1950). Properly prepared Raney nickel and zinc–chromium oxide also have been frequently used for ester hydrogenolysis (Adkins, 1954). A variety of good catalysts for both batch and fixed-bed operations can be obtained from various catalyst manufacturers. These catalysts are usually used at high temperatures and pressures (Adkins, 1954). The vigorous conditions promote various side reactions such as hydrogenolysis of the alcohol (Adkins and Billica, 1948a), occasional aromatic ring reduction, pyrolysis (Lewis *et al.*, 1969), and, in nitrogen compounds, *N*-alkylation by alcohol solvents. If any of these side reactions prove to be troublesome they can be circumvented by the avoidance of alcohol solvent in the latter case and the use of lower temperatures with massive amounts of catalyst (Adkins and Billica, 1948b) for other problems. Massive amounts of nickel have been used similarly for hydrogenation of amino esters to amino alcohols. In this type of reduction proper preparation of the nickel is essential (Adkins and Pavlic, 1947).

The hydrogenation of esters to alcohols is a reversible reaction, with alcohol formation favored by higher pressures, as illustrated by the following equilibrium data for the hydrogenation of *n*-octyl caprylate (Adkins and Burks, 1948):

$$C_7H_{15}CO_2C_8H_{17} + 2H_2 \xrightleftharpoons{260°C} 2C_8H_{17}OH$$

| % Ester at equilibrium | Pressure (psig) |
|---|---|
| 0.9 | 3980 |
| 80.3 | 140 |

N-Benzyl (Stoll *et al.*, 1943) and *O*-benzyl groups can be retained during hydrogenation of esters over copper chromite. For example, an industrial process for benzyl alcohol involves hydrogenolysis of methyl benzoate over copper chromite at 140°C and 250 atm in methanol–toluene (Kato *et al.*, 1972).

30% conversion, 95% selectivity

## 1.  *Hydrogenolysis of Unsaturated Esters*

Olefins are reduced very easily, whereas esters are reduced with relative difficulty; nonetheless, with special catalysts, unsaturated esters can be reduced to unsaturated alcohols in good yield. The reduction is practiced commercially. The catalysts are usually zinc or copper based and promoted with calcium, cadmium, or chromium (German patent 1,228,603). Japanese patent 69 13,124 teaches cadmium-on-alumina or silica–alumina. Cherkaev *et al.* (1968) have reviewed the area.

### 2.  Hydrogenolysis of Phosphate Esters

Benzyl and aryl esters of phosphorus-containing acids have been widely used as protecting groups. After suitable transformations, the benzyl or aryl group is removed by catalytic hydrogenolysis, freeing the acid and avoiding hydrolytic procedures. Palladium is employed invariably for hydrogenolysis of benzyl groups, and platinum is employed for aryl groups (Reithel and Claycomb, 1949; Baer and Kates, 1950; Baer *et al.*, 1952; Maley *et al.*, 1956; Kilgour and Bellou, 1958; Ballou and Fischer, 1954, 1955, 1956; Baer and Maurukas, 1952; Griffin and Burger, 1956; Ukita and Hayatsu, 1962; Westphal and Stadler, 1963; Gent *et al.*, 1970; Jung and Engel, 1975). Hydrogenolysis of benzyl esters is discussed in Chapter 15.

### III.  HYDROGENATION OF LACTONES

Lactones seem to be reduced more easily than linear esters (Liska and Salerni, 1960; Chanley and Mezzetti, 1964). Several $\delta$-lactones were reduced readily to the corresponding ether over platinum oxide in acetic acid containing a trace of perchloric acid, but this facile reduction could not be extended to $\gamma$- or $\epsilon$-lactones (Edward and Ferland, 1964).

Activation of the alkyl–oxygen bond in lactones facilitates hydrogenolysis. Dimethyl 2-methylterephthalate (**2**) was prepared in excellent yield by hydrogenolysis of **1** over palladium-on-carbon (Jurewicz and Forney, 1972). This facile reduction probably, but not necessarily, proceeded by initial cleavage of the benzyl–oxygen bond.

**1**                                                                **2**

The vinyl-activated oxygen atom in **3** does not undergo cleavage at the vinyl oxygen, but rather at the oxygen–carbonyl bond to afford an unusual product of reduction, an acyclic aldehyde (**4**) (Martin *et al.*, 1971).

$$\text{3} \xrightarrow[\substack{40 \text{ psig} \\ 25^\circ\text{C} \\ \text{EtOAc}}]{5\% \text{ Pd-on-C}} \text{4}$$

Aromatization can provide a driving force for facile lactone hydrogenolysis. The reduction can be viewed formally as a 1,6 addition of hydrogen (Chitwood *et al.*, 1971).

$$\xrightarrow[\substack{\text{EtOAc} \\ 3 \text{ atm}}]{5\% \text{ Pd-on-C}}$$

80%

A great deal of attention has been given to the hydrogenation of butyrolactone and its precursors, maleic and succinic anhydrides, to 1,4-butanediol, an important component of polyesters. High yields, 89%, have been obtained over copper chromite (Murata and Kobayashi, 1972); over a nickel–cobalt–thoria-on-kieselguhr, yields reached 98% (Yamaguchi and Kageyama, 1972).

$$\xrightarrow[100 \text{ atm}]{250^\circ\text{C}} HO(CH_2)_4OH$$

Hydrogenation of butyrolactone over a cobalt–rhenium catalyst in dioxane at 150 atm and 240°C gave tetrahydrofuran in 92% selectivity (Hirai and Miyata, 1970). Caution should be exercised in using dioxane at elevated temperatures, because it has decomposed explosively (Mozingo, 1955).

$$\longrightarrow \quad + H_2O$$

Substitution at the $\gamma$ position can radically alter the course of reduction. Hydrogenation of $\gamma$-(*n*-heptyl)butyrolactone over a variety of catalysts gave predominantly undecanoic acid (Dashunin, 1969).

$$CH_3(CH_2)_6 \text{—}\!\!\!\!\!\!\!\!\!\!\!\!\longrightarrow CH_3(CH_2)_9COOH$$

## IV.   HYDROGENATION OF ANHYDRIDES

Hydrogenation of anhydrides affords a number of products, depending on the course and depth of reduction. Aromatic anhydrides give a variety of ring-reduced products as well. The course of reduction is dependent on the substrate structure, reaction conditions, solvent (Eggelte *et al.*, 1973), and catalyst (Sicher *et al.*, 1961; McCrindle *et al.*, 1961, 1962; Austin *et al.*, 1937). Anhydrides are reduced with relative ease. McAlees and McCrindle (1969) established the order of decreasing reactivity for various carbonyls as acid chlorides > aldehydes, ketones > anhydrides > esters > carboxylic acids > amides. Reduction may proceed by a 1,2 addition of hydrogen to a carbonyl function or by hydrogenolysis of an oxygen–carbonyl bond. McAlees and McCrindle (1969) have evidence supporting the latter pathway. They believe that the products of reduction are determined by the rate ratio of recyclization of the initial intermediate, formed by cleavage, to its further reduction.

The course of hydrogenation under mild conditions of a variety of aromatic, carbocyclic and heterocyclic, and linear and cyclic anhydrides over palladium and platinum was reported by Kuhn and Butula (1968).

Hydrogenation of both maleic anhydride and succinic anhydride, as precursors of industrially important butyrolactone, tetrahydrofuran, and 1,4-butanediol, has been investigated intensively. Any of these products can be made the major component of the reduction. The equilibrium composition between butyrolactone and butanediol is determined by temperature and pressure. Low pressures (1 atm, 270°C) favor butyrolactone over butanediol (100:0.02), whereas at 300 atm the ratio is 6:94 (Miya *et al.*, 1973).

$$\text{(maleic anhydride)} \longrightarrow \text{(succinic anhydride)} \longrightarrow \text{(butyrolactone)} \rightleftharpoons HO(CH_2)_4OH$$
$$\searrow \swarrow$$
$$\text{(tetrahydrofuran)} + H_2O$$

Tetrahydrofuran has been obtained in 98% selectivity in vapor-phase hydrogenation of maleic anhydride at temperatures in excess of 230°C and 25 atm pressure over a copper–chromium–zinc-on-alumina catalyst (Miya *et al.*, 1973). Reduction of succinic anhydride at 260°C and 120 atm over a nickel–rhenium–barium-on-silica–alumina gives about 35 mole % tetrahydrofuran and 40 mole % butyrolactone (Kanetaka and Mori, 1972).

## REFERENCES

Adkins, H., *Org. React.* **8**, 1 (1954).
Adkins, H., and Billica, H. R., *J. Am. Chem. Soc.* **70**, 3118 (1948a).
Adkins, H., and Billica, H. R., *J. Am. Chem. Soc.* **70**, 3121 (1948b).
Adkins, H., and Burks, R. E., Jr., *J. Am. Chem. Soc.* **70**, 4174 (1948).
Adkins, H., and Pavlic, A. A., *J. Am. Chem. Soc.* **69**, 3039 (1947).
Adkins, H., Burgoyne, E. E., and Schneider, H. J., *J. Am. Chem. Soc.* **72**, 2626 (1950).
Astill, B. D., and Boekelheide, V., *J. Am. Chem. Soc.* **77**, 4079 (1955).
Austin, P. R., Bousquet, E. W., and Lazier, W. A., *J. Am. Chem. Soc.* **59**, 864 (1937).
Baer, E., and Kates, M., *J. Am. Chem. Soc.* **72**, 942 (1950).
Baer, E., and Maurukas, J., *J. Am. Chem. Soc.* **74**, 158 (1952).
Baer, E., Maurukas, J., Russell, M., *J. Am. Chem. Soc.* **74**, 152 (1952).
Ballou, C. E., and Fischer, H. O. L., *J. Am. Chem. Soc.* **76**, 3188 (1954).
Ballou, C. E., and Fischer, H. O. L., *J. Am. Chem. Soc.* **77**, 3329 (1955).
Ballou, C. E., and Fischer, H. O. L., *J. Am. Chem. Soc.* **78**, 1659 (1956).
Broadbent, H. S., *Ann. N.Y. Acad. Sci.* **145**, 58 (1967).
Broadbent, H. S., and Bartley, W. J., *J. Org. Chem.* **28**, 2345 (1963).
Broadbent, H. S., and Selin, T. G., *J. Org. Chem.* **28**, 2343 (1963).
Broadbent, H. S., Campbell, G. C., Bartley, W. J., and Johnson, J. H., *J. Org. Chem.* **24**, 1847 (1959).
Broadbent, H. S., Mylroie, V. L., and Dixon, W. R., *Ann. N.Y. Acad. Sci.* **172**, 194 (1970).
Campbell, J. S., and Kemball, C., *Trans. Faraday Soc.* **59**, 2583 (1963).
Carnahan, J. E., Ford, T. A., Gresham. W. F., Grisby, W. E., and Hager, G. F., *J. Am. Chem. Soc.* **77**, 3766 (1955).
Cavallito, C. J., and Haskell, T. H., *J. Am. Chem. Soc.* **66**, 1166 (1944).
Chanley, J. D., and Mezzetti, T., *J. Org. Chem.* **29**, 228 (1964).
Cherkaev, V. G., Bliznyak, N. V., and Bag, A. A., *Tr. Vses. Nauchno-Issled. Inst. Sint. Nat. Dushistykh Veshchestv* p. 234 (1968); *Chem. Abstr.* **71**, 38200 (1969).
Chitwood, J. L., Gott, P. G., and Martin, J. C., *J. Org. Chem.* **36**, 2216 (1971).
Dashunin, V. M., Maeva, R. V., and Rodionava, N. V., *Zh. Org. Khim.* **5**, 637 (1969); *Chem. Abstr.* **71**, 21657a (1969).
Denton, D. A., McQuillin, F. J., and Simpson, P. L., *Proc. Chem. Soc.* p. 297 (1964).
Edward, J. T., and Ferland, J. M., *Chem. Ind. (London)* p. 975 (1964).
Eggelte, T. A., DeKonning, H., and Huisman, H. O., *Tetrahedron* **29**, 2445 (1973).
Folkers, K., and Adkins, H., *J. Am. Chem. Soc.* **54**, 1145 (1932).
Ford, T. A., U.S. Patent 2,607,807, Aug. 19, 1952.
Gent, P. A., Gigg, R., and Warren, C. D., *Tetrahedron Lett.* p. 2575 (1970).
Griffin, B. S., and Burger, A., *J. Am. Chem. Soc.* **78**, 2336 (1956).
Guyer, A., Bieler, A., and Jaberg, K., *Helv. Chim. Acta* **30**, 39 (1947).
Guyer, A., Bieler, A., and Sommaruga, M., *Helv. Chim. Acta* **38**, 976 (1955).
Harnik, M., U.S. Patent 3,107,256, Oct. 15, 1963.
Hiari, H., and Miyata, K., J. Patent 72 42,832, Jan. 28, 1970; *Chem. Abstr.* **78**, 111105t (1973).
Jung, A., and Engel, R., *J. Org. Chem.* **40**, 244 (1975).

Jurewicz, A. T., and Forney, L. S., U.S. Patent 3,651,126, Mar. 21, 1972.

Kanetaka, J., and Mori, S., Jpn. Patent 72 34, 180, Nov. 20, 1972; *Chem. Abstr.* **78**, 111104 (1973).

Kato, K., Namikawa, K., and Watanabe, M., *Bull. Jpn. Pet. Inst.* **14**, 206 (1972).

Kilgour, G. L., and Ballou, C. E., *J. Am. Chem. Soc.* **80**, 3956 (1958).

Kuhn, R., and Butula, I., *Justus Liebigs Ann. Chem.* **718**, 50 (1968).

Lazier, W. A., and Arnold, H. R., *Org. Synth. Collect. Vol.* **2**, 142 (1953).

Levine, S. G., and Wall, M. E., *J. Am. Chem. Soc.* **81**, 2829 (1959).

Lewis, J. B., Hedrick, G. W., and Settine, R. L., *J. Chem. Eng. Data.* **14**, 401 (1969).

Liska, K. J., and Salerni, L., *J. Org. Chem.* **25**, 124 (1960).

McAlees, A. J., and McCrindle, R., *J. Chem. Soc. C* p. 2425 (1969).

McCrindle, R. Overton, K. H., and Raphael, R. A., *Proc. Chem. Soc.* p. 313 (1961).

McCrindle, R., Overton, K. H., and Raphael, R. A., *J. Chem. Soc.* p. 4799 (1962).

Maley, F., Maley, G. F., and Lardy, H. A., *J. Am. Chem. Soc.* **78**, 5303 (1956).

Martin, J. C., Brannock, K. C., Burpitt, R. D., Gott, P. G., and Hoyle, V. A., Jr., *J. Org. Chem.* **36**, 2211 (1971).

Miya, B., Hoshino, F., and Ono, T., *Am. Chem. Soc., Div. Pet. Chem., Dallas, Tex. 1973; Chem. Abs.* **82**, 43112g (1975).

Mozingo, R., *Org. Synth., Collect. Vol.* **3**, 181 (1955).

Murata, A., and Kobayashi, H., Ger Patent 2, 043,349, Mar. 9, 1972; *Chem. Abstr.* **76**, 126378s (1972).

Peterson, P. E., and Casey, C., *J. Org. Chem.* **29**, 2325 (1964).

Rapala, R. T., Lavagnino, E. R., Shepard, E. R., and Farkas, E., *J. Am. Chem. Soc.* **79**, 3770 (1957).

Reithel, F. J., and Claycomb, C. K., *J. Am. Chem. Soc.* **71**, 3669 (1949).

Riener, T. W., *J. Am. Chem. Soc.* **71**, 1130 (1949).

Sakuragi, T., *J. Org. Chem.* **23**, 129 (1958).

Sayles, D. C., and Degering, E. F., *J. Am. Chem. Soc.* **71**, 3161 (1949).

Schreyer, R. C., U.S. Patent 2,862,977, Dec. 2, 1958.

Shriner, R. L., and Anderson, J., *J. Am. Chem. Soc.* **61**, 2705 (1939).

Shuikin, N. I., Bel'skii, I. F., and Shostakovskii, V. M., *Dokl. Akad. Nauk SSSR* **139**, 634 (1961).

Sicher, J., Sipos, F., and Jonas, J., *Collect. Czech. Chem. Commun.* **26**, 262 (1961).

Stoll, A., Peyer, J., and Hoffman, A., *Helv. Chim. Acta* **26**, 929 (1943).

Ukita, T., and Hayatsu, H., *J. Am. Chem. Soc.* **84**, 1879 (1962).

Westphal, O., and Stadler, R., *Angew Chem., Int. Ed. Engl.* **2**, 327 1963).

Yamaguchi, M., and Kageyama, Y., Ger Patent 2,144,316, Mar. 16, 1972; *Chem. Abstr.* **76**, 126380m (1972).

*Chapter* 5

# Hydrogenation of Aldehydes

Aldehydes are reduced readily to the corresponding alcohols. There is considerable difference in behavior between aliphatic and aromatic aldehydes, and they are conveniently discussed separately. Unsaturated aldehydes are an important class of compounds the reduction of which poses particular problems that vary depending on whether the aldehyde is aliphatic or a vinylogue of an aromatic aldehyde, such as cinnamaldehyde. The course of hydrogenation of aldehydes, especially unsaturated aldehydes, can be enormously influenced by various additives.

## I. SATURATED ALIPHATIC ALDEHYDES

Aliphatic aldehydes can be reduced readily over a variety of catalysts to the corresponding alcohol with little danger of overhydrogenation to the hydrocarbon. Platinum oxide has been widely used in laboratory practice for decades. It is often quickly deactivated, but reductions in which small amounts of certain salts, such as ferrous chloride or stannous chloride (Maxted and Akhtar, 1959), are added, proceed rapidly to completion. Early workers (Carothers and Adams, 1923, 1925; Tuley and Adams, 1925; Adams and Garvey, 1926) assumed that the additive functioned by preventing reduction of active platinum oxide to an inactive platinum of lower oxidation state, since the activity of deactivated platinum could be restored by shaking the solution with air. Inactivation is accompanied by coagulation of the catalyst, and later workers assumed that the function of promoters is to prevent coagulation merely by increasing the hydrogenation rate (Baltzly, 1976); coagulation seems to be especially severe in inherently slow reductions. Perhaps the promoter serves several functions. Platinum-on-carbon and ruthenium-on-carbon, as well as platinum oxide, are promoted by stannous chloride, the most effective of many salts tested (Rylander and Kaplan, 1961). This additive is effective at about one atom of tin per atom of noble metal; larger ratios tend to poison the catalyst.

Ruthenium is an excellent catalyst for the hydrogenation of aliphatic aldehydes in both continuous and batch processing. Ruthenium has been used to convert polysaccharides directly to polyhydric alcohols in high yields through a combination of acid hydrolysis and hydrogenation at 160°–170°C and 450–750 psig (Balandin et al., 1959; Sharkov, 1963). The sensitive intermediate aldehydes are reduced as rapidly as they are formed. Ruthenium-on-carbon is an excellent catalyst for reduction of dextrose to sorbital (Boyers, 1959). In the past this reduction has always been carried out industrially with nickel (Fedor et al., 1960; Phillips, 1963; Wright, 1974), but new plants may find ruthenium more advantageous. Ruthenium is especially advantageous in reductions of carbohydrates in aqueous media since water is a very strong and unique promoter for ruthenium hydrogenations.

Aldehydes with a strong tendency to exist in the enol form may undergo hydrogenolysis of the oxygen function. The reaction has been used to introduce a methyl function, as in the reduction of 2-formyl-5α-androstane-3,17-dione over palladium-on-carbon in ethanol–hydrochloric acid to afford 2α-methyl-5α-androstane-3,17-dione in 65% yield (deRuggieri, 1965). One might guess that platinum would be more efficient than palladium for hydrogenolysis of enols.

## A. Homogeneous Catalysts

Aldehydes have been reduced over a variety of homogeneous catalysts by hydrogen and by hydrogen transfer from appropriate donors (Kwiatek et al., 1962; Coffey, 1967; Tanaka et al., 1973; Schrock and Osborn, 1970; Malunowicz and Tyrlik, 1974; Sasson et al., 1974; Imai et al., 1976). These catalysts are all of theoretical interest, but as yet they seem to offer no particular advantage in synthetic applications.

## B. Acetals and Ethers

Hydrogenation of aldehydes in methanol or ethanol is apt to be accompanied by the formation of ethers through hydrogenolysis of intermediate acetals and hemiacetals. Palladium in the presence of hydrogen strongly promotes acetal formation even in basic media (Millman and Smith, 1977). Added or contaminating iron in platinum oxide may also promote acetal (Carothers and Adams, 1924) and ether formation (Lindgren, 1950). Reduction of aldehydes in alcohols containing dry hydrogen chloride over platinum oxide is the basis for a general synthesis of ethers (Verzele *et al.*, 1963). Ether formation can be prevented by the use of ruthenium catalysts and, of course, by avoiding primary alcohols as solvents.

## II.  UNSATURATED ALIPHATIC ALDEHYDES

Unsaturated aldehydes can be reduced at either or both points of unsaturation. Usually, with an appropriate choice of conditions, the saturated aldehyde, unsaturated alcohol, or saturated alcohol can each be obtained in high yield. Structure plays an important role in determining the relative difficulty of obtaining each of these compounds. As might be expected, steric hindrance around the double bond makes preferential reduction of the carbonyl group comparatively easier.

### A.  Saturated Aldehydes

There is usually very little difficulty in reducing an unsaturated aldehyde, conjugated or not, to the corresponding saturated aldehyde, provided that the double bond is not too hindered. Palladium is the preferred catalyst for this task, being excellent for saturation of double bonds and relatively poor for reduction of aliphatic aldehydes (Rylander and Himelstein, 1964). The reduction usually comes to a stop spontaneously, with the saturated aldehyde obtained in excellent yield. However, as hindrance around the double bond increases the carbonyl may be reduced preferentially, as illustrated by the preparation of 6-hydroxymethyl-3-enol ethers derived from various 6-formyl-$\Delta^{4,6}$ steroids in reduction over palladium in methanol containing sodium acetate (Burn *et al.*, 1964).

### B.  Unsaturated Alcohols

Unsaturated alcohols can be formed from unsaturated aldehydes, frequently in excellent yield, through preferential saturation of the carbonyl group. The yield of unsaturated alcohol may be diminished by subsequent hydrogenation of the double bond, hydrogenolysis of the allylic oxygen, and isomerization of the allylic alcohol to the saturated aldehyde (Simonik and Beranek, 1972).

Various modified platinum catalysts have been used for decades to reduce $\alpha,\beta$-unsaturated aldehydes to unsaturated alcohols. The earliest of these seems to have been unsupported platinum–zinc–iron (Tuley and Adams, 1925). A recent patent teaches the use of this catalyst combination, as exemplified by the reduction of $\beta$-methylcrotonaldehyde (Ichikawa *et al.*,

1977a):

$$
\begin{array}{c}
H_3C \\
\phantom{H_3}C{=}CHCHO \\
H_3C
\end{array}
\xrightarrow[\substack{0.08 \text{ gm } Zn(OAc)_2 \\ 450 \text{ psig}}]{\substack{0.5 \text{ gm } PtO_2 \\ 0.37 \text{ gm } FeSO_4}}
\begin{array}{c}
H_3C \\
\phantom{H_3}C{=}CHCH_2OH \\
H_3C
\end{array}
$$

93%

$$
+ \quad
\begin{array}{c}
H_3C \\
\phantom{H_3}CHCH_2CH_2OH \\
H_3C
\end{array}
+ \quad
\begin{array}{c}
H_3C \\
\phantom{H_3}CHCH_2CHO \\
H_3C
\end{array}
$$

3.5%                   1.5%

The same modifiers have been adapted to supported platinum. The support used is important; satisfactory yields of unsaturated alcohol were obtained with carbon and calcium carbonate, but not with barium sulfate or alumina (Rylander *et al.*, 1963). The solvent also influences the reduction; catalysts deactivated before the reduction was complete could be regenerated by shaking the reaction mixture with air when the solvent was ethanol, but not when hexane was used. Excellent results have been obtained with un-supported platinum modified by cobalt. The level of cobalt can be varied widely and should be increased with more difficult selectivity problems (Steiner, 1976). A catalyst derived from 250 mg platinum oxide and 15 mg cobalt acetate tetrahydrate reduced $\beta$-methylcrotonaldehyde in *n*-butanol to the unsaturated alcohol in 91% yield. Reduction of 6-acetoxy-4-methyl-hexa-2,4-dien-1-al gives the diunsaturated alcohol in 95% yield:

$$
\underset{O}{CH_3\overset{\|}{C}OCH_2CH{=}}\underset{CH_3}{C{=}}CH{=}CHCHO
\xrightarrow[\substack{600 \text{ mg } Co(OAc)_3 \cdot 4H_2O \\ 2000 \text{ ml } CH_3OH \\ 25°C, 1 \text{ atm}}]{5 \text{ gm } PtO_2}
\underset{O}{CH_3\overset{\|}{C}OCH_2CH{=}}\underset{CH_3}{C{=}}CH{=}CHCH_2OH
$$

71 gm                                                                 95%

Platinum oxide modified by nickel and iron is effective for reducing citral to citronellol in 96% yield at 10 atm and 70°C (Ichikawa *et al.*, 1977b). The unconjugated double bond remains, whereas the conjugated one is reduced selectively. Without iron present the yield of citronellol is only 83% (Ichikawa and Teizo, 1977). DeSimone and Gradeff (1977) used Raney nickel promoted by chromium for this reduction, which then proceeds largely with saturation of the conjugated double bond before carbonyl reduction.

96%

Raney nickel, whether or not modified by various salts, is not suitable for selective reduction of conjugated unsaturated aldehydes to unsaturated alcohols, but Raney cobalt is. The selectivity of Raney cobalt depends on the temperature at which it is developed (50°C is optimal). The selectivity is further enhanced by the addition of ferrous chloride or other salts (Hotta and Kubomatsu, 1969). The promoter was thought to function by decreasing the amount of strongly adsorbed hydrogen relative to the amount of weakly adsorbed hydrogen; strongly adsorbed hydrogen is thought to react more readily with the carbon–carbon double bond (Takeuchi and Asano, 1963). Promoters increase the rate constant for formation of the unsaturated alcohol, without affecting the rate constants for saturated aldehyde and saturated alcohol (Hotta and Kubomatsu, 1971, 1972, 1973).

$$CH_3CH_2CH\text{=}CCHO \xrightarrow[\substack{0.5\ mmole\ FeCl_2\\ isopropanol}]{1.0\ gm\ Co} CH_3CH_2CH\text{=}CCH_2OH$$
$$\underset{CH_3}{|} \qquad\qquad\qquad\qquad \underset{CH_3}{|}$$

$$89\%$$

Acrolein, crotonaldehyde, and cinnamaldehyde were reduced to the corresponding unsaturated alcohols in high yield over reduced osmium-on-carbon or osmium-on-alumina (Rylander and Steele, 1969), but the system is apt to be erratic. Allyl alcohol is obtained by vapor-phase hydrogenation of acrolein over supported rhenium (Vanderspurt, 1977a). Selectivity is improved sharply (to ca. 60%) by addition of carbon monoxide or carbon disulfide to the process stream (Vanderspurt, 1977b). An unusual catalyst for vapor-phase hydrogenation of acrolein is a silver–cadmium–zinc alloy, preferrably supported (Vanderspurt, 1978).

$$CH_2CHCHO \xrightarrow[\substack{490\ psig\\ 160°C\\ 7\text{-}9\ sec}]{Cd\text{-}Ag\text{-}Zn} CH_2\text{=}CHCH_2OH$$
$$100\%\ conversion,$$
$$70\%\ yield$$

Excellent yields of unsaturated alcohols can be obtained from prereduced rhenium heptoxide modified by addition of pyridine. A suitable catalyst is obtained by treating rhenium heptoxide (5 gm) is 150 ml purified dioxane with hydrogen at 2000 psig for 4 hr at 120°C (Pascoe and Stenberg, 1978). The yield of unsaturated alcohol declines only slightly with increasing temperature up to 140°C, beyond which the yield declines sharply. The same reduction can be carried out over 5% iridium-on-carbon to afford crotyl alcohol in 90% yield (Khidelkel et al., 1970) and over platinum–iron–zinc-on-carbon in 89% yield (Rylander et al., 1963).

$$CH_3CH\text{=}CHCHO \xrightarrow[\substack{0.6\ ml\ pyridine\\ 2000\ psig\\ 80°C}]{0.6\ gm\ prereduced\ Re_2O_7} CH_3CH\text{=}CHCH_2OH$$
$$0.2\ mole \qquad\qquad\qquad\qquad\qquad 87\%\ yield$$

## III.  AROMATIC ALDEHYDES

Aromatic and aliphatic aldehydes differ appreciably in behavior on reduction. For instance, aromatic aldehydes are reduced very easily over palladium and slowly over ruthenium, whereas the reverse is true for aliphatic aldehydes. Also, the alcohol derived from aliphatic aldehydes is resistant to further reduction under most conditions, whereas benzyl alcohols undergo hydrogenolysis readily. Reduction of aromatic aldehydes to the hydrocarbon proceeds stepwise (Meschke and Hartung, 1960) through the alcohol, which can usually be obtained in high yield.

Palladium is the preferred catalyst for most reductions of aromatic aldehydes; it is excellent both for reduction to the alcohol and for hydrogenolysis of the alcohol to the hydrocarbon. Excellent yields of the benzyl alcohol can be obtained without difficulty in neutral solvent when hydrogen absorption is limited to 1 mole. If the yield is less than quantitative, it can be improved by the use of lower catalyst loadings and/or traces of various basic inhibitors, such as tertiary amines and alkali hydroxides. Formation of the hydrocarbon, on the other hand, is favored by increased catalyst loadings and traces of strong acids such as sulfuric, perchloric, or hydrochloric acid. The inhibiting effect of amines and the promoting effect of acid on the hydrogenolysis of benzyl alcohols is illustrated by the reduction of 2,5-dimethyl-6-nitrobenzaldehyde. This compound could not be reduced to 2,3,6-trimethylaniline in neutral medium, even under high pressure, but when the intermediate methylolaniline was converted to its hydrochloride salt hydrogenolysis proceeded without difficulty (Sato et al., 1968).

Platinum, rhodium, ruthenium, and nickel have all been used successfully in the reduction of aromatic aldehydes, but in general they are not preferred to palladium unless some selectivity problem is involved. Platinum is preferred to palladium in the reduction of halo aromatic aldehydes when the halogen is to be preserved (Carothers and Adams, 1924; Gardner and McDonnell, 1941). Ruthenium is the preferred metal when the aromatic ring is to be reduced with preservation of the alcohol function (Howk, 1949; Ponomarev and Chegolya, 1962).

## IV.  AROMATIC UNSATURATED ALDEHYDES

Reduction of aromatic $\alpha,\beta$-unsaturated aldehydes is attended by a number of complications. Molecules such as cinnamaldehyde are vinylogues of benzaldehyde, and the aldehyde function behaves toward catalysts as if it were an aromatic aldehyde. The carbonyl function is reduced easily and competes with saturation of the carbon–carbon double bond. Products found in the hydrogenation of cinnamaldehyde include hydrocinnamaldehyde, cinnamyl alcohol, phenylpropanol, phenylpropene, and phenylpropane (Straus and Grindel, 1924). The product composition has been found to depend on the method of preparation of the catalyst, the solvent, metal supports, various additives (Rylander *et al.*, 1963; Rylander and Himelstein, 1964), as well as the amount of catalyst (Csuros, 1947, 1948; Csuros *et al.*, 1946). Despite the complexity, the reaction is controlled fairly easily.

### A.  Hydrocinnamaldehydes

Some hydrocinnamaldehydes can be formed without difficulty from cinnamaldehydes in high yield over palladium catalysts, either because of the intrinsic nature of the structure or because of its contained impurities (Skita, 1915; Bogert and Powell, 1931). Low yields and impure products arise from reduction of the carbonyl function before the double bond. Two good ways of circumventing this difficulty involve the use of catalyst modifiers.

Potassium salts of weak acids, such as potassium acetate or potassium carbonate, are especially effective in preventing the formation of cinnamyl alcohols and hydrocinnamyl alcohols. A reduction carried out with 20.2 gm *p-tert*-butyl-$\alpha$-methylcinnamaldehyde, 1.2 gm 5% palladium-on-alumina, and 0.044–0.053 gm potassium acetate at 100°C and 60 psig afforded only 1.2% of the undesired *p-tert*-butyl-$\alpha$-methyldihydrocinnamic alcohol. Sodium and lithium salts of weak acids are less effective promoters than are potassium salts, and potassium salts of stronger acids, such as potassium chloride, are less effective than potassium salts of weak acids (Dunkel *et al.*, 1970). Processing can be effected continuously (Arrigo *et al.*, 1970).

Another effective additive is ferrous sulfate used with palladium-on-carbon. Only hydrocinnamaldehyde is obtained and at a rapid rate. The selectivity is reproduced easily, but the iron–palladium ratios for maximal selectivity varied with catalyst among the catalysts tested from 1:0.9 to 1:1.3 (Rylander and Himelstein, 1964). It is interesting to compare these results with those of Tuley and Adams (1925), who, using platinum catalysts, found iron additives to give cinnamyl alcohol on adsorption of 1 equivalent of hydrogen and not hydrocinnamaldehyde.

Reduction of p-tert-butyl-α-methylcinnamaldehyde to the hydrocinnamaldehyde was effected in 86% yield by the use of complexed cobalt carbonyl catalysts under oxo conditions (Kogami et al., 1972). Higher yields can be obtained by the use of dicobalt octacarbonyl in the presence of various amines. The yields depend on both the base strength of the amine and its steric hindrance. With diisopropylamine in isopropanol at 107°C and 65 atm, $H_2:CO = 12:1$, the yield of the hydrocinnamaldehyde was 96.4% (Kogami and Kumanotani, 1973).

Rhodium compounds, such as $RhCl(\phi_3P)_3$, have been used to produce quantitative yields of hydrocinnamaldehyde. However, $Rh_2Cl_2(CO)_4$ under similar conditions gives cinnamyl alcohol in 94% yield (Mizoroki et al., 1977). Under oxo conditions $RhCl(CO)(\phi_3P)_2$ reduced double bonds in α,β-unsaturated carbonyl compounds preferentially (Ucciani et al., 1976).

$$\phi CH=CHCH_2OH \xleftarrow[80\text{ atm}]{C_6H_6,\ (C_2H_5)_3N} \phi CH=CHCHO \xrightarrow[80\text{ atm}]{C_6H_6,\ (C_2H_5)_3N} \phi CH_2CH_2CHO$$

$$\begin{array}{ccc}
94\% & \begin{array}{c} H_2:CO = 1:1 \\ 60°C \\ Rh_2Cl_2(CO)_2 \end{array} & \begin{array}{c} H_2:CO = 1:1 \\ 90°C \\ RhCl(\phi_3P)_3 \end{array} \qquad 100\%
\end{array}$$

## B.  Cinnamyl Alcohols

The carbonyl function in cinnamaldehydes is hydrogenated more easily than that in aliphatic unsaturated aldehydes but, nonetheless, hydrocinnamaldehydes are the major primary products of reduction over most catalysts. Selective hydrogenation of the carbonyl function in cinnamaldehydes has been achieved over a variety of special catalysts. Perhaps the earliest of these was unsupported platinum–iron–zinc (Tuley and Adams, 1925). This combination of metals has also been used with carbon and calcium carbonate as supports (Rylander et al., 1963). Yields are above 90%. Reduction of cinnamaldehyde over 5% iridium-on-carbon (Khidekel et al., 1970) or over 5% osmium-on-carbon (Rylander and Steele, 1969) at elevated pressures has given high yields of cinnamyl alcohol. At 2000 psig and temperatures less than 120°C, reduction of cinnamaldehyde and α-methylcinnamaldehyde over prereduced rhenium heptoxide inhibited by pyridine affords the corresponding unsaturated alcohols in nearly quantitative yields (Pascoe and Stenberg, 1979).

Cinnamaldehyde is reduced by hydrogen–carbon monoxide (1:1) over rhodium trichloride in benzene containing N-methylpyrrolidine at 90°C and 80 atm to afford cinnamyl alcohol in 83% yield. N-Methylpyrrolidine was the most satisfactory of several tertiary amines tested. α-Methylcinnamaldehyde is reduced similarly to the unsaturated alcohol (Meguro et al., 1975).

## REFERENCES

Adams, R., and Garvey, B. S., J. Am. Chem. Soc. **48**, 477 (1926).
Arrigo, J. T., Christensen, N. J., Chrysler, R. L., and Sparks, A. K., U.S. Patent 3,520,935, July 21, 1970
Balandin, A. A., Vasyunina, N. A., Chepigo, S. V., and Barysheva, G. S., Dokl. Akad. Nauk SSSR **128**, 941 (1959).
Baltzly, R., J. Org. Chem. **41**, 933 (1976).
Bogert, M. T., and Powell, G., J. Am. Chem. Soc. **53**, 2747 (1931).
Boyers, G. G., U.S. Patent 2,868,847, Jan. 13, 1959.
Burn, D., Cooley, G., Davies, M. T., Ducker, J. W., Ellis, B., Feather, P., Hiscock, A. K., Kirk, D. N., Leftwick, A. P., Petrow, V., and Williamson, D. M., Tetrahedron **20**, 597 (1964).
Carothers, W. H., and Adams, R., J. Am. Chem. Soc. **45**, 1071 (1923).
Carothers, W. H., and Adams, R., J. Am. Chem. Soc. **46**, 1675 (1924).
Carothers, W. H., and Adams, R., J. Am. Chem. Soc. **47**, 1047 (1925).
Coffey, R. S., Chem. Commun. p. 923 (1967).
Csuros, Z., Muegy. Kozl. Budapest p. 110 (1947).
Csuros, Z., Magy. Kem. Lapja **3**, 29 (1948).
Csuros, Z., Zech, K., and Geczy, I., Hung. Acta Chim. **1**, 1 (1946).
deRuggieri, P., U.S. Patent 3,207,752, Sept. 21, 1965.
DeSimone, R. S., and Gradeff, P. S., U.S. Patent 4,029,709, June 14, 1977.
Dunkel, M., Eckhardt, D. J., and Stern, A., U.S. Patent 3,520,934, July 21, 1970.
Fedor, W. S., Millar, J., and Accola, A. J., Jr., Ind. Eng. Chem. **52**, 282 (1960).
Gardner, J. H., and McDonnell, T. F., J. Am. Chem. Soc. **63**, 2279 (1941).
Hotta, K., and Kubomatsu, T., Bull. Chem. Soc. Jpn. **42**, 1447 (1969).
Hotta, K., and Kubomatsu, T., Bull. Chem. Soc. Jpn. **44**, 1348 (1971).
Hotta, K., and Kubomatsu, T., Bull. Chem. Soc. Jpn. **45**, 3118 (1972).
Hotta, K., and Kubomatsu, T., Bull. Chem. Soc. Jpn. **46**, 3566 (1973).
Howk, B. W., U.S. Patent 2,487,054, Nov. 8, 1949.
Ichikawa, Y., and Teizo, Y., Jpn. Patent 77 12,106, Jan. 29, 1977; Chem. Abstr. **87**, 52778a (1977).
Ichikawa, Y., Suzuki, M., and Sawaki, T., Jpn. Patent 77 84,193, July 13, 1977; Chem. Abstr. **87**, 200794v (1977a).
Ichikawa, Y., Yamaji, T., and Sawaki, T., Jpn. Patent 77 46,008, Apr. 12, 1977; Chem. Abstr. **87**, 136052x (1977b).
Imai, H., Nishiguchi, T., and Fukuzumi, K., J. Org. Chem. **41**, 665 (1976).
Khidekel, M. L., Bakhanova, E. N., Astakhova, A. S., Brikenshtein, K. A., Savchenko, V. I., Monakhova, I. S., and Dorokhov, V. G., Izv. Akad. Nauk SSSR, Ser. Khim. p. 499 (1970).
Kogami, K., and Kumanotani, J., Bull. Chem. Soc. Jpn. **46**, 3562 (1973).
Kogami, K., Takahashi, O., and Kumanotani, J., Bull. Chem. Soc. Jpn. **45**, 604 (1972).
Kwiatek, J., Mador, I. L., and Seyler, J. K., J. Am. Chem. Soc. **84**, 304 (1962).
Lindgren, B. O., Acta Chem. Scand. **4**, 1365 (1950); Chem. Abstr. **45**, 6602 (1951).
Malunowicz, E., and Tyrlik, S., J. Organomet. Chem. **72**, 269 (1974).
Maxted, E. B., and Akhtar, S., J. Chem. Soc. p. 3130 (1959).
Meguro, S., Mizoroki, T., and Ozaki, A., Chem. Lett. p. 943 (1975).

Meschke, R. W., and Hartung, W. H., *J. Org. Chem.* **25**, 137 (1960).
Millman, W. S., and Smith, G. V., *in* "Catalysis in Organic Syntheses, 1977," (G. V. Smith, ed.), p. 48. Academic Press, New York, 1977.
Mizoroki, T., Seki, K., Meguro, S., and Ozaki, A., *Bull. Chem. Soc. Jpn.* **50**, 2148 (1977).
Pascoe, W. E., and Stenberg, J. F., *in* "Catalysis in Organic Syntheses, 1978," (W. H. Jones, ed.) Academic Press, New York, 1979.
Phillips, M. A., *Br. Chem. Eng.* **8**, 767 (1963).
Ponomarev, A. A., and Chegolya, A. S., *Dokl. Akad. Nauk. SSSR* **145**, 812 (1962).
Rylander, P. N., and Himelstein, N., *Engelhard Ind., Tech. Bull.* **4**, 131 (1964).
Rylander, P. N., and Kaplan, J., *Engelhard Ind., Tech. Bull.* **2**, 48 (1961).
Rylander, P. N., and Steele, D. R., *Tetrahedron Lett.* p. 1579 (1969).
Rylander, P. N., Himelstein, N., and Kilroy, M., *Engelhard Ind., Tech. Bull.* **4**, 49 (1963).
Sasson, Y., Albin, P., and Blum, J., *Tetrahedron Lett.* p. 833 (1974).
Sato, K., Fujima, Y., and Yamada, A., *Bull. Chem. Soc. Jpn.* **41**, 442 (1968).
Schrock, R. R., and Osborn, J. A., *Chem. Commun.* p. 567 (1970).
Sharkov, V. I., *Angew. Chem., Int. Ed. Engl.* **2**, 405 (1963).
Simonik, J., and Beranek, L., *Collect. Czech. Chem. Commun.* **37**, 353 (1972).
Skita, A., *Ber. Dtsch. Chem. Ges.* **48**, (1915).
Steiner, K., U.S. Patent 3,953,524, Apr. 27, 1976.
Straus, F., and Grindel, H., *Justus Liebigs Ann. Chem.* **439**, 276 (1924).
Takeuchi, T., and Asano, T., *Z. Phys. Chem. (Frankfurt am Main)* **36**, 118 (1963).
Tanaka, M., Watanabe, Y., Mitsudo, T., Iwane, H., and Takagi, Y., *Chem. Lett.* p. 239 (1973).
Tuley, W. F., and Adams, R., *J. Am. Chem. Soc.* **47**, 3061 (1925).
Ucciani, E., Lai, R., and Tanguy, L., *C. R. Acad. Sci., Ser. C* **283**, 17 (1976); *Chem. Abstr.* **86** 16257a (1977).
Vanderspurt, T. H., U.S. Patent 4,048,110. Sept. 13, 1977a.
Vanderspurt, T. H., U.S. Patent 4,020,116, Apr. 26, 1977b.
Vanderspurt, T. H., U.S. Patent 4,072,727, Feb. 7, 1978.
Verzele, M., Acke, M., and Anteunis, M., *J. Chem. Soc.* p. 5598 (1963).
Wright, L. W., *Chemtech.* p. 42 (1974).

*Chapter* 6

# Hydrogenation of Ketones

Catalysts in ketone hydrogenation vary widely in activity. Their relative effectiveness depends greatly on whether the ketone is aromatic or aliphatic. Overhydrogenation with loss of the oxygen function occurs rarely in non-activated aliphatic ketones but, in aromatic ketones, hydrogenolysis as well as ring saturation may be important side reactions. For these reasons aliphatic and aromatic ketones are best considered separately.

## I. ALIPHATIC KETONES

### A. Catalysts

Palladium, platinum, rhodium, ruthenium, iridium, osmium, copper chromite, and nickel catalysts have been used successfully in ketone hydrogenation. Palladium, excellent for the reduction of aromatic ketones, is relatively ineffective in the hydrogenation of nonactivated aliphatic ketones and is not generally recommended for the latter purpose. Its low activity for aliphatic carbonyl reduction often makes it, however, a catalyst of choice for selective saturation of double bonds in unsaturated carbonyl compounds. Osmium makes a sluggish catalyst for ketone hydrogenation; it has been used to minimize ring saturation and hydrogenolysis in carbonyl-containing aromatic molecules such as phenylacetone (Rylander and Hasbrouck, 1969, 1971). Ruthenium, commonly used at elevated pressures, has given some excellent results in the hydrogenation of aliphatic ketones (Raasch, 1958; Rapala and Farkas, 1958; Zirkle *et al.*, 1961; Hasek *et al.*, 1961; Hasek and Martin, 1963). Long induction periods may be encountered if ruthenium is used at low pressures (Smissman *et al.*, 1964). Rhodium functions effectively under mild conditions and has given excellent results even though other catalysts failed (Adams *et al.*, 1960, 1961; Nair and Adams, 1961). Platinum catalysts, especially platinum oxide, have been used by many workers.

Iridium catalysts are used rarely. In the hydrogenation of 2,5-hexanedione to 2,5-dimethyltetrahydrofuran, iridium proved to be 5–50 times more active than any other platinum metal catalyst (Rylander and Steele, 1967), but this exceptional activity is not general. Nickel catalysts are often used under vigorous conditions (Adkins and Connor, 1931; Howe and Haas, 1946; Hurd and Perletz, 1946) but, if large amounts of catalyst are used, mild conditions can be employed (Ruzicka *et al.*, 1945; Campbell and Hunt, 1950; Blance and Gibson, 1954; Meinwald and Frauenglass, 1960; Harnik, 1963; Ebersole and Chang, 1973). Copper chromite always requires vigorous conditions (Fuson and Corse, 1945; Hey and Musgrave, 1949; Jacobs and Goerner, 1956).

In an examination of characteristic properties of catalysts influencing the hydrogenation of ketones, Simoniková *et al.* (1973) concluded that differences in catalysts are more pronounced in the adsorption step than in the reaction step. The strength of adsorption of homologous ketones rises in the series acetone, 2-butanone, 4-methyl-2-pentanone, 3-methyl-2-butanone, and 3,3-dimethyl-2-butanone, following increasing electropositivity of the alkyl groups; the ordering of the influence of ketone structure on rate is opposite to its ordering for adsorbability (Simoniková *et al.*, 1973).

## B. Solvents

Solvents have a marked effect on the rate of hydrogenation of ketones (Breitner *et al.*, 1959). The postulate that hydrogen, ketone, and solvent all compete for the same catalyst sites has been used successfully to account for various solvent-related phenomena (Kishida and Teranishi, 1968). The more strongly the solvent is adsorbed, the greater is its fractional coverage of catalyst sites, and the less the fractional coverage by ketone and hydrogen, the surface concentrations of which determine the rate of reaction. Consequently, the rate should be highest in weakly adsorbed solvents. In the hydrogenation of acetone over Raney nickel, the rates decrease with solvent in the order *n*-hexane, cyclohexane, methanol, isopropanol. There is also a predictable dependence of rate on acetone concentration, with the rate curve passing through a maximal concentration, which varies with the solvent. It follows from this treatment that under certain conditions the apparent activation energy can be zero and even negative, as is sometimes observed (Maxted and Moon, 1935). The above scheme of competitive adsorption would not apply and would be invalid if there were interaction of solvent and substrate in solution or among adsorbed species.

Small quantities of acid (Scheutte and Thomas, 1930; Shriner and Witte, 1941; Babcock and Fieser, 1952; Tschudi and Schinz, 1952; Roy and Wheeler, 1963; Roberts, 1965) or base (Lieber and Smith, 1936; Delepine and Horeau, 1937; Adkins and Billica, 1948; Forrest and Tucker, 1948; Hennion and Watson, 1958; Takagi, 1963; Freifelder et al., 1964) may have pronounced effects on the reduction of a ketone, altering the rate or product or both. Metal oxide and metal black catalysts especially are apt to contain sufficient occluded alkali remaining from their preparation to give results different from those obtained with alkali-free catalysts. Usually acid or alkali has promoting effects on ketone reduction, but inhibition has also been noted for each. Acid promotion lends support to the postulation that protonated ketones are intermediates in the hydrogenation (Brewster, 1954). The function of alkali in promotion is often obscure. It may remove inhibiting acid, increase the enol content, decrease hydrogen availability at the catalyst surface, increase heterolytic splitting of hydrogen, or change the product-determining step by altering the adsorbability of the substrate (Mitsui et al., 1973). In summary, so many examples of promotion by alkali or acid are known that it would seem worthwhile to explore the effects of pH in optimizing a reaction despite the lack of an adequate rationale.

### Ketal and Ether Formation

A special situation exists when methanol or ethanol is used as solvent for ketone hydrogenation. In the presence of hydrogen and palladium, ketals may be formed rapidly, and these in turn may undergo hydrogenolysis to the ether. Hydrogenation of $5\alpha$-cholestan-3-one in methanol over pre-reduced palladium hydroxide gives 91% $3\beta$-methoxy- and 7% $3\alpha$-methoxy-$5\alpha$-cholestane with small amounts of $5\alpha$- and $5\beta$-cholestane. Under the same conditions, the products from $5\beta$-cholestan-3-one are 92% $3\beta$-methoxy- and 4% $3\alpha$-methoxy-$5\beta$-cholestane together with 4% $5\beta$-cholestan-3-ol (Nishimura et al., 1967).

TABLE I

Hydrogenation of 4-Methylcyclohexanone in Ethanol[a]

| Catalyst | Product |
|---|---|
| Palladium | 96.6% 1-ethoxy-4-methylcyclohexane, 89.8% cis |
| Platinum | 74.7% 4-methylcyclohexanol, 83.3% cis |
| Rhodium | 70.1% 4-methylcyclohexanol, 83.0% cis, plus ketal |
| Ruthenium | 100% 4-methylcyclohexanol, 69.0% cis |

[a] Data of Nishimura et al. (1967). Used with permission.

This ether-forming reaction was examined by the use of 4-methylcyclo-hexanone in ethanol and prereduced palladium, ruthenium, and rhodium hydroxides and prereduced platinum oxide, as catalyst, with the results summarized in Table I (Nishimura *et al.*, 1967).

Clearly, palladium is the preferred catalytic metal when the ether is desired, and ruthenium is preferred for obtaining the alcohol. Hydrogenolysis of ketals is the basis of certain ether syntheses (Howard and Brown, 1961; Verzele *et al.*, 1963).

## C.  Hydrogenolysis

Hydrogenolysis of nonactivated ketones has been observed, but it is mostly a trival side reaction (Carothers and Adams, 1925; Foresti, 1937, 1939; Koizumi, 1940) unless special structural features are present, such as those found in the cyclitol series (Posternak, 1941; Post and Anderson, 1962). Hydrogenolysis usually is promoted by acids and inhibited by bases, but the reverse trend has been noted in certain keto acids (Dauben and Adams, 1948).

Reiff and Aaron (1967) employed catalytic hydrogenolysis as a successful alternative to the Wolff–Kishner or Clemmensen reduction for a number of azabicyclic ketones. Hydrogenolysis of the ketone relative to reduction to the carbinol is favored by increased acid and catalyst concentrations, although sensitivity of the reaction to these parameters varies from one system to another. The hydrogenolysis does not go through an intermediate carbinol and seems to depend on the proximity of an ammonium function.

Extensive hydrogenolysis is apt to occur in the reduction of compounds such as $\beta$-keto esters (Walker, 1958), $\beta$-ketoamides (Julian *et al.*, 1935), and $\beta$-diketones (Patrikeev and Liberman, 1948; Walker, 1958; Rylander and Steele, 1965), all of which show a tendency to enolize. In certain cases the ability to enolize has been shown to be a prerequiste for hydrogenolysis (Julian *et al.*, 1935). Hydrogenolysis is also likely to occur readily with 1,4-diketones, such as acoric acid (Birch *et al.*, 1964), or 1,5-diketones (Shuikin and Vasilevskaya, 1964), which cyclize on hydrogenation to form tetrahydrofurans and tetrahydropyrans. The extent of hydrogenolysis depends on the substrate structure, catalyst (Misani *et al.*, 1956; Burton and Stevens, 1963; Burton *et al.*, 1964), amount of catalyst (Jeanloz *et al.*, 1947), and solvent (Orchin and Butz, 1943; Smith, 1953).

## 1. Keto Esters

Tables II–V, which give data on the hydrogenation of acetoacetic ester and 2-acetylbutyrolactone, support the generality that the extent of hydrogenolysis and stereochemistry, wherever applicable, may depend to a large degree on the catalyst, solvent, and substrate structure (Rylander and Starrick, 1966). Hydrogenolysis can be sharply curtailed by the addition of one atom of iron, as iron chloride, or zinc, as zinc acetate, per atom of catalytic metal. Reasons for the observed ordering of solvents and catalysts

TABLE II

Effect of Type of Platinum Catalyst and Additives on Hydrogenation of Acetoacetic Ester[a]

| Catalyst (mg) | Additive | % Ethyl butyrate | % Ethyl $\beta$-hydroxybutyrate |
|---|---|---|---|
| Pt black (200) | None | 90 | 10 |
| Pt black (200) | FeCl$_3$ | 0 | 100 |
| PtO$_2$ (200) | None | 0 | 100 |
| Pt black (200) (prereduced PtO$_2$) | None | 0 | 100 |
| 5% Pt/C (500) | None | 12 | 88 |
| 5% Pt/C (500) | Zn(OAc)$_2$ | 0 | 100 |

[a] All hydrogenations were carried out at room temperature and atmospheric pressure in water until they stopped spontaneously. No starting material remained.

TABLE III

Effect of Solvent on Hydrogenation of Ethyl Acetoacetate over Platinum Black and Platinum-on-Carbon[a]

| Solvent | % Enol (18°C) | % Ethylbutyrate | | % Ethyl $\beta$-hydroxybutyrate | |
|---|---|---|---|---|---|
| | | Pt black | 5% Pt/C | Pt black | 5% Pt/C |
| Water | 0.40 | 90 | 12 | 10 | 88 |
| Hexane | 46.4 | 92 | 38 | 8 | 62 |
| Acetic acid | 5.7 | 95 | 19 | 5 | 81 |
| Ethyl ether | 27.1 | 47 | [b] | 53 | [b] |
| Tetrahydrofuran | | 52 | 0 | 48 | 100 |

[a] All experiments were carried out at room temperature and atmospheric pressure.
[b] No reduction.

TABLE IV

**Effect of Metal on Product in Hydrogenation of 2-Acetylbutyrolactone**[a]

| | | Composition of product | | |
| | | % Hydroxylactone | | |
| Catalyst | % 2-Ethylbutyrolactone | $A_N$ | $A_B$ | $A_N/A_B$ |
|---|---|---|---|---|
| 5% Rh/C | 63.0 | 12.0 | 25.0 | 0.48 |
| 5% Pt/C | 39.4 | 42.4 | 18.2 | 2.32 |
| 5% Pd/C | 5.3 | 31.4 | 63.3 | 0.49 |
| 5% Ru/C | 0 | 40.0 | 60.0 | 0.67 |
| 5% Ir/C | 0 | 56.4 | 43.6 | 1.29 |
| 5% Ir/Al$_2$O$_3$ | 0 | 60.0 | 40.0 | 1.50 |
| Pt black | 39.0 | 55.5 | 5.5 | 10.1 |
| Ir black | 3.7 | 51.8 | 44.5 | 1.16 |
| Rh black | 8.0 | 35.4 | 56.6 | 0.62 |

[a] Each experiment was carried out at room temperature and atmospheric pressure in water until absorption ceased spontaneously.

TABLE V

**Effect of Solvent on Hydrogenation of 2-Acetylbutyrolactone over Platinum-on-carbon**[a]

| | | Composition of product | | |
| | | % Hydroxylactone | | |
| Solvent | % 2-Ethylbutyrolactone | $A_N$ | $A_B$ | $A_N/A_B$ |
|---|---|---|---|---|
| Methanol | 60.8 | 29.5 | 9.7 | 3.02 |
| Propanol | 60.0 | 25.1 | 14.9 | 1.68 |
| Isopropanol | 50.1 | 32.5 | 17.4 | 1.87 |
| Water | 39.4 | 42.4 | 18.2 | 2.32 |
| Ethyl acetate | 55.3 | 28.2 | 16.5 | 1.71 |
| Hexane, 83%; acetic acid, 17% | 25.0 | 47.7 | 27.3 | 1.75 |
| Chloroform | 35.0 | 55.0 | 10.0 | 5.5 |
| Tetrahydrofuran | 58 | 30 | 12 | 2.5 |
| Acetic acid | 52 | 36 | 12 | 3.0 |

[a] Each experiment was carried out at room temperature and atmospheric pressure.

do not readily emerge from these data; it must suffice at present to note that large changes can be effected. The hydroxylacetones are designated $A_N$ and $A_B$ (referring to narrow and broad hydroxyl absorption bands in the infrared spectrum), but their configuration was not firmly established. Ethylbutyrol-actone and ethyl butyrate do not arise from hydrogenolysis of the corresponding hydroxy compounds.

## 2. Diketones

An interesting means of selectively converting diketones to monoketones is by catalytic hydrogenolysis of intermediate unsaturated oxyphosphoranes in benzene, cyclohexane, or ethyl acetate. Since diketones are available from monoketones, the hydrogenolysis is also a potential solution for the difficult synthetic problem of ketone transposition (Stephenson and Falk, 1976). Hydrogenolysis is discussed further in Section I,E.

## D. Unsaturated Ketones

Generally, hydrogenation of simple unsaturated aliphatic ketones proceeds with preferential saturation of the olefinic function. In some molecules, such as mesityl oxide, this selectivity is maintained regardless of catalyst (Breitner et al., 1959), but in others the catalyst may have a marked effect on selectivity (Albertson, 1952; Phillips and Mentha, 1956; Freifelder and Stone, 1961; Koelsch and Ostercamp, 1961). The selective hydrogenation of 2 provides an illustration of the differences sometimes shown between palladium and platinum. Over palladium-on-carbon the double bond was selectively reduced to afford 3 in high yield, whereas over platinum oxide the double bond remained intact during reduction of the less hindered ketone to give 1 in 73% yield. The yield of 3 was sensitive to the palladium catalyst used. Fresh catalyst invariably led to overreduction, but this could be eliminated by the addition of small amounts of triethylamine; with suitable conditions the yields were nearly quantitative (France et al., 1969).

Kaye and Matthews (1964) reported preferential reduction of carbonyls in an unsaturated dione system over ruthenium and preferential saturation of the double bonds over palladium. One might gather from these limited examples that selective hydrogenation of the carbonyl function in unsaturated ketones can best be achieved over either platinum or ruthenium, and the double bond preferentially hydrogenated over palladium, if indeed the course of the reduction can be changed by a change in catalyst. Deuterium exchange experiments suggest that double-bond saturation in some $\alpha,\beta$-unsaturated ketones proceeds through enols obtained by 1,4 addition of hydrogen (Friedlin *et al.*, 1973).

Rhodium-on-alumina is effective in reducing both the carbonyl and double bond (**4**) in 6-allyl-1,3-dimethyl-5-hydroxyuracil (**5**), whereas only the allylic bond (**6**) is removed over palladium after absorption of 1 mole of hydrogen. The hydroxyl group can be removed by condensation with 1-phenyl-5-chlorotetrazole followed by hydrogenolysis over palladium (**7**). This appears to be the first example of the replacement of a 5-hydroxyl group of a pyrimidine by hydrogen (Otter *et al.*, 1971).

## E.  Selective Hydrogenation of Diketones

Diketones can be reduced to hydroxy ketones (Blomquist and Wolinsky, 1955) and to diols. The yield of the latter compounds depends in large measure on the extent to which hydrogenolysis can be avoided. Selectivity to the hydroxy ketone varies greatly with catalyst and with solvent, as illustrated by the data of Table VI on the hydrogenation of biacetyl and acetylacetone (Rylander and Vaflor, 1975). Diol is not found in the initial stages of reduction, suggesting that the less than 100% yields of hydroxy ketone arise partly through hydrogenolysis and partly through competition of hydroxy ketone with diketone for catalyst sites (thermodynamic selectivity). With attention to catalyst and solvent, excellent yields of either hydroxy ketone or diol can be obtained. Excellent yields of hydroxybutanone can also be obtained by hydrogenation of biacetyl over palladium-on-carbon at 105°–125°C (Skibina et al., 1976).

TABLE VI

Hydrogenation of Diketones[a]

| Catalyst | Solvent | Maximal % hydroxy ketone | Maximal % diol |
|---|---|---|---|
| | Biacetyl | | |
| 5% Pd-on-C | Cyclohexane | 99 | 99 |
| 5% Rh-on-C | Cyclohexane | 92 | 97 |
| 5% Pt-on-C | Cyclohexane | 88 | 97 |
| 5% Ru-on-C | Cyclohexane | 63 | 74 |
| | Acetylacetone | | |
| 5% Pd-on-C | Cyclohexane | 86 | 99 |
| | Isopropanol | 91 | 99 |
| 5% Rh-on-C | Cyclohexane | 60 | 68 |
| | Isopropanol | 68 | 77 |
| 5% Pt-on-C | Cyclohexane | 27 | 92 |
| | Isopropanol | 58 | 97 |
| 5% Ru-on-C | Cyclohexane | 35 | 99 |
| | Isopropanol | 59 | 79 |
| 5% Pd-on-Al$_2$O$_3$ | Methanol | 95 | — |
| | Isopropanol | 90 | — |
| | Cyclohexane | 86 | — |
| | Tert-Butanol | 78 | — |
| | Tetrahydrofuran | 76 | — |

[a] Pressure 500 psig; temperature 25°C.

In the hydrogenation of tetramethyl-1,3-cyclobutanedione to the diol, ruthenium proved to be an excellent catalyst and more effective than palladium, platinum, or rhodium (Hasek *et al.*, 1961). Hydrogenation of certain cyclic diketones may proceed with ring closure either through transannular aldolization (Buchanan *et al.*, 1963) or by the establishment of a new carbon–carbon bond through interaction of both carbonyl functions to give a bicyclic diol (Cope and Kagan, 1958). Hydrogenation of $\delta$-diketones affords tetrahydropyrans and, through aldolization reactions, cyclohexanone derivatives (Shuikin and Vasilevskaya, 1964). Hydrogenation of $\gamma$-diketones affords tetrahydrofurans in yields that depend on the catalyst, solvent, and substrate structure (Rylander and Vaflor, 1975).

Hydrogenation of 5,5-dimethyl-1,3-cyclohexadione (**8**) over nickel in methanol at 2250 psig and 80°C gave the *cis*-diol **9** in 94% yield, whereas hydrogenation of 2,2-dimethyl-1,3-cyclohexadione (**10**) gave the trans isomer **11** (Mironov *et al.*, 1973). The high yield of diol under these conditions contrasts with the extensive hydrogenolysis that occurs on reduction of **8** over platinum under mild conditions (Liberman and Kazanskii, 1946).

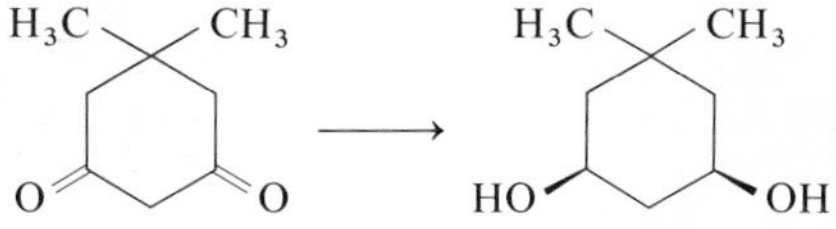

8 &rarr; 9

10 &rarr; 11

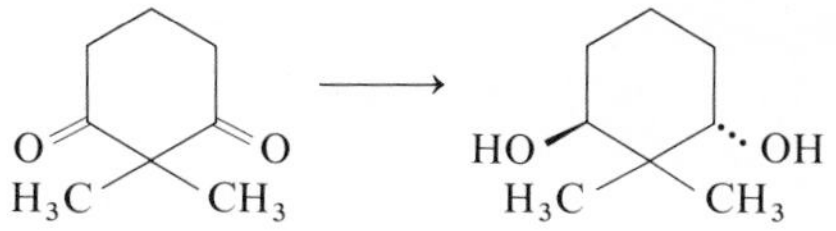

## F. Aldol Condensations

A variety of products can be derived by aldol condensations of ketones followed by hydrogenation. Under favorable conditions both reactions can be carried out at the same time (Imai *et al.*, 1971).

$$2CH_3CCH_3 \xrightarrow[\substack{\text{Pd-on-Y, zeolite} \\ \text{600 psig}}]{200°C} CH_3CCH_2C(CH_3)_2 + H_2O$$

Other workers kept the acidic function and the hydrogenation catalyst separate, for example, by mixing zeolite with palladium-on-carbon. The yield of methyl isobutyl ketone was 96% (Takagi *et al.*, 1970; Takagi and Murakami, 1970).

## G.  Stereochemistry

Over the years various investigators have devised rules for predicting the stereochemical outcome of ketone hydrogenation with some success (Barton, 1953; Wicker, 1956; Findlay, 1959), but exceptions abound. The number of parameters affecting stereochemistry is such that it is almost impossible to make any simple encompassing statement. Some idea of the variation in stereochemistry induced by varying reaction parameters can be gained from studies of hydrogenation of substituted cyclohexanones (Anziani and Cornubert, 1945; Wicker, 1956, 1957; Eliel and Lukach, 1957; Hückel and Hubele, 1958; Hückel and Maier, 1958; Rylander and Steele, 1963).

### 1.  Catalysts

The catalytic metal greatly influences the result, as indicated in Table VII, which gives data on the hydrogenation of 4-*tert*-butylcyclohexanone (Takagi *et al.*, 1973).

TABLE VII

Hydrogenation of 4-tert-Butylcyclohexanone
in Cyclohexane[a]

| Catalyst | Ratio of *cis*- to *trans*-4-*tert*-butylcyclohexanol |
|---|---|
| Rhodium black | 2.1 |
| Palladium black | 1.6 |
| Ruthenium black | 1.5 |
| Osmium black | 1.3 |
| Iridium black | 0.91 |
| Platinum black | 0.38 |

[a] Data of Takagi *et al.* (1973). Used with permission.

These results were accounted for by the assumption of two different adsorption modes on the catalyst. Mode B leading to the *trans*-alcohol is preferred for palladium, rhodium, ruthenium, and osmium. The assumption of these modes of adsorption has been elaborated to also account for results of deuteration. Deuteration over ruthenium, osmium, iridium, and platinum gives alcohols containing only C-1 deuterium, whereas over palladium and rhodium about 50% of the atoms at C-2 and C-6 are deuterium and, interestingly, with equal amounts in the axial and equatorial positions. Enol forms of the ketone are not considered important intermediates in these deuterations (Takagi *et al.*, 1973).

The mean number of deuterium atoms incorporated into the cyclohexanol molecule falls in the sequence Rh > Pd > Ru > Pt $\simeq$ Ir $\simeq$ Os, which is similar to the order established by Bond and Wells (1964) or Wells and Wilson (1967) for olefin isomerization. The similarity in metal orderings for deuteration and isomerization suggests a similar mechanistic sequence for the two reactions (Takagi et al., 1973).

Table VIII gives stereochemical data relating to the alcohols formed on hydrogenation of various alkylcyclohexanones. Rhodium-on-carbon is generally the most effective and platinum oxide the least effective in producing axial alcohols. It is at present uncertain how far this generality can be extended. Not shown in the table is the effect of alkali; it tends to increase the percentage of axial alcohol obtained over nickel but to decrease the percentage of axial alcohol when used with rhodium, palladium, or platinum. The effect of alkali removal from platinum oxide, shown on the last line, is impressive. Stereochemistry is also related to hydrogen availability at the catalyst surface. Mechanisms inferred from these findings have been discussed at length (Mitsui et al., 1973). Most phenomena are accounted for by considering whether the stereochemistry of the product is determined mainly by the configuration of the ketone in its initial adsorption state or by its configuration at the transition state.

The system rhodium in isopropanol–hydrochloric acid or in tetrahydrofuran–hydrochloric acid has been claimed to be one of the best means of producing axial alcohols from unhindered ketones by hydrogenation (Nishimura et al., 1977b). For instance, hydrogenation of 5α-cholestan-3-one

**TABLE VIII**

**Percent Axial Alcohol Produced in Hydrogenation of Alkylcyclohexanones in Ethanol[a]**

| Catalyst | Compound | | | | | | |
|---|---|---|---|---|---|---|---|
|  | 2-$CH_3$ | 3-$CH_3$ | 4-$CH_3$ | 4-*tert*-$C_4H_9$ | 3-*tert*-$C_4H_9$ | 3-$C_6H_5$ | 2-$C_6H_{11}$ |
| $PtO_2$ | 42 | 23 | 34 | 38 | 41 | 20 | 65 |
| RaNi | 63 | 52 | 65 | 74 | 64 | 60 | 88 |
| Ni boride | 72 | 65 | 77 | 85 | 56 | — | 88 |
| Rh-on-C | 70 | 88 | 84 | 94 | 94 | — | 87 |
| Pd-on-C | 39 | 40 | 44 | 44 | 45 | — | — |
| Pt black[b] | 80 | 44 | 53 | 57 | 52 | 42 | 92 |

[a] Data of Mitsui *et al.* (1973). Reprinted with permission of Pergamon Press.
[b] Platinum oxide reduced by hydrogen and washed well with water.

over platinum in *tert*-butyl alcohol gives predominantly the equatorial alcohol 5α-cholestan-3β-ol, whereas hydrogenation over rhodium in iso-propyl alcohol–hydrochloric acid affords the axial alcohol 5α-cholestan-3α-ol in high yield (Nishimura *et al.*, 1975). Steroidal 3,17- and 3,20-diones are selectively hydrogenated at C-3 to give the corresponding 3-axial hy-droxy ketones in excellent yield (Nishimura *et al.*, 1977a). But exceptions exist. The methoxy function in 2- and 4-methoxycyclohexanones shows exceptional orientating properties, especially over iridium and platinum, and with these ketones highest yields of axial alcohol are obtained over alkali-free platinum in *tert*-butyl alcohol (Table IX).

**TABLE IX**

| Catalyst | Solvent | Cis/trans |
|---|---|---|
| Platinum black | *tert*-Butanol | 22 |
| Rhodium black | Isopropanol–HCl | 6.7 |

Results from the hydrogenation of 4-methylcyclohexanone support the generality regarding rhodium as a preferred catalyst for the formation of axial alcohol (Table X).

Hydrogenation of 5α-cholestan-3-one with palladium gives an equatorial 3β-ol, whereas reduction of 5β-cholestan-3-one under the same conditions affords an axial 3β-ol with high stereoselectivity in alcohols or in tetra-

TABLE X

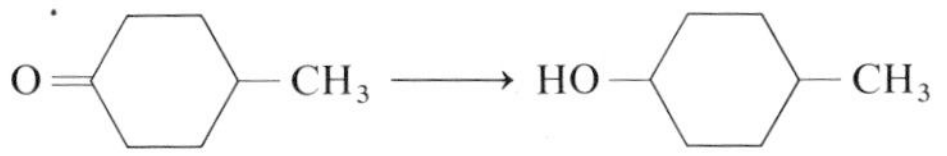

$$O = \hspace \hspace - CH_3 \longrightarrow HO - \hspace - CH_3$$

| Catalyst | Solvent | Cis/trans |
|----------|---------|-----------|
| Platinum black | *tert*-Butanol | 3.5 |
| Rhodium black | Isopropanol–HCl | 11 |

hydrofuran. With platinum the results depend on the solvent; with rhodium in ethanol both ketones afford the axial alcohols (Nishimura *et al.*, 1977a).

## 2. Solvents

Solvents may have a large influence on the stereochemistry of ketone reduction, as illustrated by the data in Table XI. The effect of solvent cannot be divorced from the effect of catalyst, and even closely related catalysts such as platinum black prepared by reduction of platinum hydroxide with hydrogen in water (Nishimura *et al.*, 1975) and platinum oxide (Mitsui *et al.*, 1973) give quite different results. However, if the platinum oxide is prereduced and carefully washed free of alkali it then gives substantially the same results as the catalyst from platinum hydroxide (Nishimura *et al.*, 1975, 1977a).

The increasing cis–trans ratios with the series ethanol < isopropanol < *tert*-butanol also holds for 4-methylcyclohexanone and 2- and 4-methoxy-cyclohexanones. The results of platinum in *tert*-butanol are striking for the

TABLE XI

**Effect of Solvent and Catalyst on Cis/Trans Ratios of Alcohols Produced from 2-Methylcyclohexanone**

| Catalyst | Solvent | Cis/trans |
|----------|---------|-----------|
| Platinum oxide[a] | Ethanol | 1.3 |
| | Acetic acid | 0.6 |
| | Dioxane | 1.6 |
| | Hexane | 0.6 |
| Reduced platinum hydroxide[b] | Ethanol | 3.5 |
| | Isopropanol | 6.7 |
| | *tert*-Butanol | 13 |

[a] Data of Mitsui *et al.* (1973). Used with permission of Pergamon Press.

[b] Data of Nishimura *et al.* (1975). Used with permission.

methoxy compounds, giving cis–trans ratios of 67 and 22 for the 2- and 4-methoxycyclohexanone, respectively (Nishimura et al., 1975).

### 3.  Temperature

Some interesting effects of moderate temperature changes were reported by Solodar (1976) in studies on the selective hydrogenation of menthone and isomenthone. Hydrogenation of isomenthone (12) over 6% ruthenium-on-carbon in neutral or basic conditions produced isomenthol (13) in 87–94% selectivity as long as the temperature was 150°C, but at 105°C neoisomenthol (14) predominated. In other experiments 150°C temperatures completely suppressed or greatly reduced the rate of reaction, whereas at 105°C reduction proceeded smoothly. Platinum was generally the most selective under all conditions for converting isomenthone to neoisomenthol (14).

As a working generality, Wicker (1957) earlier proposed that the more stable isomer is favored by higher temperatures during hydrogenation of phenols and cyclohexanones.

### 4.  Exo Addition

Addition of hydrogen to certain bridged polycyclic ketones follows an exo addition rule to give endo-hydroxyl. Hydrogenation of 1-azabicyclo-[2.2.1]heptan-3-one (Spry and Aaron, 1969) or 1-azabicyclo[3.2.1]octan-3-one (Thill and Aaron, 1968) occurs from the less hindered exo side of the molecule to give the corresponding endo-hydroxyl isomer in high yield. Sodium in ethanol reduction gives the thermodynamically controlled, more stable equatorial isomer. The hydrogenation results are analogous to those obtained with norcamphor, which also affords the corresponding endo-hydroxyl isomer (Komppa and Beckmann, 1934; Alder and Stein, 1936; Alder et al., 1956).

## 5. Amino Ketones

The stereochemistry of reduction of amino ketones may depend on whether there is a free pair of electrons on the nitrogen atom (Kugita and May, 1961; May and Kugita, 1961; May *et al.*, 1961; Saito and May, 1961; Mokotoff, 1968). For instance, the free base **17** was converted to the $\alpha$ isomer **18** by catalytic hydrogenation, whereas the $\beta$ isomer **15** was produced from the methyl quaternary salt **16** under comparable conditions (House *et al.*, 1963). These results could be attributed to steric factors, to prior complexing of the amine function on the catalyst surface, or to assistance of hydride transfer by participation of the amine function.

15          16, R = CH$_3$          18
            17, R = —

However, in the azabicyclooctanone ($n = 1$), azabicyclononanone ($n = 2$), and azabicyclodecanone ($n = 3$) systems the stereochemistry of reduction is determined mainly by the bulk and conformation of the polymethylene chain, with the effect of solvent and salt formation of secondary importance (House *et al.*,) (Table XII).

The stereochemical results of hydrogenation of 1-, 2-, and 3-ketoquinolizidines over palladium, platinum, rhodium, and ruthenium in several solvents were reported (Rader *et al.*, 1964). Strikingly large differences in the percentages of axial alcohol as a function of catalyst and solvent were found. Some of the results were attributed to the ability of the free electrons on the bridgehead nitrogen to bond with the catalyst surface and thus produce an "anchor effect."

Hydrogenation of 1,5-dimethyl-3-piperidone hydrobromide in aqueous potassium hydroxide over platinum oxide gave the axial alcohol *trans*-1,5-dimethyl-3-piperidol, whereas in water the cis equatorial isomer was formed (Mistryukov and Katvalyan, 1969).

(19)

**TABLE XII[a]**

| | A | | | | B | |
|---|---|---|---|---|---|---|
| $n$ | Free base (%) | Quaternary salt (%) | Solvent | $n$ | Free base (%) | Quaternary salt (%) |
| 1 | 98 | — | Isopropanol | 1 | 2 | — |
| 1 | 81 | 85 | Acetic acid | 1 | 19 | 15 |
| 2 | 38 | — | Isopropanol | 2 | 62 | — |
| 2 | 4 | 44 | Acetic acid | 2 | 96 | 56 |
| 3 | 1 | — | Isopropanol | 3 | 99 | — |
| 3 | 3 | < 1 | Acetic acid | 3 | 97 | > 99 |

[a] Data of House *et al.* (1963). Used with permission.

Reduction of 2-aminopropiophenones (**19**) by either hydrogen and palladium-on-carbon or sodium borohydride proceeds stereospecifically to afford *erythro*-2-amino-1-phenylpropanols (**20**), whereas reduction of the 2-acetamido derivatives gave a mixture of *threo*- and *erythro*-2-acetamido-1-phenyl-1-propanols. The authors proposed a cyclic intermediate to account for the stereospecificity (Matsumoto *et al.*, 1962).

**19**, R = H, OC$_2$H$_5$                                                                 **20**

### 6. Homogeneous Iridium-Catalyzed Reductions

An effective method for generating high-purity axial alcohols is by iridium-catalyzed hydrogen transfer with phosphite esters or phosphorus acid as reducing agents (Haddad *et al.*, 1964; Browne and Kirk, 1969; Henbest and Mitchell, 1970; Fanta and Erman, 1971).

Ketone + phosphite + water $\longrightarrow$ *sec*-alcohol + phosphate

Iridium is usually more effective as a catalyst than platinum, rhodium, ruthenium, palladium, cobalt, or iron. In special cases tris(triphenylphosphine) rhodium chloride proved to be even more selective than iridium (Orr *et al.*, 1970). By way of illustration, the axial alcohol *trans*-9,10,*cis*-8,9-*H*-8-hydroxy-2-methyldecahydroisoquinoline (**23**) was prepared in 98% yield by reduction of *trans*-2-methyldecahydroisoquinol-8-one (**22**) with trimethyl phosphite and chloroiridic acid as catalyst. The equatorial alcohol **21** is formed on sodium reduction of the ketone (Mathison and Morgan, 1974).

### 7. Asymmetric Hydrogenation

Asymmetric hydrogenation of carbonyl functions by catalysts modified with optically active compounds, such as α-amino or α-hydroxy acids, has occasioned a long series of papers, for this type of reduction is of both theoretical and practical interest. Under favorable conditions optical yields as high as 87% have been realized (Orito *et al.*, 1976). The maximal stereoselectivity obtained is a function of conversion, tending to decrease as conversion increases (Gross and Rys, 1974). Optical yield is sensitive to the modifier, substrate structure, water, solvent, pH, and catalyst among other variables. A leading reference to this area is Tanabe *et al.* (1973). There has also been success with homogeneous asymmetric hydrogenation catalysts (Ohgo *et al.*, 1971). An effective chiral synthesis of *R*(−)-pantolactone was achieved by asymmetric hydrogenation of α-keto-β,β-dimethyl-γ-butyrolactone with BPPM–rhodium complex (Achiwa *et al.*, 1977).

## II.  AROMATIC KETONES

Aromatic ketones can be reduced to either the corresponding aromatic alcohol or alkyl aromatic with little difficulty and usually in excellent yield. The alcohol is an intermediate in the formation of the alkyl aromatic, although there is evidence that some hydrogenolysis of the oxygen function may occur without intervention of the carbinol (Isagulyants et al., 1972). The rate of reduction and strength of adsorption are influenced by ring substitutents. In a series of competitive experiments, the relative strengths of adsorption of acetophenone and a number of ring-substituted acetophenones were measured over palladium-on-carbon. All substituted acetophenones studied, except p-diacetylbenzene, are adsorbed more weakly than acetophenone. The surprising conclusion was reached that methyl-, methoxy-, amino-, and hydroxy-substituted acetophenones differ only slightly in strength of adsorption; the latter three substituents, despite having lone electron pairs, behave as an alkyl group of corresponding size (van Bekkum et al., 1969). This finding is especially interesting in view of the "anchor effect" often invoke for these substituents to explain preferred orientation. Electronic substituent effects on the rate of hydrogenation of acetophenone are correlated by a Hammett–Yukawa (Yukawa et al., 1966) equation with a 0.7 $\rho$ value. Electron-withdrawing substituents accelerate and electron-donating substituents retard the rate of hydrogenation. Electron-donating substituents stabilize the conjugated adsorbed state relative to acetophenone and hence increase the energy difference between the adsorbed state and the transition state of hydrogenation, which consequently retards the rate. The reverse is true for electron-withdrawing substituents (van Bekkum et al., 1969).

2-Alkyl substituents in acetophenones cause marked inhibition in the rate of hydrogenation, an effect attributed mainly to steric factors. In contrast, the rate of hydrogenation is accelerated by $\omega$ substituents in alkyl phenyl ketones; for instance, the rate for pivalophenone is three times that for acetophenone. This effect is attributed to the difficulty of flat adsorption and to a relatively lower conjugation energy in the adsorbed state of the more substituted ketone. This explanation is supported by the observation that increased branching retards hydrogenation in cyclohexyl alkyl ketones (van Bekkum et al., 1969).

### A.  Hydrogenolysis

Hydrogenolysis of a ketonic group adjacent to an aromatic nucleus occurs readily (Hartung and Simonoff, 1953; Berger et al., 1970). Yields are generally high, and the necessary conditions are mild. The procedure is usually better than chemical reductions such as the Clemmensen or Wolff–Kishner reduction (Eisenbraun *et al.*, 1971). Generally palladium makes the most useful catalyst for hydrogenolysis of aromatic ketones in that it combines a good activity for hydrogenation and hydrogenolysis of the carbonyl with low activity for hydrogenation of the aromatic nucleus (Levine and Temin, 1957). Nonetheless, it may be necessary to interrupt the reduction at the theoretical absorption of hydrogen to prevent a slow ring saturation (Jensen, 1960). Other metals such as copper–chromite, platinum, and nickel have been used successfully for hydrogenolysis, but palladium is by far the most frequently used catalyst (Hartung and Simonoff, 1953).

Hydrogenolysis is favored by elevated temperatures (Koo, 1953; Walker, 1956, 1958) and by acids (Lock and Walter, 1944; Kindler *et al.*, 1953; Johnson *et al.*, 1957; Wilt and Schneider, 1961). Polar solvents are generally used, and excellent results have been obtained with lower alcohols, acetic acid, and acetic acid containing sulfuric or perchloric acid. The beneficial effects of acid are illustrated by the reduction of the hydrogenolysis-resistant diketone **25**. Attempted hydrogenolysis with palladium-on-carbon in neutral media over a wide pressure range afforded only the hemiketal **24**, but when the reduction of either **24** or **25** was carried out in glacial acetic acid containing concentrated sulfuric acid, 7,7′-dicarbomethoxy-*syn,cis*-truxane (**26**) was obtained in good yield (Anastassiou and Griffin, 1968).

In reductions in which the starting ketone is sensitive to acids, the acid is best added after the carbinol stage has been reached (Murphy, 1961).

Brieger and Fu (1976) reduced aromatic aldehydes and ketones to the corresponding hydrocarbons in good yield by catalytic transfer reduction using cyclohexene or limonene as hydrogen donor, palladium-on-carbon as catalyst, and a Lewis acid promoter, such as ferric chloride, aluminum chloride, or even water. Good yields are obtained by simply refluxing the mixture for several hours.

### Ketone Derivatives

Carbonyl compounds are often isolated or purified as derivatives, such as oximes, semicarbazones, phenylhydrazones, or 2,4-dinitrophenylhydrazones. Burnham and Eisenbraun (1971) examined the direct hydrogenolysis of these derivatives to ascertain whether they could be converted directly to the hydrocarbon without prior hydrolysis to the ketone. They concluded that 2,4-dinitrophenylhydrazones were the most suitable compounds for successful hydrogenolyses over palladium-on-carbon catalysts.

## B.   Reduction to the Aromatic Carbinol

Reduction of an aromatic ketone to the corresponding carbinol usually occurs readily. Less than perfect yields arise mainly through hydrogenolysis of the carbinol or ring saturation. Hydrogenolysis can be minimized by interrupting the reduction after the theoretical absorption of hydrogen, by avoiding acidic media, and by using organic bases (Kindler et al., 1957; Hiskey and Northrop, 1961; Cohen et al., 1963) or alkali.

Table XIII shows the pronounced effect of acidity on the maximal yield of phenylethanol obtainable by hydrogenation of acetophenone over 5% palladium-on-carbon in methanol under ambient conditions (Rylander and Hasbrouck, 1968). The presence of small amounts of acetic or hydrochloric acid sharply lowers the yield of phenylethanol. In methanol without additive the yield was 90%, although other workers (Freifelder et al., 1964) obtained 99.9 + % under similar conditions. The lower yield was attributed to a slight residual acidity remaining in the catalyst and, in fact, when the reduction was carried out with 0.008 mole sodium hydroxide per mole acetophenone, a quantitative yield of phenylethanol was obtained (Table XIII ).

**TABLE XIII**

**Hydrogenation of Acetophenone over 5% Palladium-on-carbon in Methanol**

| Promoter | Mole promoter per mole acetophenone | Maximal % phenylethanol |
|---|---|---|
| None | — | 90 |
| Acetic acid | 0.20 | 60 |
| Hydrochloric acid | 0.014 | 76 |
| Sodium hydroxide | 0.008 | 100 |

The influence of solvent on the reduction of aromatic ketones is further illustrated by the hydrogenation of isoflavone (**27**), which can be quantitatively converted to either isoflavanone (**28**) or a mixture of isomeric isoflavan-4-ols (**29**) over palladium-on-carbon. In dioxane the reduction selectively affords isoflavanone whereas in protic solvents, ethanol, aqueous ethanol, or aqueous acetone, intermediate hydrogenation affords about equal amounts of isoflavanone and isoflavan-4-ols. Isoflavane (**30**) is formed in the presence of hydrochloric or acetic acid in ethanol. In strong base a complex degradation occurs (Szabo and Antal, 1973).

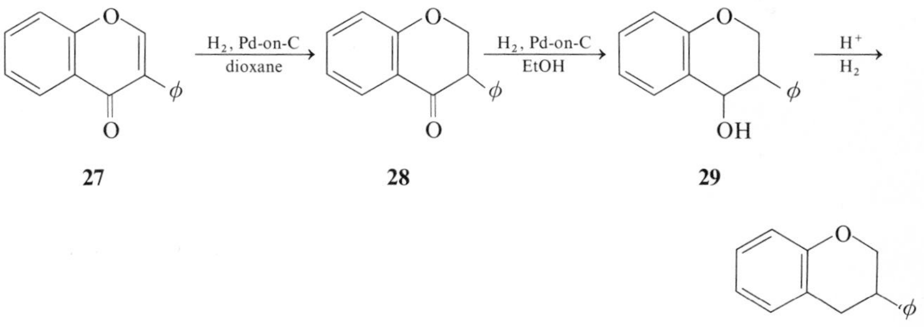

Catalysts

Palladium is the most frequently used catalyst for hydrogenation of aromatic ketones to either the corresponding carbinol or hydrocarbon. Its activity for ring saturation is relatively low and that for ketone reduction high (Breitner *et al.*, 1959). In a comparison of catalysts for the hydrogenation of acetophenone to phenylethanol, palladium-on-carbon was considered the catalyst of choice (Freifelder *et al.*, 1964). Palladium-on-carbon was also the catalyst of choice for the conversion of 2-acetylpyridine to 2-(1-hydroxyethyl)pyridine, but in the reduction of the 3- and 4-acetylpyridines complications with palladium ensued including partial ring reduction and pinacol formation. The best yield of 4-(1-hydroxyethyl)pyridine (80%) was obtained with platinum oxide (Freifelder, 1964).

C.   Saturated Carbinols and Ketones

Catalysts have an important influence on the course of aromatic ketone reduction; by the proper choice of metal, high yields of a variety of products can be obtained. The following scheme, in which acetophenone is used as a model substrate, illustrates the means by which various types of products can be formed in excellent yield (Rylander and Hasbrouck, 1969):

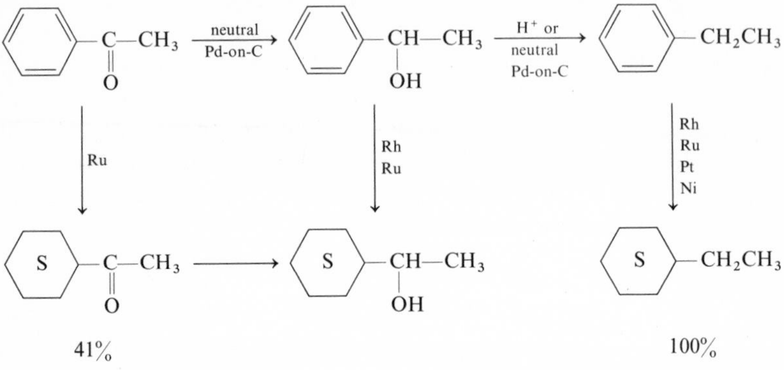

Facile preparation of the fully saturated deoxygenated material, ethyl-cyclohexane, requires oxygen hydrogenolysis before ring saturation; the saturated carbinol is stable towards hydrogenolysis under most conditions. Cyclohexylethanol can be obtained in high yield from either phenylethanol or methylcyclohexyl ketone or from acetophenone via both these products as well as intermediates such as methyl cyclohexenyl ketone and methyl-cyclohexenylcarbinol (Taya *et al.*, 1968). Ruthenium, without solvent, is superior to rhodium for the preparation of cyclohexylethanol from either acetophenone or phenylethanol because there is less hydrogenolysis. However, oxygen hydrogenolysis is sensitive to solvent, and with the appropriate solvent excellent yields of cyclohexylethanol can be obtained using rhodium catalysts (Rylander and Hasbrouck, 1968). Reduction of acetophenone over rhodium-on-alumina offers only a fair route to methyl cyclohexyl ketone; the best yield obtained was 41% in *tert*-butanol (Table XIV).

Alumina is a better support than carbon for rhodium in the hydrogenation of acetophenone to cyclohexylethanol; in various solvents the yields with alumina were 15–25 percentage points higher than with carbon (Rylander and Hasbrouck, 1968). Rhodium (75%)–platinum (25%) oxide (Nishimura, 1960) in methanol gave cyclohexylethanol in 95% yield. This catalyst has been recommended when hydrogenolyses of various sorts are to be avoided.

Hydrogenolysis decreases as pressure increases in the reduction of aceto-phenone in ethanol over rhodium-on-alumina (Freidlin *et al.*, 1970). The yield of methylcyclohexylcarbinol at 150° C increases from 85 to 93% by increasing the pressure from 50 to 300 atm. (Barinov and Mushenko, 1973).

The extent of hydrogenolysis of aromatic ketones that occurs during saturation of the aromatic ring also depends on ring substituents, as illus-trated by the hydrogenation of 4-chromanones containing electron-with-

**TABLE XIV**

**Hydrogenation of Acetophenone in Various Solvents over 5%
Rhodium-on-alumina[a]**

| Solvent | Maximal % phenylethanol | Maximal % cyclohexylethanol | Maximal % methyl cyclohexyl ketone |
|---|---|---|---|
| Methanol | 71 | 85 | 29 |
| Isopropanol | 45 | 99+ | 36 |
| *tert*-Butanol | 41 | 99+ | 41 |
| Ethyl acetate | 62 | 99+ | 35 |
| Dioxane | 47 | 66 | 24 |
| Acetone[b] | 59 | 99+ | 35 |

[a] Temperature 25°C; atmospheric pressure.
[b] Acetone was converted to isopropanol as the reaction progressed.

drawing substituents (Hirsch and Schwartzkopf, 1973, 1974). One might
surmise that substituents in general would increase hydrogenolysis of these
oxygen functions inasmuch as substituents would tend to increase the life
of intermediate vinylic and allylic double bonds.

| | |
|---|---|
| $R_1=R_2=H$ | 95 |
| $R_1=H; R_2=CO_2CH_3$ | 60 |
| $R_1=CO_2CH_3; R_2=H$ | 45 |

| | | |
|---|---|---|
| $R_1=R_2=H$ | — | 5 |
| $R_1=H; R_2=CO_2CH_3$ | 25 | 15 |
| $R_1=CO_2CH_3; R_2=H$ | 35 | 20 |

Maintenance of a benzylic oxygen during ring saturation proved to be of
value in the synthesis of *d*-phyllocladene from abietic acid. The authors
(Shimagaki and Tahara, 1975) noted that the scheme has application in the
syntheses of other natural products as well.

TABLE XV[a]

A

B                    C

|              | Conversion | % in products | | |
| Catalyst | (%) | A | B | C |
|---|---|---|---|---|
| RaNi | 96 | 52 | 37 | 11 |
| RaNi[b] | 91 | 64 | 36 | — |
| 5% Pd-on-C | 98 | 29 | 61 | 10 |
| 5% Pd-on-BaSO$_4$ | 94 | 23 | 52 | 25 |

[a] Data of Hanaya (1970). Used with permission.
[b] 0.002 mole sodium hydroxide.

**TABLE XVI[a]**

| Catalyst | % Erythro | % Threo | Catalyst |
|---|---|---|---|
| RaNi | 67 | 33 | |
| PtO$_2$ | 81 | 19 | |
| | 79 | 21 | Pd-on-C |
| | 75 | 25 | Pd-on-C, NaOH |
| | 63 | 37 | RaNi |
| | 73 | 27 | RaNi, NaOH |
| | 82 | 18 | PtO$_2$ |
| | 91 | 9 | PtO$_2$, NaOH |

[a] Data of Sohma and Mitsui (1969, 1970). Used with permission.

COOCH$_3$     $\xrightarrow[\text{EtOH, H}_2]{\text{RuO}_2}$     COOCH$_3$     $\xrightarrow[\text{C}_5\text{H}_5\text{N}]{\text{CrO}_3}$

COOCH$_3$    O      H   H   OH    COOCH$_3$

H   H   COOCH$_3$   O    COOCH$_3$    $\longrightarrow$

## D.  Stereochemistry

Hanaya (1966a,b, 1967) examined the stereochemistry of the hydrogenation of 2-phenyl-1-tetralone, 2-acetoxy-1-tetralone, 2-methyl-1-tetralone,
and 3-methyl-4-chromanone and arrived at the following conclusions
concerning the directing effects of substituents alpha to the carbonyl
(Hanaya, 1970). When the substituent is sterically bulky, such as phenyl or
acetoxyl groups, the molecule is adsorbed and hydrogenated on the catalyst
surface on the side opposite the substituent, irrespective of the catalyst.
The nature of the catalyst gains in importance when the substituent is small
(Table XV). Nickel is most influenced by catalyst hindrance, whereas palladium tends to give thermodynamically more stable products. Nickel is
considered to have a larger affinity for the oxygen atom than palladium
(Mitsui *et al.*, 1966) and, therefore, the stereochemistry over nickel is controlled mainly by the first adsorption step of the molecule. The stereochemistry over palladium, on the other hand, with less affinity for the oxygen
atom is controlled not by the adsorption step, but by the step in which
hydrogen is transferred from the catalyst surface to the molecule (Hanaya,
1970)

In the hydrogenation of **31** the cis isomer (**32**) predominates regardless
of the catalyst. Its formation is accounted for by the steric effects of the ester
function, which is orientated quasi-equatorially (Hanaya, 1969).

Considerable variation in stereochemical results can be achieved by the
use of appropriate catalysts in the hydrogenation of the aromatic ketones
1,3-diphenylpropane-1,2-dione  and  2-hydroxy-1,3-diphenylpropan-1-one
(Sohma and Mitsui, 1969) (Table XVI). Even larger variations can be
achieved in the more complex reduction of benzalacetophenone oxide, a
precursor of these compounds, due to the possibility of opening the oxirane

| | **31** | **32** | **33** | |
|---|---|---|---|---|
| RaNi | | 73% | 9% | 18% |
| Pd-on-C | | 72% | 3% | 25% |

ring with either inversion or retention of configuration (Sohma and Mitsui, 1970). The use of sodium hydroxide induces a change in the erythro content, but the direction of change depends on the catalyst.

## REFERENCES

Achiwa, K., Kogure, T., and Ojima, I., *Tetrahedron Lett.* p. 4431 (1977).
Adams, R., Miyano, S., and Fles, D., *J. Am. Chem. Soc.* **82**, 1466 (1960).
Adams. R., Miyano, S., and Nair, M. D., *J. Am. Chem. Soc.* **83**, 3323 (1961).
Adkins, H., and Billica, H. R., *J. Am. Chem. Soc.* **70**, 695 (1948).
Adkins, H., and Connor, R., *J. Am. Chem. Soc.* **53**, 1091 (1931).
Albertson, N. F., *J. Am. Chem. Soc.* **74**, 249 (1952).
Alder, K., and Stein, G., *Justus Liebigs Ann. Chem.* **525**, 183 (1936).
Alder, K., Wirtz, H., and Koppelberg, H., *Justus Liebigs Ann. Chem.* **601**, 138 (1956).
Anastassiou, A. G., and Griffin, G. W., *J. Org. Chem.* **33**, 3441 (1968).
Anziani, P., and Cornubert, R., *Bull. Soc. Chim. Fr.* **12**, 359 (1945).
Babcock, J. C., and Fieser, L. F., *J. Am. Chem. Soc.* **74**, 5472 (1952).
Barinov, N. S., and Mushenko, D. V., *Zh. Prikl. Khim.* (*Leningrad*) **46**, 940 (1973); *Chem. Abstr.* **79**, 4674b (1973).
Barton, D. H. R., *J. Chem. Soc.* p. 1027 (1953).
Berger, J. G., Teller, S. R., and Pachter, I. J., *J. Org. Chem.* **35**, 3122 (1970).
Birch, A. J., Hochstein, F. A., Quartey, J. A. K., and Turnbull, J. P., *J. Chem. Soc.* p. 2923 (1964).
Blance, R.B., and Gibson, D. T., *J. Chem. Soc.* p. 2487 (1954).
Blomquist, A. T., and Wolinsky, J., *J. Am. Chem. Soc.* **77**, 5423 (1955).
Bond, G. C., and Wells, P. B., *Adv. Catal.* **15**, 206 (1964).
Breitner, E., Roginski, E., and Rylander, P. N., *J. Org. Chem.* **24**, 1855 (1959).
Brewster, J. H., *J. Am. Chem. Soc.* **76**, 6361 (1954).
Brieger, G., and Fu, T-H., *J. Chem. Soc. Chem. Commun.* p. 757 (1976).
Browne, P. A., and Kirk, D. N., *J. Chem. Soc. C* p. 1653 (1969).
Buchanan, G. L., Hamilton, J. G., and Raphael, R. A., *J. Chem. Soc.* p. 4606 (1963).
Burnham, J. W., and Eisenbraun, E. J., *J. Org. Chem.* **36**, 737 (1971).
Burton, J. S., and Stevens, R., *J. Chem. Soc.* p. 4382 (1963).
Burton, J. S., Elvidge, J. A., and Stevens, R., *J. Chem. Soc.* p. 3816 (1964).
Campbell, N. R., and Hunt, J., *J. Chem. Soc.* p. 1379 (1950).
Carothers, W. H., and Adams, R., *J. Am. Chem. Soc.* **47**, 1047 (1925).
Cohen, S., Thom, E., and Bendich, A., *J. Org. Chem.* **28**, 1379 (1963).

Cope, A. C., and Kagan, F., *J. Am. Chem. Soc.* **80**, 5499 (1958).

Dauben, W. G., and Adams, R. E., *J. Am. Chem. Soc.* **70**, 1759 (1948).

Delepine, M., and Horeau, A., *Bull. Soc. Chim. Fr.* p. 31 (1937).

Ebersole, R. C., and Chang, F. C., *J. Org. Chem.* **38**, 2713 (1973).

Eisenbraun, E. J., Hinman, C. W., Springer, J. M., Burnham, J. W., Chou, T. S., Flanagan, P. W., and Hamming, M. C., *J. Org. Chem.* **36**, 2480 (1971).

Eliel, E. L., and Lukach, C. A., *J. Am. Chem. Soc.* **79**, 5986 (1957).

Fanta, W. I., and Erman, W. F., *J. Org. Chem.* **36**, 358 (1971).

Findlay, S. P., *J. Org. Chem.* **24**, 1540 (1959).

Foresti, B., *Ann. Chim. Appl.* **27**, 359 (1937).

Foresti, B., *Soc. Ital. Prog. Sci., Atti Riun., 27th, Bologna* **5**, 346 (1939).

Forrest, J., and Tucker, S. H., *J. Chem. Soc.* p. 1137 (1948).

France, D. J., Hand, J. J., and Los, M., *Tetrahedron* **25**, 4011 (1969).

Freidlin, L. K., Borunova, N. V., Gvinter, L. I., Daniélova, S. S., and Badakh, R. N., *Izv. Akad. Nauk SSSR, Ser. Khim.* (8) p. 1692 (1970).

Freidlin, L. K., Gvinter, L. I., Zamureenko, V. A., Sukhobok, A. A., and Kustanovich, I. M., *Zh. Org. Khim.* **9**, 3 (1973); *Chem. Abstr.* **78**, 110230 (1973).

Freifelder, M., *J. Org. Chem.* **29**, 2895 (1964).

Freifelder, M., and Stone, G. R., *J. Org. Chem.* **26**, 3805 (1961).

Freifelder, M., Anderson, T., Ng, Y. H., and Papendick, V., *J. Pharm. Sci.* **53**, 967 (1964).

Fuson, R. C., and Corse, J., *J. Am. Chem. Soc.* **67**, 2054 (1945).

Gross, L. H., and Rys, P., *J. Org. Chem.* **39**, 2429 (1974).

Haddad, Y. M. Y., Henbest, H. B., Husbands, J., and Mitchell, T. R. B., *Proc. Chem. Soc.* p. 361 (1964).

Hanaya, K., *Nippon Kagaku Zasshi* **87**, 745 (1966a).

Hanaya, K., *Nippon Kagaku Zasshi* **87**, 991 (1966b).

Hanaya, K., *Bull. Chem. Soc. Jpn.* **40**, 1884 (1967).

Hanaya, K., *Nippon Kagaku Zasshi* **90**(3), 314 (1969); *Chem. Abstr.* **71**, 12873r (1969).

Hanaya, K., *Bull. Chem. Soc. Jpn.* **43**, 442 (1970).

Harnik, M., *J. Org. Chem.* **28**, 3386 (1963).

Hartung, W. H., and Simonoff, R., *Org. React.* **7**, 263 (1953).

Hasek, R. H., and Martin, J. C., *J. Org. Chem.* **28**, 1468 (1963).

Hasek, R. H., Elam, E. U., Martin, J. C., and Nations, R. G., *J. Org. Chem.* **26**, 700 (1961).

Henbest, H. B., and Mitchell, T. R. B., *J. Chem. Soc. C* p. 785 (1970).

Hennion, G. P., and Watson, E. J., *J. Org. Chem.* **23**, 656 (1958).

Hey, D. D., and Musgrave, O. C., *J. Chem. Soc.* p. 3156 (1949).

Hirsch, J. A., and Schwartzkopf, G., *J. Org. Chem.* **38**, 3534 (1973).

Hirsch, J. A., and Schwartzkopf, G., *J. Org. Chem.* **39**, 2040 (1974).

Hiskey, R. G., and Northrop, R. C., *J. Am. Chem. Soc.* **83**, 4798 (1961).

House, H. O., Muller, H. C., Pitt, C. G., and Wickham, P. P., *J. Org. Chem.* **28**, 2407 (1963).

Howard, W. L., and Brown, J. H., Jr., *J. Org. Chem.* **26**, 1026 (1961).

Howe, A. P., and Haas, H. B., *Ind. Eng Chem.* **38**, 251 (1946).

Hückle, W., and Hubele, A., *Justus Liebigs Ann. Chem.* **613**, 27 (1958).

Hückel, W., and Maier, M., *Justus Liebigs Ann. Chem.* **616**, 46 (1958).

Hurd, C. D., and Perletz, P., *J. Am. Chem. Soc.* **68**, 38 (1946).

Imai, T., Mitsuda, Y., Ebisawa, H., Kametaka, T., and Minoura, T., Jpn. Patent 71 02,643, Jan 22, 1971; *Chem. Abstr.* **74**, 111569q (1971).

Isagulyants, G. V., Borunova, N. V., Freidlin, L. K., Daniélova, S. S., and Kovalenko, L. I., *Izv. Akad. Nauk SSSR, Ser. Khim* (7) p. 1554 (1972); *Chem. Abstr.* **77**, 151257e (1972).

Jacobs, R. L., and Goerner, G. L., *J. Org. Chem.* **21**, 837 (1956).

Jeanloz, R., Prins, D. A., and Euw, J. V., *Helv. Chim. Acta* **30**, 374 (1947).
Jensen, F. R., *J. Org. Chem.* **25**, 269 (1960).
Johnson, W. S., Christiansen, R. G., and Ireland, R. E., *J. Am. Chem. Soc.* **79**, 1995 (1957).
Julian, P. L., Pikl, J., and Wantz, F. E., *J. Am. Chem. Soc.* **57**, 2026 (1935).
Kaye, I. A., and Matthews, R. S., *J. Org. Chem.* **29**, 1341 (1964).
Kindler, K., Oelschläger, H., and Henrich, P., *Chem. Ber.* **86**, 501 (1953).
Kindler, K., Günther, H., Helling, H., and Sussner, E., *Justus Liebigs Ann. Chem.* **605**, 200 (1957).
Kishida, S., and Teranishi, S., *J. Catal.* **12**, 90 (1968).
Koelsch, C. F., and Ostercamp, D. L., *J. Org. Chem.* **26**, 1104 (1961).
Koizumi, M., *Sci. Pap. Inst. Phys. Chem. Res. Jpn.* **37**, 414 (1940).
Komppa, G., and Beckmann, S., *Justus Liebigs Ann. Chem.* **512**, 172 (1934).
Koo, J., *J. Am. Chem. Soc.* **75**, 720 (1953).
Kugita, H., and May, E. L., *J. Org. Chem.* **26**, 1954 (1961).
Levine, M., and Temin, S. C., *J. Org. Chem.* **22**, 85 (1957).
Liberman, A. L., and Kazanskii, B. A., *Bull. Acad. Sci. URSS, Cl. Sci. Chim.* p. 77 (1946); *Chem. Abstr.* **42**, 5861 (1948).
Lieber, E., and Smith, G. B. L., *J. Am. Chem. Soc.* **58**, 1417 (1936).
Lock, G., and Walter, E., *Chem. Ber.* **77**, 286 (1944).
Mathison, I. W., and Morgan, P. H., *J. Org. Chem.* **39**, 3210 (1974).
Matsumoto, T., Nishida, T., and Shirahama, H., *J. Org. Chem.* **27**, 79 (1962).
Maxted, E. B., and Moon, C. H., *J. Chem. Soc.* p. 1190 (1935).
May, E. L., and Kugita, H., *J. Org. Chem.* **26**, 188 (1961).
May, E. L., Kugita, H., and Ager, J. H., *J. Org. Chem.* **26**, 1621 (1961).
Meinwald, J., and Frauenglass, E., *J. Am. Chem. Soc.* **82**, 5235 (1960).
Mironov, V. A., Fedorovich, A. D., and Akhrem, A. A., *Izv. Akad. Nauk SSSR, Ser. Khim.* (6) p. 1288 (1973); *Chem. Abstr.* **79**, 91646k (1973).
Misani, F., Speers, J., and Lyon, A. M., *J. Am. Chem. Soc.* **78**, 2801 (1956).
Mistryukov, E. A., and Katvalyan, G. T., *Izv. Akad. Nauk SSSR, Ser. Khim.* No. 3, p. 702 (1969); *Chem. Abstr.* **71**, 38159n (1969).
Mitsui, S., Senda, Y., and Saito, H., *Bull. Chem. Soc. Jpn.* **39**, 694 (1966).
Mitsui, S., Saito, H., Yamashita, Y., Kaminaga, M., and Senda, Y., *Tetrahedron* **29**, 1531 (1973).
Mokotoff, M., *J. Org. Chem.* **33**, 3557 (1968).
Murphy, J. G., *J. Org. Chem.* **26**, 3104 (1961).
Nair, M. D., and Adams, R., *J. Org. Chem.* **27**, 3059 (1961).
Nishimura, S., *Bull. Chem. Soc. Jpn.* **33**, 566 (1960).
Nishimura, S., Itaya, T., and Shiota, M., *Chem. Commun.* p. 422 (1967).
Nishimura, S., Katagiri, M., and Kunikata, Y., *Chem. Lett.* p. 1235 (1975).
Nishimura, S., Ishige, M., and Shiota, M., *Chem. Lett.* p. 963 (1977a).
Nishimura, S., Ishige, M., and Shiota, M., *Chem. Lett.* p. 535 (1977b).
Ohgo, Y., Takeuchi, S., and Yoshimura, J., *Bull. Chem. Soc. Jpn.* **44**, 583 (1971).
Orchin, M., and Butz, L. W., *J. Am. Chem. Soc.* **65**, 2296 (1943).
Orito, Y., Niwa, S., and Imai, S., *Yuki Gosei Kagaku Shi* **34**(4), 236 (1976); *Chem. Abstr.* **85**, 123295n (1976).
Orr, J. C., Mersereau, M., and Sanford, A., *Chem. Commun.* p. 162 (1970).
Otter, B. A., Taube, A., and Fox, J. J., *J. Org. Chem.* **36**, 1251 (1971).
Patrikeev, V. V., and Liberman, A. L., *Dokl. Akad. Nauk SSSR* **62**, 97 (1948).
Phillips, A. P., and Mentha, J., *J. Am. Chem. Soc.* **78**, 140 (1956).
Post, G. G., and Anderson, L., *J. Am. Chem. Soc.* **84**, 471 (1962).
Posternak, T., *Helv. Chim. Acta* **24**, 1045 (1941).

Raasch, M. S., U.S. Patent 2,824,888 (1958).
Rader, C. P., Wicks, G. E., Jr., Young, R. L., Jr., and Aaron, H. S., *J. Org. Chem.* **29**, 2252 (1964).
Rapala, R. T., and Farkas, E., *J. Am. Chem. Soc.* **80**, 1008 (1958).
Reiff, L. P., and Aaron, H. S., *Tetrahedron Lett.* p. 2329 (1967).
Roberts, D. R., *J. Org. Chem.* **30**, 4375 (1965).
Roy, S. K., and Wheeler, D. M. S., *J. Chem. Soc.* p. 2155 (1963).
Ruzicka, L., Plattner, P. A., and Wild, H., *Helv. Chim. Acta* **28**, 395 (1945).
Rylander, P. N., and Hasbrouck, L., *Engelhard Ind. Tech. Bull.* **8**, 148 (1968).
Rylander, P. N., and Hasbrouck, L., *Engelhard Ind., Tech. Bull.* **10**, 50 (1969).
Rylander, P. N., and Hasbrouck, L., U.S. Patent 3,557,223, Jan. 19, 1971.
Rylander, P. N., and Starrick, S., *Engelhard Ind., Tech. Bull.* **7**, 106 (1966).
Rylander, P. N., and Steele, D. R., *Engelhard Ind., Tech. Bull.* **3**, 125 (1963).
Rylander, P. N., and Steele, D. R., *Engelhard Ind., Tech. Bull.* **5**, 113 (1965).
Rylander, P. N., and Steele, D. R., cited in Rylander, P. N., "Catalytic Hydrogenation over Platinum Metals," p. 267. Academic Press, New York, 1967.
Rylander, P. N., and Vaflor, X., *Am. Chem. Soc., Cent. Reg. Meet. Morgantown, W. Va. Pap.* No. 12 1975.
Saito, S., and May, E. L., *J. Org. Chem.* **26**, 4536 (1961).
Scheutte, H. A., and Thomas, R. W., *J. Am. Chem. Soc.* **52**, 3010 (1930).
Shimagaki, M., and Tahara, A., *Tetrahedron Lett.* p. 1715 (1975).
Shriner, R. L., and Witte, M., *J. Am. Chem. Soc.* **63**, 2134 (1941).
Shuikin, N. I., and Vasilevskaya, G. K., *Izv. Akad. Nauk SSSR, Ser. Khim.* (3) p. 557 (1964).
Simoniková, J., Ralkova, A., and Kochloefl, K., *J. Catal.* **29**, 412 (1973).
Skibina, E. M., Levin, Y. M., Ioffe, I. I., and Artamonov, P. A., *Zh. Prikl. Khim.* (*Leningrad*) **49**(7), 1554 (1976); *Chem. Abstr.* **85**, 123287m (1976).
Smissman, E. E., Muren, J. F., and Dahle, N. A., *J. Org Chem.* **29**, 3517 (1964).
Smith, H., *J. Chem. Soc.* p. 803 (1953).
Sohma, A., and Mitsui, S., *Bull. Chem. Soc. Jpn.* **42**, 1451 (1969).
Sohma, A., and Mitsui, S., *Bull. Chem. Soc. Jpn.* **43**, 448 (1970).
Solodar, J., *J. Org. Chem.* **41**, 3461 (1976).
Spry, D. O., and Aaron, H. S., *J. Org. Chem.* **34**, 3674 (1969).
Stephenson, L. M., and Falk, L. C., *J. Org. Chem.* **41**, 2928 (1976).
Szabó, V., and Antal, E., *Tetrahedron Lett.* p. 1659 (1973).
Takagi, K., and Murakami, M., Ger. Patent 1,936,203, Feb. 12, 1970; *Chem. Abstr.* **72**, 110803h (1970).
Takagi, K., Murakami, M., and Iketani, K., Ger. Patent 1,951,276, Apr. 16, 1970; *Chem. Abstr.* **73**, 14201k (1970).
Takagi, Y., *Sci. Pap. Inst. Phys. Chem. Res. Jpn.* **57**, 105 (1963).
Takagi, Y., Tetratani, S., and Tanaka, K., *in* "Catalysis" (J. W. Hightower, ed.), p. 757. Am. Elsevier, New York, 1973.
Tanabe, T., Okuda, K., and Izumi, Y., *Bull. Chem. Soc. Jpn.* **46**, 514 (1973).
Taya, K., Hiramoto, M., and Hirota, K., *Sci. Pap. Inst. Phys. Chem. Res.* (*Jpn.*) **62**(4), 145 (1968).
Thill, B. P., and Aaron, H. S., *J. Org. Chem.* **33**, 4376 (1968).
Tschudi, G., and Schinz, H., *Helv. Chim. Acta* **35**, 1230 (1952).
van Bekkum, H., Kieboom, A. P. G., and van de Putte, K. J. G., *Rec. Trav. Chim. Pays-Bas* **88**, 52 (1969).
Verzele, M., Acke, M., and Anteunis, M., *J. Chem. Soc.* p. 5598 (1963).
Walker, G. N., *J. Am. Chem. Soc.* **78**, 3201 (1956).
Walker, G. N., *J. Org. Chem.* **23**, 133 (1958).

Wells, P.B., and Wilson, G. R., *J. Catal.* **9**, 70 (1967).
Wicker, R. J., *J. Chem. Soc.* p. 2165 (1956).
Wicker, R. J., *J. Chem. Soc.* p. 3299 (1957).
Wilt, J. W., and Schneider, C. A., *J. Org. Chem.* **26**, 4196 (1961).
Yukawa, Y., Tsuno, Y., and Sawada, M., *Bull. Chem. Soc. Jpn.* **39**, 2274 (1966).
Zirkle, C. L., Gerns, F. R., Pavloff, A. M., and Burger, A., *J. Org. Chem.* **26**, 395 (1961).

# Chapter 7

# Hydrogenation of
# Nitro Compounds

Aromatic nitro groups are hydrogenated easily, usually in quantitative yields, to the corresponding amines. With special conditions intermediate aromatic hydroxylamines can be made the major product of reduction. Aliphatic nitro functions are reduced much more slowly, but nonetheless easily, to afford mainly amines, hydroxylamines, or oximes depending on the conditions. Many hydrogenations require either maintanence or reduction of other functional groups as well and, at times, interactions of these functions with products of the nitro group reduction. These reactions are discussed below.

Nitro functions are reduced so readily that mass transfer resistance of hydrogen from gas phase to bulk liquid and from liquid to catalyst particles may play an important role in determining the overall rate (Yao and Emmett, 1959, 1961a,b; Malpani and Chandalia, 1973; Roberts, 1976). A practical consequence of this is that it is easy to overestimate the amount of catalyst required to complete a reaction in a given time. Another consequence is that the rate may appear to be erroneously insensitive to substrate structure (Hernandez and Nord, 1947, 1948; Yao and Emmett, 1959). Furthermore, mass transfer resistances may affect selectivity.

The reduction of a nitro function is highly exothermic, and consideration must be given to heat removal to keep within a tolerable temperature rise. Most nitro functions can be reduced under ambient conditions; in industrial processes, however, elevated temperatures and pressures are usually used to minimize catalyst usage and maximize output.

## I.  CATALYSTS

Supported and unsupported palladium, platinum, and nickel are excellent catalysts for the hydrogenation of nitro functions. Rhodium is also effective, but a special requirement would be necessary to justify its use. Ruthenium does not seem to offer any particular advantage, unless one wants to saturate the ring as well in a single step (Whitman, 1952). The catalyst of choice for any particular reduction depends largely on other functional groups present and on the products required.

Platinum group metal sulfides are used commercially in the reduction of selected aromatic nitro compounds. These catalysts are much less sensitive to poisoning by sulfur compounds than are the nonsulfided counterparts. In comparative test experiments using nitrobenzene with and without added phenyl disulfide, palladium- or platinum-on-carbon was severely poisoned by the sulfur compound, unsupported palladium sulfide was apparently unaffected, and platinum sulfide-on-carbon gave a slower but still acceptable rate (Greenfield, 1967).

Aromatic nitro compounds can be reduced to anilines by hydrogen transfer by the use of ruthenium or rhodium trichlorides as catalyst and indoline, tetrahydroquinoline, piperidine, or pyrrolidine as hydrogen donor (Imai et al., 1977). Nitrosobenzene, but not phenylhydroxylamine, was detected as an intermediate, which suggests that some system of this sort may be a key to the yet unsolved problem of aromatic nitroso synthesis by catalytic hydrogenation.

## II.  SOLVENTS

A large number of solvents have been employed with good results, including methanol, ethanol, propanol, acetone, ethyl acetate (Adams et al., 1927), glycerol, glycol, isopropanol (Parrett and Lowy, 1926), sulfuric acid–acetic acid (Oelschläger, 1965), dilute hydrochloric and sulfuric acids (Freifelder and Robinson, 1963), benzene (Kaye and Roberts, 1951), and water (Benner and Stevenson, 1952). The list could be much extended. The choice of solvent, if one is used at all (Parrett and Lowy, 1926; British Patent 832,153 Apr. 6, 1960), would seem to depend more on the overall chemistry than on the effect of the solvent on the reduction of the nitro function per se.

## III.  PARTIAL REDUCTION OF NITRO COMPOUNDS

Catalytic hydrogenation of nitro groups may give rise, by accident or design, to partially reduced products containing nitroso, oxime, hydroxylamine, azo, or hydrazo functions.

## A.  Aromatic Hydroxylamines

Aromatic hydroxylamines are sometimes obtained under conditions in which complete reduction to the amine would have been expected. Most examples occur in structures having basic nitrogen or sulfur functions, but adventitious impurities may also have had an influence (Roblin and Winnek, 1940; Fanta, 1953; Berse et al., 1957; Schipper et al., 1961; Aeberli and Houlihan, 1967; Kauer and Sheppard, 1967; Yale, 1968: Barker and Ellis, 1970; Hodge, 1972). In other cases, hydroxylamines are obtained on interruption of the reduction (DiCarlo, 1944; Cavill and Ford, 1954; Goldkamp, 1964; Taya, 1966), but, in general, hydroxylamines usually do not accumulate in sufficient amounts to make the reduction synthetically useful. A large-scale hydrogenation of impure 3,4-dichloronitrobenzene resulted in the accumulation of so much 3,4-dichlorophenylhydroxylamine that a destructive, runaway reaction occurred, triggered by the exothermic disproportionation of the hydroxylamine at its autodecomposition temperature of around 260°C (Tong et al., 1977).

High yields of phenylhydroxylamines can be obtained by hydrogenation of nitro aromatics over platinum-on-carbon or alumina in alcoholic solvents containing small amounts of dimethyl sulfoxide (DMSO). This is an especially effective promoter, for it gives sharply increased yields of the hydroxylamine without a serious decrease in rate. A further advantage is that above a certain minimum its concentration is not critical; many other promoters poison the catalyst easily. Platinum was the most effective metal tested, due in part to its tendency to adsorb nitro aromatics preferentially. In competitive hydrogenations of a mixture of nitrobenzene and phenylhydroxylamine over platinum or palladium, nitrobenzene was reduced preferentially over platinum, whereas phenylhydroxylamine disappeared preferentially over palladium (Rylander et al., 1970, 1972).

$$H_3C \overline{\phantom{x}}\!\!\left\langle \phantom{xx} \right\rangle\!\!\overline{\phantom{x}} NO_2 \xrightarrow[\substack{1.5\ gm\ DMSO \\ 25°C,\ 0-50\ psig}]{\substack{1\ gm\ 5\%\ Pt\text{-}on\text{-}C \\ CH_3OH,\ 2H_2}} H_3C \overline{\phantom{x}}\!\!\left\langle \phantom{xx} \right\rangle\!\!\overline{\phantom{x}} NHOH$$

25 gm

## B.  Rearrangement

When hydrogenation of nitro aromatics is carried out in acidic solution, the intermediate N-phenylhydroxylamines undergo rearrangement. p-Fluoroaniline was obtained in 61% yield by reduction of nitrobenzene over platinum oxide in anhydrous hydrogen fluoride at 55 psig and 50°C (Fidler et al., 1961), and chloroanilines arose by hydrogenation in hydrochloric acid (Bradbury, 1966). Commerical syntheses of p-aminophenol involve the reduction of intermediate phenylhydroxylamine over platinum catalysts in dilute sulfuric acid (Spiegler, 1956; Benner, 1965, 1968; Benwell, 1970). The yield is sensitive to acid concentration, catalyst, temperature, hydrogen pressure, mode of addition, solvents, and additives, such as, water-soluble quaternary ammonium compounds. Early workers thought that satisfactory yields could be obtained only at subatmospheric hydrogen pressure, but this restriction is no longer necessary. Dimethyl sulfoxide has been used as an additive to decrease the rate of hydrogenation of phenylhydroxylamine and permit acid-catalyzed rearrangement to compete favorably with reduction to aniline (Rylander et al., 1970, 1973).

## C.  Cyclization

Intermediate aromatic hydroxylamines formed by hydrogenation of aromatic nitro compounds may be trapped by cyclization with suitably disposed carbonyl functions. The extent of trapping depends, among other things, on the electrophilicity of the carbonyl function and on the solvent. In catalytic hydrogenation over platinum oxide of 2-nitro-2'-carboxybiphenyl and its carboxy derivatives in ethanol containing mineral acid, trapping abilities decreased in the order carbamoyl > carbomethoxy > carboxy; without mineral acid the order is reversed. The ordering of trapping functions with mineral acid is in agreement with the postulate that the carbamoyl group is the most basic group of the three examined, and consequently its conjugate acid the most electrophilic (Muth and Beers, 1969). Marked solvent effects are also observed. Reduction of 2-nitro-2'-carbamoylbiphenyl (2) over platinum in tetrahydrofuran (THF) afforded N-hydroxyphenanthridone (1) in 70% yield, whereas, in ethanol, phenanthridone (3) was obtained in 91% yield. The authors suggested that these results are attributable to the more extensive hydrogen bonding of the intermediate hydroxylamine function, making it less available for trapping (Muth et al., 1967).

A convenient synthesis of 1-hydroxy-2-indolinones (Davis *et al.*, 1973) and 1-hydroxycarbostyrils (Davis *et al.*, 1964, 1970, 1972) involves partial hydrogenation of appropriate nitro compounds and spontaneous cyclization of intermediate hydroxylamines. These syntheses are illustrated by the hydrogenation of *o*-nitrophenylglycine. Reduction of the hydrochloride salt in dilute hydrochloric acid over 5% platinum-on-carbon affords 3-amino-1-hydroxy-2-indolinone hydrochloride in 62% yield. Reduction of the free base, on the other hand, in neutral solution over palladium gives *o*-aminophenylglycine, which spontaneously cyclizes to 3-amino-2-indolinone when the solution is made acidic.

Certain other functions, as well as carbonyls, can interact with intermediate hydroxylamines. Hydrogenation of 2-nitrophenylpyruvic acid oxime (or semicarbazone or phenylhydrazone) over either palladium or platinum affords a mixture of 1-hydroxyindole-2-carboxylic acid and indole-2-carboxylic acid. Attack on the carboxylic acid group apparently is inhibited by its ionization through reaction with ammonia produced in the reaction (Baxter and Swan, 1967).

A route to 1-hydroxybenzimidazoles and imidazopyridines is by hydrogenation of the corresponding 2-nitroanilide. Cyclodehydration probably occurs at the hydroxylamine stage through attack by the nucleophilic nitrogen on the carbonyl. Cyclization to the 1-hydroxy compound is favored by acids, presumably through protonation of the carbonyl oxygen and formation of a full positive charge on the carbon atom. Substituents on the benzo ring that lower this positive charge density hinder formation of the $N$-hydroxyimidazole and favor formation of the deoxy compound, and, conversely, substituents that increase the positive charge enhance the probability of $N$-hydroxyimidazole formation. The use of polar solvents, such as ethanol, maximizes yields of $N$-hydroxyimidazoles, whereas nonpolar, aprotic solvents favor 2-aminoanilides (Doherty and Fuhr, 1973).

Basic substituents in the nitroanilide or bases, such as triethylamine, added to a neutral substrate cause the formation of azo compounds as the principal product. If a basic substituent is present, 1 mole of acid is required for the formation of any $N$-hydroxy compound.

In dinitroanilides, one nitro group can be reduced preferentially.

80%

Cyclization of intermediate hydroxylamines with neighboring functional groups may lead to compounds that are not in themselves cyclic but are derived from cyclic intermediates. Cohen and Gray (1972) reported an extremely facile cleavage of a *tert*-benzamide, which was deduced to occur through nucleophilic assistance of an intermediate neighboring hydroxyl-amine group. When a different batch of palladium-on-carbon was used, the proposed hydroxylamine intermediate could be isolated in fair yield.

The reaction scheme is similar to that used by Musso and Schroeder (1965) to account for the formation of o-aminobenzamide on hydrogenation of o-nitrobenzonitrile. Heavy-water experiments established that the amide oxygen was originally in the nitro function (Moll *et al.*, 1963).

In the hydrogenation of (**4**) the product depends on whether platinum or palladium is used (Schroeder *et al.*, 1965). The results are in accord with expectations; platinum is expected to produce more hydroxylamine than is palladium.

**4**

## D. Aliphatic Nitro Compounds

Aliphatic hydroxylamines have been obtained, often in excellent yield, by hydrogenation of the corresponding aliphatic nitro compounds over palladium black (Kahr and Zimmerman, 1957), palladium-on-alumina (Joris and Vitrone, 1958), palladium-on-carbon (McWhorter, 1965), palladium-on-calcium carbonate (Meister and Franke, 1959), and, in fixed-bed processing, over 0.5% palladium-on-8-mesh alumina. Oximes can be derived from aliphatic nitro compounds either by oxidation of the hydroxyl-amine formed in the hydrogenation (Meister and Franke, 1959) or by the use in hydrogenation of palladium (Foster and Kirby, 1961) or platinum (Weise, 1954) catalysts inhibited by heavy metals.

Knifton (1976b) found Group IB metal salts in alkylpolyamine solvents to be excellent catalysts for the selective hydrogenation of nitroalkanes to oximes. Favorable aspects of these solvent systems are that they are non-acidic, Nef-type hydrolysis thus being avoided, nonaqueous, for better solubility, and basic, favoring formation of the more reactive nitroalkane anion. Nitrocyclohexane is converted to cyclohexanone oxime in 93% yield with cuprous chloride as catalyst at 90°C and 35 atm hydrogen pressure in ethylenediamine as solvent. The apparent order of activity for Group IB and IIB metal salts is $Cu(I) \approx Cu(II) > Ag(I) \gg Hg(II)$.

The rationale for hydrogenation to the oxime calls for heterolytic splitting of hydrogen by solvated copper(I) to form a hydride complex followed by nucleophilic displacement, or addition, by the nitroalkane anion to the coordination sphere of the copper complex. Partial deoxygenation of the nitroalkane anion through hydride attack via a quasi-cyclic transition state is believed to be the slow, rate-determining step. Knifton (1973) also used carbon monoxide as a reducing agent in the homogeneous copper-catalyzed reduction of nitro compounds to oximes.

## IV.   REDUCTION OF ALIPHATIC NITRO COMPOUNDS

Aliphatic nitro compounds are not as easily hydrogenated as aromatic nitro compounds. Many successful reductions are characterized by the use of high catalyst loadings (Baer and Fischer, 1960; Barker, 1964), vigorous conditions (Lieber and Smith, 1936), or relatively lengthy reaction times (Controulis *et al.*, 1949; Iffland and Cassis, 1952; Nielsen, 1962). Nonetheless, the reductions, which provide a convenient source of many complex amines, usually can be made to proceed to completion without difficulty.

### A.   Amino Sugars

Reduction of aliphatic nitro compounds has provided a convenient route to various amino sugars, as illustrated by the following selected examples.

A synthesis of 2,3-diamino sugars was based on amination of 2,3-unsaturated 3-nitroglycosides followed by hydrogenation to the diamino stage (Baer and Neilson, 1967; Baer and Ong, 1969). In an example of the hydrogenation step, methyl-2-(2-carboxyanilino)-2,3-dideoxy-3-nitro-$\beta$-D-glucopyranoside (5) undergoes both hydrogenolytic dearylation (Kuhn and Haas, 1958) and reduction of the nitro function to afford 6 when treated with hydrogen and palladium-on-barium sulfate (Kuhn and Haas, 1955) in dilute hydrochloric acid (Baer and Kienzle, 1969).

Excellent yields of 3-acetamido-3-deoxy-1,2,5,6-di-$O$-isopropylidene-$\alpha$-D-allofuranose (8) were obtained by hydrogenation under ambient conditions of 7 over prereduced 10% palladium-on-carbon in methanol containing 1 equivalent of acetic acid (Takamoto et al., 1973). The nitro compound was prepared readily by low-temperature oxidation of the 3-oximino compound with trifluoroperacetic acid (Takamoto et al., 1971).

Benzoyl migration accompanied hydrogenation of methyl-2-$O$-benzoyl-4,6-$O$-benzylidene-3-$C$-nitromethyl-$\alpha$-D-allopyranoside (10) over Raney nickel to afford methyl-4,6-$O$-benzylidene-3-$C$-benzamidomethyl-$\alpha$-D-allopyranoside (11) in 65% yield. On the other hand, palladium-on-carbon promoted hydrogenolysis of the benzyl–oxygen bond, affording methyl-2-$O$-benzoyl-3-$C$-nitromethyl-$\alpha$-D-allopyranoside (9) (Yoshimura et al., 1973).

9            10            11

## B.  Amino Alcohols

Convenient syntheses of amino alcohols involve base-catalyzed condensation of a carbonyl compound with an alkyl nitro compound, followed by hydrogenation of the nitro function. For instance, facile reduction of the nitro function in 2,6-diphenyl-4-(nitromethyl)-4-piperidinol occurs readily over platinum oxide in acetic acid to afford 2,6-diphenyl-4-aminomethyl-4-piperidinol in 92% yield (Overberger *et al.*, 1970).

Nitro alcohols can be converted rapidly to the corresponding amino alcohol over palladium catalysts even when the alcohol is benzylic. The amino group stabilizes the alcohol against hydrogenolysis (Marquardt and Edwards, 1972), and the reduction ceases spontaneously at the amino alcohol stage. Loss of the hydroxy can be achieved by reduction in strong acid media (Rosenmund and Kung, 1942) or through conversion to the chloride and subsequent hydrogenolysis (Marquardt and Edwards, 1972).

## V.  UNSATURATED NITRO COMPOUNDS

Unsaturated, nonconjugated, aliphatic nitro compounds can usually be reduced with ease to the corresponding saturated nitro compounds over either palladium or platinum catalysts (deMauny, 1940; Sowden and Fischer, 1947; Roberts *et al.*, 1954; Freeman, 1960). On the other hand, when the competition is between an aromatic nitro function and an olefin, selective reduction of the nitro function to afford an unsaturated amino compound is the more likely course (Adams *et al.*, 1927; Fieser and Joshel, 1940; Blout and Silverman, 1944; Roger, 1947; Benoit and Marinopoulos, 1950; Boekelheide and Michels, 1952; Baker and Jordaan, 1965). Hydrogenation of the stilbene **12** over palladium reduced only the nitro function to afford **13**,

12                                      13

but in the more extended system **(14)** the olefins were reduced as well **(15)**, perhaps via 1,4 addition (Baker and Lourens, 1968).

14                                      15

Sulfided noble metal catalysts have been used effectively to reduce unsaturated aromatic nitro compounds to the unsaturated amines (Braden *et al.*, 1977).

### Conjugated Nitroolefins

Hydrogenation of conjugated nitroolefins may yield a variety of products through partial and complete reduction as well as through reductive hydrolysis. Reduction of $\beta$-nitrostyrenes **(16)** in neutral media results largely in coupled products that are thought to be dimeric **(17)** (Sonn and Schellenberg, 1917; Kohler and Drake, 1923). However, the hydrogenation products of $\beta$-nitrostyrene, after the absorption of 1 mole of hydrogen,

16                                      17

showed in a mass spectrometer mass fragments larger than dimers (Rylander, 1970).

In strong acid solutions, such as sulfuric acid–acetic acid (Kindler et al., 1934; Dyumaev and Belostotskaya, 1962; Green, 1962) or acetic acid–hydrochloric acid (Daly et al., 1961), β-nitrostyrenes are reduced smoothly to the saturated amine. Dilute aqueous hydrochloric acid has also been used effectively. The yields of recrystallized hydrochloride salt are generally high (Brossi et al., 1968; Wagner et al., 1971).

Oximes can be obtained in excellent yield by hydrogenation of aromatic α,β-unsaturated nitro compounds. Nearly quantitative yields of the oximes of substituted phenylacetaldehydes are obtained by reduction of β-nitrostyrenes in pyridine over palladium-on-carbon (Reichert and Koch, 1935; Reichert and Marquardt, 1950). The same technique may be applied to aliphatic α,β-unsaturated nitro compounds, but the yields are only about 60% (Seifert and Condit, 1963). Somewhat higher yields of oximes can be obtained by reduction in acidic media over palladium catalysts (Seifert and Condit, 1963; Freidlin et al., 1970).

A convenient synthesis of 2-nitrophenylacetaldehyde oximes involves catalytic hydrogenation of 2-β-dinitrostyrenes over 5% rhodium-on-alumina in ethanol–acetic acid–ethyl acetate (Baxter and Swan, 1967).

## VI. HALONITRO AROMATICS

Hydrogenation of halonitro aromatics to the corresponding haloanilines was at one time a formidable problem. Now the reaction is practiced industrially on a large scale with little difficulty. The extent of dehalogenation may depend on the halogen, structure of the molecule, catalyst support (Rylander et al., 1965), metal, amount of catalyst (Freifelder et al., 1961), solvent (Weizmann, 1949), and reaction conditions (Kindler et al., 1953) but, despite this complexity, dehalogenation can usually be kept to low levels.

## A.  Catalysts

Platinum seems to be the best catalyst for minimizing dehalogenation combined with a fast rate of reduction of the nitro function. It can be used effectively with (Kosak, 1970a) or without inhibitors. Addition of nickel(II) and chromium(III) ions to platinum catalysts substantially increases their activity (Kosak, 1970b). Excellent results have been obtained over sulfides of palladium, platinum, rhodium, ruthenium (Greenfield, 1967), or rhenium (Broadbent, 1967), but vigorous conditions must be used. For example, a mixture of 103.5 gm 2,5-dichloronitrobenzene, 230 ml methanol, and 3.0 gm 5% platinum sulfide-on-carbon (manufactured by Engelhard Industries, Newark, New Jersey, Keith and Bair, 1966) was reduced at 85°C and 500–800 psig for 1.25 hr. The yield of 2,5-dichloroaniline was 99.5% (Dovell and Greenfield, 1965, 1967).

Palladium is the catalyst of choice for the dehalogenation of halo aromatics but, with suitable inhibitors, the hydrogenolysis tendencies of palladium can be completely suppressed.

## B.  Inhibitors

Bases of various kinds are often used stoichiometrically to help effect catalytic dehydrohalogenation, yet, paradoxically, the same bases in lesser amounts can be used to good advantage in inhibiting loss of halogen during reduction of halonitrobenzenes. For instance, magnesium oxide or hydroxide (Spiegler, 1963), calcium hydroxide (Dietzler and Keil, 1962), and sodium acetate (German Patent 1,159,956) have all been used effectively as dehalogenation inhibitors.

Organic bases can also be very useful inhibitors, but their effectiveness is highly sensitive to structure. Only 0.5% dehalogenation of $p$-chloronitrobenzene in reduction to chloroaniline over platinum-on-carbon at 95°C and 400–500 psig occurred when morpholine (1% based on substrate) was present. Piperazine was also effective, but the open-chain analogues ethanolamine and ethylenediamine were catalyst poisons. In the morpholine series, effectiveness as a dehalogenation inhibitor decreases with increasing steric hindrance in the order morpholine > $N$-methylmorpholine > $N$-ethylmorpholine > 3,5-dimethylmorpholine (Kosak, 1970a). Triphenylphosphite is also an effective dehalogenation inhibitor (Craig et al., 1969).

Kosak (1977) described a remarkable set of inhibitors that allow hydrogenation of chloronitro, bromonitro, and iodonitro aromatics with very little hydrogenolysis of the halogen. The inhibitors are of the group

$$X-PH-OH$$
$$\|$$
$$O$$

where X = H, OH, $C_{1-12}$ alkyl, or phenyl. Excellent results can be obtained even when palladium, a catalyst often used for achieving hydrogenolysis, is employed.

In general, loss of halogen during reduction of halonitro aromatics is favored by increasing temperature and retarded by increasing pressure. In the morpholine-inhibited system, the effects of pressure are pronounced, especially when large amounts of morpholine are employed. At pressures under 50 psig, complete displacement of the halogen by morpholine occurs with no attendant hydrogenation. At 100 psig the product is a mixture of 43% p-chloroaniline, 53% p-(N-morpholino)aniline, and 6% aniline, whereas at 200 psig the product is 95% p-chloroaniline (Kosak, 1970a).

## VII.   DINITRO COMPOUNDS

Hydrogenation of dinitro compounds, which are unable to interact intra-molecularly, to the corresponding diamines presents no problem. Dinitro-toluene is hydrogenated industrially in very large volumes over either nickel or palladium-on-carbon to diaminotoluene, which is subsequently converted to toluene diisocyanate. Selective hydrogenation of one of two nitro groups is more difficult, but some outstanding successes have been recorded.

The selective reduction of 2-(4-nitrophenyl)-4(5)-nitroimidazole (18) to 2-(4-aminophenyl)-4(5)-nitroimidazole (19), was achieved in 86–89% yield by careful attention to reaction conditions (Jones et al., 1973).

The yield of 19 was found to depend on temperature, pressure, solvent, agitation, and catalyst. Palladium-on-carbon proved to be more selective than palladium or platinum oxide, platinum-on-carbon, ruthenium-on-carbon, rhodium-on-carbon, Raney nickel, or Raney cobalt. Raney nickel was surprisingly inactive, and Raney cobalt was nonselective, although it had been selective in partial reduction of other dinitro compounds (Jones et al., 1969). Concentrated aqueous ammonia was the solvent of choice; poor selectivity was found when aqueous alkali or anhydrous alcoholic ammonia as well as a number of other polar and nonpolar solvents were used. Agitation was important; poor results were obtained when significant amounts of starting material dissolved either by agitation or warming. The reaction is sensitive to pressure, and all successful experiments were made at 40 psig. The yield was lower (75%) at 20 psig, and at atmospheric pressure there was no hydrogen uptake at all. Effective temperatures were 20°–35°C; reactions carried out either above or below this range were nonselective.

The exacting reaction conditions required for selective hydrogenation arise apparently through the need to meet two requirements. The imidazole ring should exist as its resonance-stabilized anion, in order to inhibit hydrogenation of the imidazole nitro function, and the concentration of solubilized reactant should be limited, in order to prevent the formation of dimeric species. The reaction mixture begins as a heavy slurry and ends as a clear solution. Concentrated ammonia is an effective solvent, whereas anhydrous alcoholic ammonia is not, for the concentration of substrate is too great in the latter (Jones *et al.*, 1973).

Raney copper proved to be uniquely selective among a large number of catalysts for the hydrogenation of 2,4-dinitroalkylbenzenes to the corresponding 4-amino-2-nitroalkylbenzenes. Extraordinarily large catalyst loadings (50% based on substrate) are required, but this is of little economic consequence because the catalyst can be used repeatedly; both activity and selectivity increase with each reuse of the catalyst up to six reuses. Improved catalyst performance on reuse was believed to be due to the formation of the diamino compound. It was possible to rapidly condition substandard catalysts by using them to reduce a small amount of the dinitro compound to the diamine (Jones *et al.*, 1969).

81–92%

Selectivity for reduction of the 4-nitro group increases with steric hindrance of the alkyl substituent. Electronic effects are also operative; reduction of 2,4-dinitroanisole gave only 2,4-diaminoanisole. On the contrary, reduction of 2,4-dinitro-1-(*N*-piperidyl)benzene was extraordinarily selective. The result is attributed both to the steric hindrance of the piperidyl group and to the steric inhibition of resonance, which drastically reduces the electron-releasing ability of the piperidyl nitrogen (Jones *et al.*, 1969).

$$> 99\%$$

A useful selective hydrogenation of dinitro- and trinitroanilines to afford substituted nitro-o-phenylenediamines was developed by Miesel *et al.* (1976). The hydrogen source was hydrazine in the presence of ruthenium-on-carbon catalyst. This catalyst proved to be more effective than palladium-, platinum-, or rhodium-on-carbon, ruthenium-on-alumina, or Raney nickel.

$$R = 5\text{-}NH_2, 5\text{-}CF_3, 6\text{-}CF_3, 6\text{-}Cl$$

As an example, 370 gm 85% hydrazine hydrate (5% excess) was added dropwise to 1000 gm 2,6-dinitro-4-trifluoromethylaniline in 12 liters of ethanol at 60°C containing 25 gm 5% ruthenium-on-carbon. The solution temperature was allowed to rise to reflux. When the exotherm had ceased, the solution was refluxed an additional hour. After filtration and recrystallization, 600 gm (68% yield) of 3-nitro-5-trifluoromethyl-o-phenylenediamine was obtained.

Simple alkyl substitution on the amino group does not affect the selectivity of diamine formation. Even when the amino nitrogen forms part of an imidazole ring, the ortho nitro group is selectively reduced. However, nitroanilides cyclize to N-hydroxybenzimidazoles by capture of the intermediate hydroxylamine.

Lyle and LaMattina (1974) were able to selectively reduce certain 2,6-dinitroanilines over palladium-on-carbon in 1,2-dimethoxyethane–chloroform as solvent.

## Homogeneous Catalysts

Various ruthenium complexes, such as dichlorotris(triphenylphosphine) ruthenium(II) have been found to be effective catalysts for the hydrogenation of nitroparaffins (Knifton, 1976b) and nitro aromatics (Knifton, 1976a) to the corresponding amines. Of particular interest is the ability of these catalysts to selectively hydrogenate only one of two nitro functions.

This high selectivity is not found with heterogeneous catalysts. The excellent yield of p-nitroaniline obtained with the homogeneous catalyst is attributed to electron withdrawal by the second nitro group in p-dinitrobenzene, ensuring preferential bonding with the dinitro compound (Knifton, 1976b). High selectivity was also found in the reduction of 2,4-dinitrotoluene; the more hindered nitro group is reduced preferentially.

## VIII.  CYCLIZATIONS

The interaction of products of nitro group reduction, i.e., amines, oximes and hydroxylamines, with other suitably placed functions provides a convenient route to a variety of nitrogen heterocyclic compounds. Several examples of cyclizations involving amines are given below; cyclizations involving partially reduced nitro functions are discussed in Section III,C.

## A.  Nitrocarboxylic Acids and Nitro Esters

The reduction of a nitro function in a molecule containing a suitably placed carboxylic acid (Munk and Schultz, 1952; Walker, 1955; Masamune *et al.*, 1964; Kupchan and Merianos, 1968) or ester (Leonard and Beck, 1948; Baer and Achmatowicz, 1964a,b) can easily lead to lactams. The tendency toward cyclization is sometimes so high that it cannot be prevented (Gensler *et al.*, 1975); in other cases, cyclization occurs on subsequent treatment. A short and convenient route to 2-substituted indole-4-carboxylic acids involves hydrogenation of a 5-nitroisocoumarin over palladium-on-carbon in ethanol followed, without isolation, by hydrolysis and acid-catalyzed cyclization (Ames and Ribeiro, 1976). Success depends on hydrogenation of the nitro function in preference to the double bond.

R = CH$_3$, C$_6$H$_5$         R = CH$_3$, 62%
                               R = C$_6$H$_5$, 46%

A serviceable synthesis of certain pyrrolidones involves Michael addition of nitromethane to substituted methyl acrylates followed by hydrogenation and cyclization to the lactam. Reactive keto functions can be protected via ketalization (Gutsche and Zandstra, 1974). Noble metal catalysts would probably obviate the need for the vigorous hydrogenation conditions used.

24 gm

Hydrogenation of ethyl α-cyano-3-nitro-2-pyridine acetate (**20**) over 10% palladium-on-carbon in 95% ethanol affords ethyl 3-amino-α-cyano-2-pyridine acetate (**21**) quantitatively. The latter compound was converted to 3-cyano-4-azaoxindole (**22**) by refluxing xylene (Finch *et al.*, 1972).

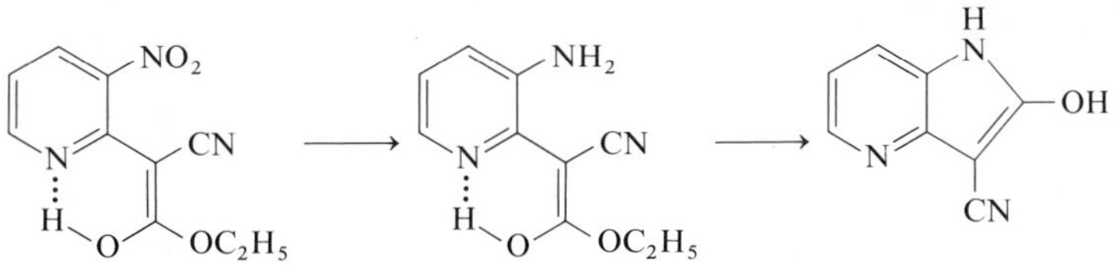

20    21    22

If reductions of α-cyano-o-nitro compounds are carried out so that the cyano group is also reduced, cyclization may occur between the imino and amino functions (Walker, 1965). Reduction of **23** over palladium-on-carbon gave indole-3-carboxamides (**24**) (Bordais and Germain, 1970). Partially reduced nitro functions cyclize, giving amine oxide derivatives (Bauer, 1938; Cheeseman, 1963).

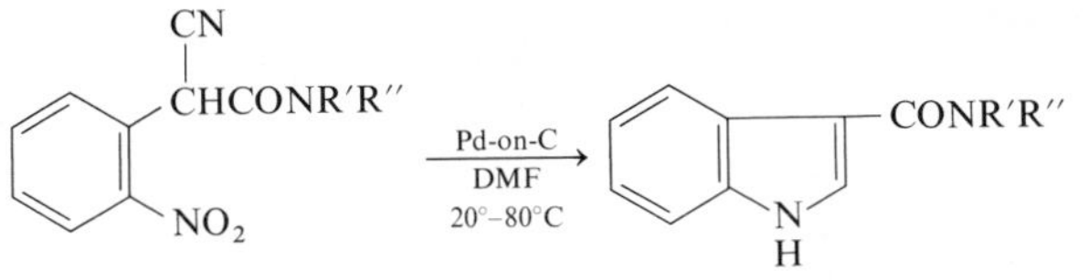

CN
CHCONR′R″
NO₂
$\xrightarrow[\substack{DMF \\ 20°-80°C}]{Pd-on-C}$
CONR′R″

23    24

The direction of ring closure may be determined by subtle factors. For instance, hydrogenation of ethyl 2-oxo-3-(2′-methoxy-5′-nitro-4′-pyridyl) butyrate (**25**) over palladium-on-carbon in ethanol affords exclusively ethyl 3-methyl-5-methoxy-6-azaindole-2-carboxylate (**26**),

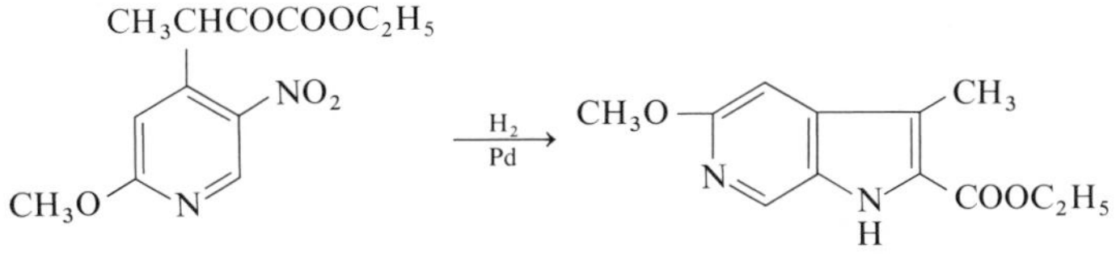

$CH_3CHCOCOOC_2H_5$
NO₂
$CH_3O$
$\xrightarrow[Pd]{H_2}$
$CH_3O$
$CH_3$
$COOC_2H_5$

25    26

whereas the similar compound **27** affords exclusively a naphthyridinone (**28**). These differences were rationalized through consideration of the influence of substituents on the enol–keto equilibrium (Frydman et al., 1973).

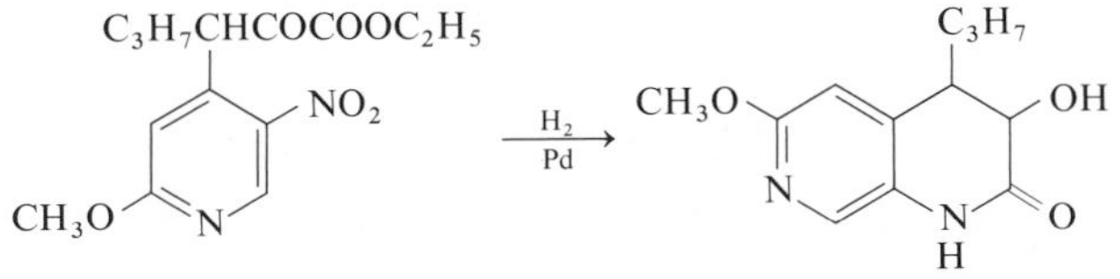

$C_3H_7CHCOCOOC_2H_5$
NO₂
$CH_3O$
$\xrightarrow[Pd]{H_2}$
$CH_3O$
$C_3H_7$
OH
O

27    28

## B.  Nitroaldehydes and Nitroketones

A variety of heterocyclic compounds can be prepared through reductive cyclization of nitroaldehydes and nitroketones. For instance, hydrogenation of 1-(2-nitrobenzyl)pyrrol-2-aldehyde over palladium-on-carbon in ethanol–ethyl acetate affords 10,11-dihydro-5H-pyrrolo[1,2-a]benzo-[e]-1,4-diazepine (Marino et al., 1969).

Similarly, hydrogenation of **30** gives **31**. Under slightly different conditions **30** forms **29**, perhaps by hydrogenolysis of **31**. Apparently, dehalogenation over palladium is unimportant (Topliss et al., 1967).

**29**                              **30**                              **31**

Indoles are formed in excellent yield by reduction of o-nitrobenzyl ketones over palladium-on-carbon (Augustine et al., 1973).

The reaction also provides a convenient entry to 4-azaindoles. Hydrogenation of ethyl 3-nitropyridyl-2-pyruvate over 9% palladium-on-carbon in ethanol affords ethyl 4-azaindole-2-carboxylate, quantitatively (Yakhontov et al., 1969).

Reductive cyclization of nitro ketones has been applied effectively to the synthesis of potential folic acid antogonists. Ring closure occurred spontaneously (Elliott et al., 1968).

Hydrogenation of aliphatic $\gamma$-nitro ketones over platinum oxide or Raney nickel gives pyrrolidines in good yields (Kloetzel, 1947). A mixture of pyrazines and piperazines is formed on hydrogenation of $\alpha$-nitro ketones in either batch or continuous processing (Ellis, 1969). The reaction was demonstrated with a variety of catalysts.

## C.   Nitroamides

Imidazoles can be obtained by reductive cyclization of 2-nitroacetanilides (Neale et al., 1969). Cyclization and complete reduction of the nitro function may depend on the solvent employed (Schulenberg and Archer, 1965).

## D.   Dinitro Compounds

A general way into the indole system involves reductive cyclization of a dinitrostyrene (Huebner et al., 1953; Heacock et al., 1963). For instance, hydrogenation of **32** was rapid and afforded the indole **33** in 50% yield (Benington et al., 1960).

However, hydrogenation of α-2-dinitrochalcone (34) over 5% palladium-on-carbon in acetic acid–ethanol affords 3-amino-2-phenylquinoline (35) as the only identifiable species. No indolic compound was found (Augustine *et al.*, 1973).

A mixture of cinnolines and diaminobiphenyls is formed by hydrogenation of 2,2′-dinitrobiphenyls. Cinnolines are not precursors of the diamines, and it is believed that the two types of compounds arise because the dinitro compound can be adsorbed in either a trans configuration, making interaction of the functional groups difficult, or in a cis configuration, favoring interaction and cinnoline formation (Ross and Kuntz, 1952; Blood and Noller, 1957).

## REFERENCES

Adams, R., Cohen, F. L., and Rees, C. W., *J. Am. Chem. Soc.* **49**, 1093 (1927).
Aeberli, P., and Houlihan, W. J., *J. Org. Chem.* **32**, 3211 (1967).
Ames, D. E., and Ribeiro, O., *J. Chem. Soc., Perkin Trans. I* p. 1074 (1976).
Augustine, R. L., Gustavsen, A. J., Wanat, S. F., Pattison, I. C., Houghton, K. S., and Koletar, G., *J. Org. Chem.* **38**, 3004 (1973).
Baer, H. H., and Achmatowicz, B., *J. Org. Chem.* **29**, 3180 (1964a).
Baer, H. H., and Achmatowicz, B., *Angew. Chem., Int. Ed. Engl.* **3**, 224 (1964b).
Baer, H. H., and Fischer, H. O. L., *J. Am. Chem. Soc.* **82**, 3709 (1960).
Baer, H. H., and Kienzle, F., *J. Org. Chem.* **34**, 3848 (1969).
Baer, H. H., and Neilson, T., *J. Org. Chem.* **32**, 1068 (1967).
Baer, H. H., and Ong, K. S., *J. Org. Chem.* **34**, 560 (1969).
Baker, B. R., and Jordaan, J. H., *J. Med. Chem.* **8**, 35 (1965).
Baker, B. R., and Lourens, G. J., *J. Med. Chem.* **11**, 34 (1968).
Barker, R., *J. Org. Chem.* **29**, 869 (1964).
Barker, G., and Ellis, G. P., *J. Chem. Soc. C* p. 2230 (1970).
Bauer, K. H., *Ber. Dtsch. Chem. Ges. B* **71**, 2226 (1938).
Baxter, I., and Swan, G. A., *J. Chem. Soc. C* p. 2446 (1967).
Benington, F., Morin, R. D., and Clark, L. C., Jr., *J. Org. Chem.* **25**, 1542 (1960).
Benner, R. G., U.S. Patent 3,383,416, July 8, 1965.
Benner, R. G., U.S. Patent 3,383,416, May 14, 1968.
Benner, R. G., and Stevenson, A. C., U.S. Patent 2,619,503, Nov. 25, 1952.
Benoit, G., and Marinopoulos, D., *Bull. Soc. Chim. Fr.* p. 829 (1950).
Benwell, N. R. W., Br. Patent 1,181,969, Feb. 18, 1970.
Berse, C., Boucher, R., and Piché, L., *J. Org. Chem.* **22**, 805 (1957).
Blood, A. E., and Noller, C. R., *J. Org. Chem.* **22**, 711 (1957).

Blout, E. K., and Silverman, D. C., *J. Am. Chem. Soc.* **66**, 1442 (1944).
Boekelheide, V., and Michels, A. P., *J. Am. Chem. Soc.* **74**, 256 (1952).
Bordais, J., and Germain, C., *Tetrahedron Lett.* p. 195 (1970).
Bradbury, W. C., U.S. Patent 3,265,735, Aug. 9, 1966.
Braden, R., Knupfer, H., and Hartung, S., U.S. Patent 4,002,673, Jan. 11, 1977.
Broadbent, H. S., *Ann. N.Y. Acad. Sci.* **145**, 58 (1967).
Brossi, A., Van Burik, J., and Teitel, S., *Helv. Chim. Acta* **51**, 1978 (1968).
Cavill, G. W., and Ford, D. L., *J. Chem. Soc.* p. 565 (1954).
Cheeseman, G. W. H., *Adv. Heterocycl. Chem.* **2**, 218 (1963).
Cohen, T., and Gray, W. F., *J. Org. Chem.* **37**, 741 (1972).
Controulis, J., Rebstock, M. C., and Crooks, H. M., Jr., *J. Am. Chem. Soc.* **71**, 2463 (1949).
Craig, W. C., Davis, G. J., and Shull, P. O., U.S. Patent 3,474,144, Oct. 21, 1969.
Daly, J., Horner, L., and Witkop, B., *J. Am. Chem. Soc.* **83**, 4787 (1961).
Davis, A. L., Choun, O. H. P., Cook, D. E., and McCord, T. J., *J. Med. Chem.* **7**, 632 (1964).
Davis, A. L., Hughes, J. W., Hance, R. L., Gault, V. L., Tommy, J., and McCord, T. J., *J. Med. Chem.* **13**, 549 (1970).
Davis, A. L., Smith, D. R., Foyt, D. C., Black, J. L., and McCord, T. J., *J. Med. Chem.* **15**, 325 (1972).
Davis, A. L., Smith, D. R., and McCord, T. J., *J. Med. Chem.* **16**, 1043 (1973).
deMauny, H. C., *Bull. Soc. Chim. Fr.* **7**, 133 (1940).
DiCarlo, F. J., *J. Am. Chem. Soc.* **66**, 1420 (1944).
Dietzler, A. J., and Keil, T. R., U.S. Patent 3,051,753, Aug. 28, 1962.
Doherty, G. O., and Fuhr, K. H., *Ann. N.Y. Acad. Sci.* **214**, 221 (1973).
Dovell, F. S., and Greenfield, H., *J. Am. Chem. Soc.* **87**, 2767 (1965).
Dovell, F. S., and Greenfield, H., U.S. Patent 3,350,450, Oct. 31, 1967.
Dyumaev, K. M., and Belostotskaya, I. S., *Zh. Obshch. Khim.* **32**, 2661 (1962).
Elliott, R. D., Temple, C., Jr., and Montgomery, J. A., *J. Org. Chem.* **33**, 533 (1968).
Ellis, A. F., U.S. Patent 3,453,278, July 1, 1969.
Fanta, P. E., *J. Am. Chem. Soc.* **75**, 737 (1953).
Fidler, D. A., Logan, J. S., and Boudakian, M. M., *J. Org. Chem.* **26**, 4014 (1961).
Fieser, L. F., and Joshel, L. M., *J. Am. Chem. Soc.* **62**, 1211 (1940).
Finch, N., Robison, M. M., and Valerio, M. P., *J. Org. Chem.* **37**, 51 (1972).
Foster, R. E., and Kirby, A. F., U.S. Patent 2,967,200, Jan. 3, 1961.
Freeman, J. P., *J. Am. Chem. Soc.* **82**, 3869 (1960).
Friedlin, L. K., Litvin, E. F., and Chursina, V. M., *Katal. Vosstanov. Gidrirov Zhidk. Faze* p. 59 (1970); *Chem. Abstr.* **76**, 45665f (1972).
Freifelder, M., and Robinson, R. M., U.S. Patent 3,079,435, Feb. 26, 1963.
Freifelder, M., Martin, W. B., Stone, G. R., and Coffin, E. L., *J. Org. Chem.* **26**, 383 (1961).
Frydman, B., Buldain, G., and Repetto, J. C., *J. Org. Chem.* **38**, 1824 (1973).
Gensler, W. J., Lawless, S. F., Bluhm, A. L., and Dertouzos, H., *J. Org. Chem.* **40**, 733 (1975).
Goldkamp, A. H., U.S. Patent 3,125,589, Mar. 17, 1964.
Green, M., U.S. Patent 3,062,884, Nov. 6, 1962.
Greenfield, H., *Ann. N.Y. Acad. Sci.* **145**, 108 (1967).
Gutsche, C. D., and Zandstra, H. R., *J. Org. Chem.* **39**, 324 (1974).
Heacock, R. A., Hutzinger, O., Scott, B. D., Daley, J. W., and Witkop, B., *J. Am. Chem. Soc.* **85**, 1825 (1963).
Hernandez, L., and Nord, F. F., *Experientia* **3**, 489 (1947).
Hernandez, L., and Nord, F. F., *J. Colloid Sci.* **3**, 363 (1948).
Hodge, E. B., *J. Org. Chem.* **37**, 320 (1972).
Huebner, C. F., Troxell, H. A., and Schroeder, D. C., *J. Am. Chem. Soc.* **75**, 5887 (1953).

Iffland, D. C., and Cassis, F. A., Jr., *J. Am. Chem. Soc.* **74**, 6284 (1952).
Imai, H., Nishiguchi, T., and Fukuzumi, K., *J. Org. Chem.* **42**, 431 (1977).
Jones, W. H., Benning, W. F., Davis, P., Mulvey, D. M., Pollak, P. I., Schaeffer, J. C., Tull, R., and Weinstock, L. M., *Ann. N.Y. Acad. Sci.* **158**, 471 (1969).
Jones, W. H., Pines, S. H., and Sletzinger, M., *Ann. N.Y. Acad. Sci.* **214**, 150 (1973).
Joris, G. G., and Vitrone, J., Jr., U.S. Patent 2,829,163, Apr. 1, 1958.
Kahr, K., and Zimmerman, K., Swiss Patent 325,080, Dec. 14, 1957.
Kauer, J. C., and Sheppard, W. A., *J. Org. Chem.* **32**, 3580 (1967).
Kaye, I. A., and Roberts, I. M., *J. Am. Chem. Soc.* **73**, 4762 (1951).
Keith, C. D., and Bair, D., U.S. Patent 3,275,567, Sept. 27, (1966).
Kindler, K., Brandt, E., and Gehlhaar, E., *Justus Liebigs Ann. Chem.* **511**, 209 (1934).
Kindler, K., Oelschlager, H., and Henrich, P., *Chem. Ber.* **86**, 167 (1953).
Kloetzel, M. C., *J. Am. Chem. Soc.* **69**, 2271 (1947).
Knifton, J. F., *J. Org. Chem.* **38**, 3296 (1973).
Knifton, J. F., *J. Org. Chem.* **41**, 1200 (1976a).
Knifton, J. F., *in* "Catalysis in Organic Syntheses, 1976" (P. N. Rylander and H. Greenfield, eds.), p. 257. Academic Press, New York, 1976b.
Kohler, E. P., and Drake, N. L., *J. Am. Chem. Soc.* **45**, 1281 (1923).
Kosak, J. R., *Ann. N.Y. Acad. Sci.* **172**, 175 (1970a).
Kosak, J. R., U.S. Patent 3,546,297, Dec. 8, 1970b.
Kosak, J. R., U.S. Patent 4,020,107, Apr. 26, 1977.
Kuhn, R., and Haas, H. J., *Angew. Chem.* **67**, 785 (1955).
Kuhn, R., and Haas, H. J., *Justus Liebigs Ann. Chem.* **611**, 57 (1958).
Kupchan, S. M., and Merianos, J. J., *J. Org. Chem.* **33**, 3735 (1968).
Leonard, N. J., and Beck, K. M., *J. Am. Chem. Soc.* **70**, 2504 (1948).
Lieber, E., and Smith, G. B. L., *J. Am. Chem. Soc.* **58**, 2170 (1936).
Lyle, R. E., and La Mattina, J. L., *Synthesis* p. 726 (1974).
McWhorter, J. R., Jr., U.S. Patent 3,173,953, Mar. 16, 1965.
Malpani, P. S., and Chandalia, S. B., *Indian Chem. J.* **8**, 15 (1973).
Marino, A., DeMartino, G., Filacchioni, G., and Giuliano, R., *Farmaco, Ed. Sci.* **24**, 276 (1969).
Marquardt, F. H., and Edwards, S., *J. Org. Chem.* **37**, 1861 (1972).
Masamune, T., Takasugi, M., Suginome, H., and Yokoyama, M., *J. Org. Chem.* **29**, 681 (1964).
Meister, H., and Franke, W., U.S. Patent 2,886,596, May 12, 1959.
Miesel, J. L., O'Doherty, G. O. P., and Owen, J. M., *in* "Catalysis in Organic Syntheses, 1976," (P. N. Rylander and H. Greenfield, eds.), p. 273. Academic Press, New York, 1976.
Moll, H., Musso, H., and Schroeder, H., *Angew. Chem. Int. Ed. Engl.* **2**, 212 (1963).
Munk, M., and Schultz, H. P., *J. Am. Chem. Soc.* **74**, 3433 (1952).
Musso, H., and Schroeder, H., *Chem. Ber.* **98**, 1562 (1965).
Muth, C. W., and Beers, R. N., *Proc. W. Va. Acad. Sci.* **41**, 235 (1969).
Muth, C. W., Elkins, J. R., DeMatte, M. L., and Chiang, S. T., *J. Org. Chem.* **32**, 1106 (1967).
Neale, A. J., Davies, K. M., and Ellis, J., *Tetrahedron* **25**, 1423 (1969).
Nielsen, A. T., *J. Org. Chem.* **27**, 1998 (1962).
Oelschläger, H., *Chem. Ber.* **89**, 2025 (1965).
Overberger, C. G., Reichenthal, J., and Anselme, J. P., *J. Org. Chem.* **35**, 138 (1970).
Parrett, A. N., and Lowy, A., *J. Am. Chem. Soc.* **48**, 778 (1926).
Reichert, B., and Koch, W., *Arch. Pharm. (Weinheim, Ger.)* **273**, 265 (1935).
Reichert, B., and Marquardt, H., *Pharmazie* **5**, 10 (1950).
Roberts, G. W., *in* "Catalysis in Organic Syntheses, 1976" (P. N. Rylander and H. Greenfield, eds.), p. 1. Academic Press, New York, 1976.
Roberts, J. D., Lee, C. C., and Saunders, W. H., Jr., *J. Am. Chem. Soc.* **76**, 4501 (1954).

Roblin, R. O., Jr., and Winnek, P. S., *J. Am. Chem. Soc.* **62**, 1999 (1940).

Roger, R., *J. Chem. Soc.* p. 560 (1947).

Rosenmund, K. W., and Kung, E., *Chem. Ber.* **75**, 1850 (1942).

Ross, S. D., and Kuntz, I., *J. Am. Chem. Soc.* **74**, 1297 (1952).

Rylander, P. N., unpublished observations, Engelhard Industries Res. Lab., Newark, New Jersey, 1970.

Rylander, P. N., Kilroy, M., and Coven, V., *Engelhard Ind., Tech. Bull.* **6**, 11 (1965).

Rylander, P. N., Karpenko, I. M., and Pond, G. R., *Ann. N.Y. Acad. Sci.* **172**, 266 (1970).

Rylander, P. N., Karpenko, I. M., and Pond, G. R., U.S. Patent 3,694,509, Sept. 26, 1972.

Rylander, P. N., Karpenko, I. M., and Pond, G. R., U.S. Patent 3,715,397, Feb. 6, 1973.

Schipper, E., Chinery, E., and Nichols, J., *J. Org. Chem.* **26**, 4145 (1961).

Schroeder, H., Schwabe, U., and Musso, H., *Chem. Ber.* **98**(8), 2556 (1965).

Schulenberg, J. W., and Archer, S., *J. Org. Chem.* **30**, 1279 (1965).

Seifert, W. K., and Condit, P. C., *J. Org. Chem.* **28**, 265 (1963).

Sonn, A., and Schellenberg, A., *Chem. Ber.* **50**, 1513 (1917).

Sowden, J. C., and Fischer, H. O. L., *J. Am. Chem. Soc.* **69**, 1048 (1947).

Spiegler, L., U.S. Patent 3,073,865, Jan. 15, 1963.

Takamoto, T., Sudoh, R., and Nakagawa, T., *Tetrahedron Lett.* p. 2053 (1971).

Takamoto, T., Yokota, Y., Sudoh, R., and Nakagawa, T., *Bull. Chem. Soc. Jpn.* **46**, 1532 (1973).

Taya, K., *Chem. Commun.* p. 464 (1966).

Tong, W. R., Seagraves, R. L., and Wiederhorn, R., *Natl. Loss Prev. Symp.* A.I.Ch.E. *83rd, Houston, Tex., 1977.*

Topliss, J. G., Shapiro, E. P., and Taber, R. I., *J. Med. Chem.* **10**, 642 (1967).

Wagner, D. P., Rachlin, A. I., and Teitel, S., *Synth. Commun.* **1**, 47 (1971).

Walker, G. N., *J. Am. Chem. Soc.* **77**, 3844 (1955).

Walker, G. N., *J. Med. Chem.* **8**, 583 (1965).

Weise, J., Ger. Patent 917,426, Sept. 2, 1954.

Weizmann, A., *J. Am. Chem. Soc.* **71**, 4154 (1949).

Whitman, G. M., U.S. Patent 2,606,925, Aug. 12, 1952.

Yakhontov, L. N., Azimov, V. A., and Lapan, E. I., *Tetrahedron Lett.* p. 1909 (1969).

Yale, H. L., *J. Org. Chem.* **33**, 2382 (1968).

Yao, H. C., and Emmett, P. H., *J. Am. Chem. Soc.* **81**, 4125 (1959).

Yao, H. C., and Emmett, P. H., *J. Am. Chem. Soc.* **83**, 796 (1961a).

Yao, H. C., and Emmett, P. H., *J. Am. Chem. Soc.* **83**, 799 (1961b).

Yoshimura, J., Sato, K., Kobayashi, K., and Shin, C., *Bull. Chem. Soc. Jpn.* **46**, 1515 (1973).

# Chapter 8

# Hydrogenation of Nitriles

Catalytic hydrogenation of nitriles may give rise to a number of products including primary, secondary, and tertiary amines, imines, hydrocarbons, aldehydes, amides, and alcohols. Despite the complexity of the reaction, it usually can be controlled so as to be a very useful synthetic tool. In bifunctional molecules, nitrile reduction products may interact with other suitably disposed functions; this type of reaction has proved useful in the elaboration of various ring systems.

## I. MECHANISM

The formation of primary, secondary, and tertiary amines as well as certain hydrolysis products is usually accounted for by the assumption of an imine intermediate according to a scheme proposed over a half a century ago [Eq. (1)] (von Braun *et al.*, 1923). Minor variations have been proposed by others (Mignonac, 1920; Winans and Adkins, 1932; Juday and Adkins, 1955; Greenfield, 1967).

$$RC \equiv N \xrightarrow{\text{H}_2} RCH = NH \xrightarrow{\text{H}_2} RCH_2NH_2 \qquad (1)$$

Addition of the primary amine to the intermediate imine gives a product from which the secondary amine can be formed by hydrogenolysis [Eq. (2)]. How much of this addition reaction occurs on the catalyst surface is an open

$$RCH = NH + RCH_2NH_2 \rightleftharpoons RCH(NH_2)NHCH_2R \longrightarrow (RCH_2)_2NH + NH_3 \qquad (2)$$

question. Alternatively, the addition product may undergo elimination of ammonia to afford a Schiff's base, which then undergoes hydrogenation [Eq. (3)]. If the intermediate imine is sterically hindered it can be the major product of the reduction (Chiavarelli and Marini-Bettolo, 1956).

$$RCH(NH_2)NHCH_2R \xrightarrow{-NH_3} RCH{=}NCH_2R \xrightarrow{H_2} (RCH_2)_2NH \qquad (3)$$

Tertiary amines can be formed similarly through addition of a secondary amine to an imine followed by hydrogenolysis [Eq. (4)].

$$RCH_2NHCH_2R + RCH{=}NH \rightleftharpoons RCH(NH_2)N(CH_2R)_2$$
$$RCH(NH_2)N(CH_2R)_2 \longrightarrow (RCH_2)_3N + NH_3 \qquad (4)$$

In aqueous media, imine intermediates can undergo hydrolysis with formation of carbonyl compounds, which may remain or undergo reduction or reductive alkylation (Miyatake and Tsunoo, 1952) [Eq. (5)].

$$RCH{=}NH + H_2O \longrightarrow RCHO + NH_3 \qquad (5)$$

Nitrile reductions carried out in the presence of an amine provide an excellent way of preparing unsymmetric amines (Kindler *et al.*, 1950; Saito *et al.*, 1956), which, as Kindler and Hesse (1933) pointed out, is to be expected from reactions (1) and (2).

$$RCN + R^1NH_2 \longrightarrow RCH_2NHR^1 + NH_3 \qquad (6)$$

Reaction (1) suggests that formation of secondary and tertiary amines could be minimized if the primary amine were effectively removed from further reaction with the intermediate imine, for example, by formation of an acid salt. Indeed, many workers have employed strong acid solutions with success for exactly this purpose (Rosenmund and Pfannkuch, 1923; Carothers and Jones, 1925; Hartung, 1928; van Tamelen and Smissman, 1953; Freifelder and Ng, 1965). Similarly, the primary amine can be effectively removed by acetylation if the reaction is conducted in an anhydride solution (Carothers and Jones, 1925; McBee and Wiseman, 1950; Overberger and Mulvaney, 1959; Musso and Figge, 1962; Griffin *et al.*, 1962).

Ammonia is also effective in minimizing coupling reactions during nitrile hydrogenation (Huber, 1944). Reductions in ammoniacal alcohol solutions have been recommended as the most suitable means for producing primary amines from a variety of nitriles (Freifelder, 1960; Freifelder and Ng, 1965). Presumably, ammonia functions by competing with the primary amine for the intermediate imine [Eq. (7)].

$$RCN \longrightarrow RCH{=}NH \underset{}{\overset{NH_3}{\rightleftharpoons}} \underset{\underset{NH_2}{|}}{RCHNH_2} \longrightarrow RCH_2NH_2 + NH_3 \qquad (7)$$

Nitrile reduction with hydrazine and a catalyst also favors primary amine formation, perhaps for the same reason that the presence of ammonia does. Alternatively, the suggestion of Greenfield (1967), that bases [carbonates (Kalina and Pasek, 1969), hydroxides (Guth *et al.*, 1963; Fluchaire and Chambert, 1944), and amines] suppress the formation of coupled products by selective poisoning of the catalyst in hydrogenolysis reactions leading to these amines, may be sufficient to explain the action of ammonia.

Equations (1)–(7) account for the formation of the major product types found in nitrile hydrogenation and for some of the changes in products with various solvents. They fail, however, to account for the extraordinary differences in selectivity exhibited by various catalysts.

## II.  CATALYSTS

The products of nitrile hydrogenation depend markedly on the catalyst and on whether the nitrile is aliphatic or aromatic. In the hydrogenation of propionitrile, rhodium produces predominantly dipropylamine, whereas reduction over palladium or platinum gives high yields of tripropylamine (Rylander and Kaplan, 1960; Rylander and Steele, 1965). These results were obtained regardless of the solvent (nonreactive) or the support. Similar differences in catalyst were obtained using butyronitrile (Greenfield, 1967) and valeronitrile (Rylander et al., 1973). However, these differences are not maintained with long-chain nitriles, which give less coupled products regardless of catalyst. Nickel, nickel boride (Russell et al., 1972), and cobalt appear to be the best catalysts for converting low molecular weight aliphatic nitriles to primary amines. The yield can be made nearly quantitative by reduction in ammoniacal methanol (Greenfield, 1967). Barnett (1969)

**TABLE I**

Effects of Pressure on Hydrogenation of Valeronitrile[a] in Methanol[b]

| Catalyst | Pressure (psig) | Pentylamine (%) | Dipentylamine (%) | Tripentylamine (%) |
|---|---|---|---|---|
| 5% Rh/C | 50 | 0 | 93 | 7 |
| 5% Rh/C | 750 | 0 | 100 | 0 |
| 5% Rh/C | 1400 | 29 | 71 | 0 |
| 5% Rh/Al$_2$O$_3$ | 50 | 13 | 87 | 0 |
| 5% Rh/Al$_2$O$_3$ | 1400 | 54 | 46 | 0 |
| Rh$_2$O$_3$ | 50 | 23 | 77 | 0 |
| Rh$_2$O$_3$ | 750 | 68 | 32 | 0 |
| Rh$_2$O$_3$ | 1400 | 100 | 0 | 0 |

[a] Each hydrogenation was carried out with 2.0 ml valeronitrile, 200 mg catalyst, and 40 ml methanol at room temperature.

[b] Data of Rylander et al. (1973). Used with permission.

claims that, without base present, nickel boride or cobalt boride is the most effective catalyst for converting low molecular weight nitriles to primary amines, rhodium boride is most effective for conversion to secondary amines, and platinum boride is most effective for conversion to tertiary amines.

Coupling decreases impressively with increasing pressure (Table I; Rylander *et al.*, 1973) and presumably would continue to decrease at pressures above 1400 psig. The use of elevated pressures to diminish coupling reactions seems worthy of further exploration.

Aromatic nitriles, typified by benzonitrile, on reduction give a mixture of benzylamines, dibenzylamines, and ring-reduced products in proportions that depend on catalyst, solvent, and reaction conditions (Table II). In this reduction, supported platinum and rhodium behave much alike, with palladium distinctly different, whereas, in aliphatic nitrile reductions, palladium and platinum behave similarly, with rhodium distinctly different.

**TABLE II**

**Hydrogenation of Benzonitrile**

| Catalyst | Solvent | Temp (°C) | Pressure (psig) | Benzylamine[a] | Dibenzylamine[a] | Miscellaneous compounds[a] |
|---|---|---|---|---|---|---|
| 5% Pd-on-C[b] | Octane | 25 | 50 | 63 | 34 | 0 |
| 5% Pt-on-C[b] | Octane | 25 | 50 | 0 | 94 | 6[c] |
| 5% Rh-on-C[b] | Octane | 25 | 50 | 0 | 100 | 0 |
| 5% Pd-on-C[b] | Ethanol | 25 | 50 | 59 | 41 | 0 |
| 5% Pt-on-C[b] | Ethanol | 25 | 50 | 0 | 93 | 7[c] |
| 5% Rh-on-C[b] | Ethanol | 25 | 50 | 0 | 100 | 0 |
| 5% Pd-on-C[b] | Benzene | 25 | 50 | 34 | 63 | 3[c] |
| 5% Pt-on-C[b] | Benzene | 25 | 50 | 0 | 97 | 3[c] |
| 5% Rh-on-C[b] | Benzene | 25 | 50 | 0 | 100 | 0 |
| $PtO_2$[d] | Ethanol | 65 | 1500 | 0 | — | 100[e] |
| PdO[d] | Ethanol | 65 | 1500 | 0 | — | 100[e] |
| $RuO_2$[d] | Ethanol | 100 | 1500 | 0 | — | 100[e] |
| 7:3 Rh–$PtO_x$[d] | Ethanol | 75 | 1500 | 91 | — | 7[f] |
| 3:1 Ir–$PtO_x$[d] | Ethanol | 100 | 1500 | 42 | 58 | — |
| $Rh(OH)_3$[d] (LiOH) | Ethanol | 100 | 1500 | 90 | — | 10[f] |

[a] Area percentage of gas chromatogram.
[b] Data of Rylander *et al.* (1973). Used with permission.
[c] Toluene, an imine, and several unknowns.
[d] Data of Takagi *et al.* (1967). Used with permission.
[e] Reported only as high-boiling material.
[f] Hexahydrobenzylamine.

At elevated temperatures and pressures, platinum (Adams *et al.*, 1941), palladium (Shriner and Adams, 1924), and ruthenium (Pichler and Buffleb, 1940) oxides give no benzylamine; instead they yield only high-boiling materials, presumably coupled and reduced derivatives. However, 7:3 rhodium–platinum oxide (Nishimura, 1961), which is frequently recommended when hydrogenolysis is to be avoided, was effective in giving high yields of benzylamine. Rhodium hydroxide (Takagi *et al.*, 1965) inhibited by lithium hydroxide is also effective in giving benzylamine in high yield. Lithium hydroxide here, as in the reduction of anilines, presumably inhibits the hydrogenolysis reaction leading to coupled products.

Rhodium hydroxide and 7:3 rhodium–platinum oxide promoted by lithium hydroxide are effective catalysts for the reduction of adiponitrile, isophthalonitrile, and terephthalonitrile to the corresponding diamines in good yield; without promotion, yields are poor (Takagi *et al.*, 1967). Alkali-promoted ruthenium-on-alumina is effective in reducing adiponitrile to hexamethylenediamine in excellent yield under conditions of high pressure and temperature (e.g., 2500 psig, 150°C) and in the presence of ammonia (Brake, 1969). A superior commercial process has been developed for the manufacture of dibenzylamine by hydrogenation of benzonitrile over 5% platinum-on-carbon without solvent but with sufficient water present to prevent catalyst poisoning (Greenfield, 1976).

An interesting use of rhodium catalysts was made in the selective reduction of 4-cyanodialkylbutyraldehydes to the corresponding 5-aminopentanols. The reduction specifically requires rhodium and the presence of ammonia. The expected reductive amination of the aldehyde function does not seem to be a major reaction. As an example, 100 gm (0.8 mole) 4-cyano-2,2-dimethylbutyraldehyde, 200 ml methanol, 5 gm 5% rhodium-on-alumina, and 100 ml ammonia are hydrogenated in a stirred autoclave at 125°C and 1200 psig. The yield of 5-amino-2,2-dimethylpentanol is 68%. Coupled products increase as the reaction temperature is lowered; at 70°C the yield falls to 43%, and 28 gm of the coupled product 5,5′-iminobis(2,2-dimethylpentanol) is formed (Martin and Gott, 1970).

$$\underset{\underset{CH_3}{|}}{\overset{\overset{CH_3}{|}}{OHCC}}CH_2CH_2CN \longrightarrow \underset{\underset{CH_3}{|}}{\overset{\overset{CH_3}{|}}{HOCH_2C}}CH_2CH_2CH_2NH_2$$

Ammonia and rhodium were also used effectively in preparing branched-chain amino sugars. The sequence involves a Wittig reaction with diethyl cyanomethylphosphonate, stereoselective hydrogenation of the resulting unsaturated sugar over palladium, and reduction of the nitrile over rhodium–ammonia (Rosenthal and Baker, 1969, 1973a,b).

A facile synthesis of *o*-hydroxy- and *p*-hydroxy-substituted phenethyl-amines illustrates the use of a strong acid in preventing coupling reactions during nitrile hydrogenation (Schwartz *et al.*, 1976).

## III.   UNSYMMETRICAL AMINES

A convenient way of preparing unsymmetric amines is by hydrogenation of a nitrile in the presence of an appropriate amine. The product composi-tion depends as would be expected, on the nitrile–amine ratio and on the catalyst. Table III shows the effect of this ratio in the model system valero-nitrile–butylamine with 5% rhodium-on-carbon as catalyst (Rylander *et al.*, 1973). Large percentages of butylpentylamine form even when the amine–

TABLE III

Effect of Butylamine–Valeronitrile Ratio on Product Composition[a,b]

| Butylamine–valeronitrile (molar ratio) | Dipentylamine[c] | Butylpentylamine[c] |
|---|---|---|
| 0.434 | 36 | 64 |
| 0.650 | 25 | 75 |
| 0.860 | 19 | 81 |
| 1.30 | 12 | 88 |
| 1.70 | 5 | 95 |
| 2.20 | 7 | 93 |
| 3.5 | 5 | 95 |
| 6.6 | ca. 0 | 100 |

[a] Data of Rylander et al. (1973). Used with permission.

[b] Each experiment was carried out at room temperature and an initial pressure of 50 psig with 300 mg 5% rhodium-on-carbon, 2.0 ml valeronitrile, and butylamine.

[c] Area percentage of gas chromatogram peaks.

nitrile molar ratio is substantially less than 1. These results suggest that reduction of the intermediate imine to pentylamine is slow relative to addition of butylamine to the imine.

Product composition may vary considerably with catalyst, as indicated by the reduction of a mixture of valeronitrile and butylamine (0.77:1 molar ratio) in methanol (Table IV).

Much larger differences among catalysts are found in unsymmetric coupling reactions involving aromatic nitriles. Product distribution is also solvent dependent (Table V).

TABLE IV

Hydrogenation of Valeronitrile–Butylamine[a]

$$C_4H_9CN + C_4H_9NH_2 \xrightarrow[25°C]{50\ psig} (C_5H_{11})_2NH + C_4H_9NHC_5H_{11}$$

|  | A | B |
|---|---|---|
| Catalyst | Product A (%) | Product B (%) |
| $Rh_2O_3$ | 37 | 61 |
| 75% Rh, 25% Pt oxide | 39 | 61 |
| 5% Pt-on-C | 3 | 97 |
| 5% Pd-on-C | 7 | 93 |
| 5% Rh-on-C | 12 | 88 |
| 5% Rh-on-$Al_2O_3$ | 22 | 78 |

[a] Data of Rylander et al. (1973).

TABLE V

Hydrogenation of Benzonitrile and Butylamine[a,b]

| Catalyst | Solvent | Reaction time (hr) | Benzylamine (%) | Butylbenzylamine (%) | Dibenzylamine (%) | Miscellaneous compounds[c] |
|---|---|---|---|---|---|---|
| 5% Pd/C | Octane | 4 | 100 | 0 | 0 | 0 |
| 5% Pt/C | Octane | 7 | 0 | 92 | 0 | 8 |
| 5% Rh/C | Octane | 8 | 0 | 100 | 0 | 0 |
| 5% Pd/C | Ethanol | 2 | 83 | 0 | 0 | 17 |
| 5% Pt/C | Ethanol | 3 | 0 | 88 | 0 | 12 |
| 5% Rh/C | Ethanol | 19 | 0 | 21 | 79 | 0 |
| 5% Pd/C | Benzene | 2 | 36 | 45 | 19 | 0 |
| 5% Pt/C | Benzene | 2 | 0 | 93 | 0 | 7 |
| 5% Rh/C | Benzene | 12 | 0 | 0 | 100 | 0 |

[a] Data of Rylander et al. (1973).

[b] Each experiment was carried out at room temperature and an initial pressure of 50 psig with 2.0 gm catalyst, 50 ml solvent, 10.3 gm benzonitrile (0.1 mole), and 7.3 gm butylamine (0.1 mole).

[c] Miscellaneous compounds include toluene, an imine determined by infrared, and several unknowns.

## IV. CYCLIZATIONS

Hydrogenation of nitriles in molecules containing a second suitably disposed reactive function provides a convenient entry into a variety of ring systems through cyclization reactions involving either the resulting amine or intermediate imine. Interacting second functions have included other nitriles (Bergel et al., 1948; Schultz, 1948; Bowden and Green, 1952), amines (Trofimenko, 1963), ketones (Helberger and von Rebay, 1939; Boekelheide, 1947; Boekehlheide and Schilling, 1950; Mandell et al., 1963), esters (Winans and Adkins, 1933; Koelsch, 1943; Barr and Cook, 1945; Boekelheide et al., 1953; House et al., 1962), acids, amides, and olefins, as well as a variety of functions such as nitro (Snyder et al., 1958; Hester, 1964; Walker, 1965), oximino, and various heterocyclic rings (Boekelheide et al., 1953) that are themselves first fully or partially hydrogenated before undergoing cyclization reactions.

Cyclizations involving cyano ketones presumably go through a normal reductive alkylation reaction after the nitrile function has been preferentially reduced to an amine. Preferential reduction of the ketone group prevents cyclization from occurring except under vigorous conditions. Table VI shows that the selectivity obtained in the reduction of cyano ketones may depend markedly on the catalyst (Rylander and Wolsky, 1970).

Generalizing from these limited data, it would appear that platinum is not the catalyst of choice when cyclization is desired; conversely, it is the

**TABLE VI**

**Hydrogenation of 2-( β-Cyanoethyl)cyclohexanone in Methanol**

| | Product composition (%) | | |
|---|---|---|---|
| Catalyst | ![structure] CH$_2$CH$_2$CH=NH | ![structure] CH$_2$CH$_2$CN | ![structure] |
| 5% Pd-on-C[a] | 24 | 0 | 76 |
| 5% Pt-on-C | 0 | 100 | 0 |
| 5% Rh-on-C | 0 | 0 | 100 |

[a] Stopped spontaneously. Analysis by mass spectrometry, vapor-phase chromatography, infrared spectroscopy, and H$_2$ adsorption. Conditions ambient.

preferred catalyst for selective reduction of the ketone in cyano ketones. Rhodium, which has been rarely used in this type of reduction, appears to be very effective for cyclization. Raney nickel is relatively inactive compared to noble metals, and massive amounts of it must be used under mild conditions or more vigorous conditions must be employed.

Reductive cyclization of 2.31 gm of the ketonitrile **2** over 1 spoonful of Raney nickel at 3 atm for 24 hr affords a mixture of myosmine (**3**) and nornicotine (**4**) in 67% combined yield from pyridine-3-aldehyde (**1**). The yield of myosmine can be increased at the expense of nornicotine by shortening the hydrogenation time (Leete *et al.*, 1972).

An interesting technique for changing the stereochemistry of the ring juncture formed on reduction of a γ-ketonitrile was discovered by Robinson *et al.* (1966). Hydrogenation of either **6** or **8** over palladium gave the cis isomer **7**. Presumably **8** is an intermediate in the hydrogenation of **6**. However, if **6** is reduced in the presence of a large excess of anhydrous ammonia, the trans isomer **5** is formed. The latter course is achieved by reductive alkylation of the carbonyl to afford the equatorial cyclohexylamine, which then adds to the aldimine and undergoes hydrogenolysis to afford **5**. The use of methylamine instead of ammonia gave almost entirely the *cis-N*-methyldecahydroquinoline instead of the expected trans isomer.

5                                                                6

7                                                                8

The sterochemistry of ring closure is also affected by catalyst activity. Mandel *et al.* (1965), using a variation of the quinolizidine synthesis (Mandell *et al.*, 1963), hydrogenated the dicyano ketone **9** over 5 and 10% palladium-on-carbon in glacial acetic acid. Hydrogenation over 5% palladium-on-carbon afforded *dl*-matrine (**10**), whereas over 10% palladium-on-carbon under the same conditions the product was *dl*-leontine (**11**). Presumably, *dl*-matrine is first formed and then isomerized to *dl*-leontine.

9                              10                              11

Reductive cyclization involving ring closure with an amide is illustrated by the preparation of hexahydroimidazo[1,2-*a*]pyrazines and hexahydro-4*H*-pyrazino[1,2-*a*]pyrimidines through hydrogenation of 1-cyanomethyl-piperazin-2-ones and 2-cyanoethylpiperazin-2-ones over Raney nickel in ammoniacal methanol. The yields are excellent (Uchida and Ohta, 1973).

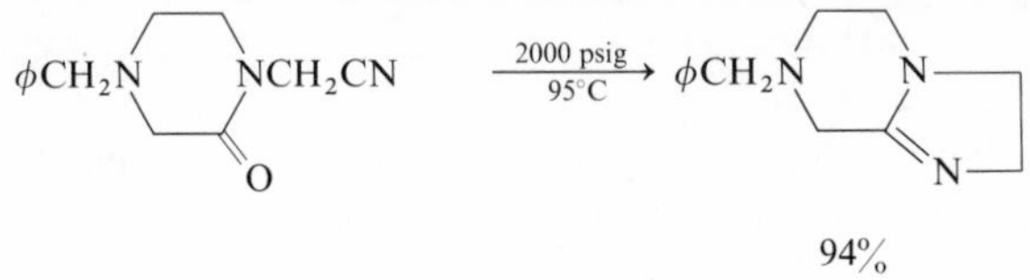

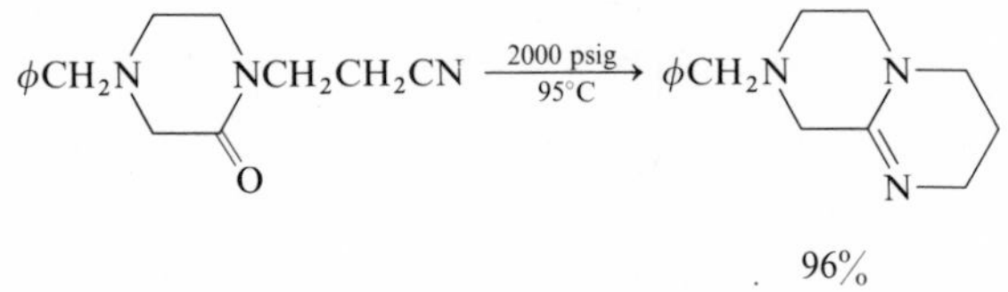

94%

96%

Hydrogenation of *o*-nitrophenylacetonitriles provides a useful synthesis of indoles. Presumably, in these reactions the nitro group is first and easily reduced to an amine followed by cyclization with the imine, which is formed by addition of the fourth mole of hydrogen to the nitrile (Snyder *et al.*, 1958). This type of cyclization was applied by Bordais and Germain (1970) to the synthesis of a variety of indole-3-carboxamides obtained in two steps from the corresponding halonitro compounds and cyanoacetamides.

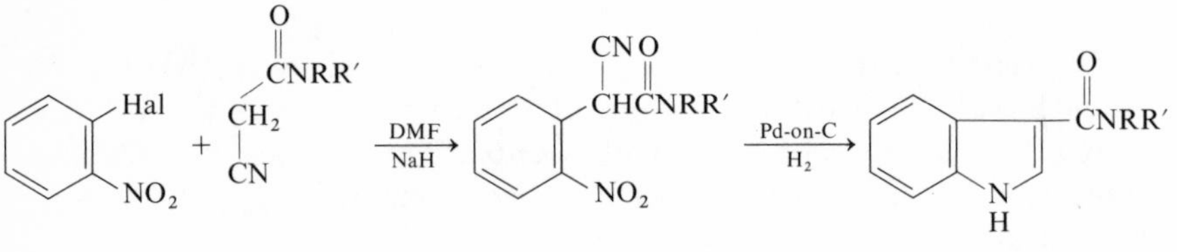

Six-membered rings result from reductive cyclization of δ-nitrocyano compounds. For instance, hydrogenation of 4-nitroindol-3-ylacetonitrile over 10% palladium-on-carbon in ethyl acetate affords 1,3,4,5-tetrahydro-pyrrole[4,3,2-*de*]quinoline (Hester, 1964).

An interesting double ring closure of **12** over W2 Raney nickel gave the tetracyclic pyridone diester **13**, a potential camptothecin intermediate, in a single step. The nitrile function is reduced to an aminomethyl group, which undergoes intramolecular enamine exchange and cyclization (Borch *et al.*, 1972). The same type of closure was achieved in 42% yield with the corresponding 3-cyanopyridine derivative.

12                                                                          13

## V.  REDUCTIVE HYDROLYSIS

Hydrogenation of nitriles in acidic media provides a good method of converting a nitrile to an aldehyde. If the aldehyde is aromatic, further reduction to the carbinol occurs readily (Miyatake and Tsunoo, 1952). Intermediate aldehydes may react with amines, forming ketimines (Snyder *et al.*, 1958) or semicarbazones, if the reduction is carried out in the presence of semicarbazide (Plieninger and Werst, 1955; Rogers, 1963).

A typical reductive hydrolysis is illustrated by the conversion of 4-amino-5-cyano-2-cyclobutylmethylpyrimidine (**14**) to 4-amino-2-cyclobutylmethyl-5-pyrimidylmethanol (17). A mixture of 18.8 gm of **14** and 3.5 gm of 10% palladium-on-carbon was hydrogenated for 4.5 hr in 25 ml of 2.0 *N* HCl at 3 atm to afford 15.0 gm of carbinol (Mizzoni *et al.*, 1970). The sequence involved reduction to the aldimine **15**, hydrolysis, and hydrogenation of the aromatic aldehyde **16**.

14                                  15

16                                  17

Tinapp (1969) examined in detail factors affecting the yield of aldehyde in the reductive hydrolysis of aromatic nitriles in strong acid over Raney nickel. Excellent results were obtained by the use of a mixture of tetrahydrofuran and water (10:1) containing a molar amount of sulfuric acid equal to the nitrile present.

$$CH_3O-\langle\ \rangle-CN \xrightarrow[\substack{H_2SO_4 \\ 1\ atm\ 25°C}]{\substack{RaNi \\ THF-H_2O}} CH_3O-\langle\ \rangle-CHO$$

with OCH₃ substituent on left ring and CH₃O substituent on right ring

91%

## VI. METHYLATION

Displacement of aromatic halogen by cyanide followed by hydrogenation and hydrogenolysis (Block and Coy, 1972) is a convenient sequence for the introduction of a methyl group (Jorgensen *et al.*, 1974).

$$CH_3O-\langle\ \rangle-O-\langle\ \rangle-CH_2CH \longrightarrow$$

with CH₃ and I substituents; NHAc and COOC₂H₅ groups

$$CH_3O-\langle\ \rangle-O-\langle\ \rangle-CH_2CH \xrightarrow[\substack{p\text{-cymene} \\ H_2}]{20\%\ Pd\text{-on-C}}$$

with CH₃ and NC substituents; NC at top; NHAc and COOC₂H₅ groups

$$CH_3O-\langle\ \rangle-O-\langle\ \rangle-CH_2CH$$

with CH₃ substituent; H₃C substituents; NHAc and COOC₂H₅ groups

## REFERENCES

Adams, R., Voorhees, V., and Shriner, R. L., *Org. Synth.* **8**, 93 (1941).
Barnett, C., *Ind. Eng. Chem., Prod. Res. Dev.* **8**, 145 (1969).
Barr, W., and Cook, J. W., *J. Chem. Soc.* p. 438 (1945).
Bergel, F., Morrison, A. L., and Rinderknecht, H., U.S. Patent 2,446,803, Aug. 10, 1948.
Block, P., Jr., and Coy, D. H., *J. Chem. Soc., Perkins. Trans. I* p. 63 (1972).
Boekelheide, V., *J. Am. Chem. Soc.* **69**, 790 (1947).
Boekelheide, V., and Schilling, W. M., *J. Am. Chem. Soc.* **72**, 712 (1950).
Boekelheide, V., Linn, W. J., O'Grady, P., and Lamborg, M., *J. Am. Chem. Soc.* **75**, 3243 (1953).
Borch, R. F., Grudzinskas, C. V., Peterson, D. A., and Weber, L. D., *J. Org. Chem.* **37**, 1141 (1972).
Bordais, J., and Germain, C., *Tetrahedron Lett.* p. 195 (1970).

Bowden, K., and Green, P. N., *J. Chem. Soc.* p. 1164 (1952).

Brake, L. D., U.S. Patent 3,471,563, Oct. 7, 1969.

Carothers, W. H., and Jones, G. A., *J. Am. Chem. Soc.* **47**, 3051 (1925).

Chivarelli, S., and Marini-Bettolo, G. B., *Gazz. Chim. Ital.* **86**, 515 (1956).

Fluchaire, M. L. A., and Chambert, F., *Bull. Soc. Chim. Fr.* p. 22 (1944).

Freifelder, M., *J. Am. Chem. Soc.* **82**, 2386 (1960).

Freifelder, M., and Ng, Y. H., *J. Pharm. Sci.* **54**, 1204 (1965).

Greenfield, H., *Ind. Eng. Chem., Prod. Res. Dev.* **6**, 142 (1967).

Greenfield, H., *Ind. Eng. Chem., Prod. Res. Dev.* **15**, (1976).

Grinffin, G. W., Basinski, J. E., and Peterson, L. I., *J. Am. Chem. Soc.* **84**, 1012 (1962).

Guth, V., Leitich, J., Specht, W., and Wessely, F., *Monatsh. Chem.* **94**, 1262 (1963).

Hartung, W. H., *J. Am. Chem. Soc.* **50**, 3370 (1928).

Helberger, J. H., and von Rebay, A., *Justus Liebigs Ann. Chem.* **539**, 187 (1939).

Hester, J. B., Jr., *J. Org. Chem.* **29**, 1158 (1964).

House, H. O., Wickham, P.P., and Müller, H. C., *J. Am. Chem. Soc.* **84**, 3139 (1962).

Huber, W., *J. Am. Chem. Soc.* **66**, 876 (1944).

Jorgensen, E. C., Murray, W. J., and Block, P. J., Jr., *J. Med. Chem.* **17**, 434 (1974).

Juday, R., and Adkins, H., *J. Am. Chem. Soc.* **77**, 4559 (1955).

Kalina, M., and Pasek, J., *Kinet. Katal.* **10**(3), 574 (1969).

Kindler, K., and Hesse, F., *Arch. Pharm.* (*Weinheim, Ger*) **271**, 439 (1933).

Kindler, K., Shrader, K., and Middelhoff, B., *Arch. Pharm.* (*Weinheim, Ger*) **283**, 184 (1950).

Koelsch, C. F., *J. Am. Chem. Soc.* **65**, 2093 (1943).

Lette, E., Chedekel, M. R., and Bodem, G. B., *J. Org. Chem.* **37**, 4465 (1972).

McBee, E. T., and Wiseman, P. A., U.S. Patent 2,515,246, July 18, 1950.

Mandell, L., Piper, J. U., and Singh, K. P., *J. Org. Chem.* **28**, 3440 (1963).

Mandell, L., Singh, K. P., Gresham, J. T., and Freeman, W. J., *J. Am. Chem. Soc.* **87**, 5234 (1965).

Martin, J. C., and Gott, P. G., U.S. Patent 3,520,932, July 21, 1970.

Mignonac, G., *C. R. Acad. Sci.* **171**, 1148 (1920).

Miyatake, K., and Tsunoo, M., *Yakugaku Zasshi* **72**, 630 (1952).

Mizzoni, R. H., Lucas, R. A., Smith, R., Boxer, J., Brown, J. E., Goble, F., Konopka, E., Gelzer, J., Szanto, J., Maplesden, D. C., and de Stevens, G., *J. Med. Chem.* **13**, 878 (1970).

Musso, H., and Figge, K., *Chem. Ber.* **95**, 1844 (1962).

Nishimura, S., *Bull. Chem. Soc. Jpn.* **34**, 1544 (1961).

Overberger, C. G., and Mulvaney, J. E., *J. Am. Chem. Soc.* **81**, 4697 (1959).

Pichler, H., and Buffleb, H., *Brennst.-Chem.* **21**, 257 (1940).

Plieninger, H., and Werst, G., *Chem. Ber.* **88**, 1956 (1955).

Robinson, M. M., Lambert, B. F., Dorfman, L., and Pierson, W. G., *J. Org. Chem.* **31**, 3220 (1966).

Rogers, A. O., U.S. Patent 3,078,274, Feb. 19, 1963.

Rosemund, K. W., and Pfannkunch, E., *Ber. Dtsch. Chem. Ges. B* **56**, 2258 (1923).

Rosenthal, A., and Baker, D. A., *Tetrahedron Lett.* p. 397 (1969).

Rosenthal, A., and Baker, D. A., *J. Org. Chem.* **38**, 193 (1973a).

Rosenthal, A., and Baker, D. A., *J. Org. Chem.* **38**, 198 (1973b).

Russell, T. W., Hoy, R. C., and Cornelius, J. E., *J. Org. Chem.* **37**, 3552 (1972).

Rylander, P. N., and Kaplan, J., *Am. Chem. Soc. Meet., New York, 1960*.

Rylander, P. N., and Steele, D. R., *Engelhard Ind., Tech. Bull.* **5**, 113 (1965).

Rylander, P. N., and Wolsky, O., unpublished observations, Engelhard Ind. Res. Lab., Newark, New Jersey, 1970.

Rylander, P. N., Hasbrouck, L., and Karpenko, I., *Ann. N.Y. Acad. Sci.* **214**, 100 (1973).

Saito, N., Tanaka, C., and Takatani, S., *Yakugaku Zasshi* **76**, 341 (1956).

Schultz, H. P., *J. Am. Chem. Soc.* **70**, 2666 (1948).

Schwartz, M. A., Zoda, M., Vishnuvajjala, B., and Mami, I., *J. Org. Chem.* **41**, 2502 (1976).

Shriner, R. L., and Adams, R., *J. Am. Chem. Soc.* **46**, 1683 (1924).

Snyder, H. R., Merica, E. P., Force, C. G., and White, E. G., *J. Am. Chem. Soc.* **80**, 4622 (1958).

Takagi, Y., Naito, T., and Nishimura, S., *Bull. Chem. Soc. Jpn.* **38**, 2119 (1965).

Takagi, Y., Nishimura, S., Taya, K., and Hirota, K., *Sci. Pap. Inst. Phys. Chem. Res. (Jpn.)* **61**, 114 (1967).

Tinapp, P., *Chem. Ber.* **102**, 2770 (1969).

Trofimenko, S., *J. Org. Chem.* **28**, 2755 (1963).

Uchida, H., and Ohta, M., *Bull. Chem. Soc. Jpn.* **46**, 3612 (1973).

van Tamelen, E. E., and Smissman, E. E., *J. Am. Chem. Soc.* **75**, 2031 (1953).

Von Braun, J., Blessing, G., and Zobel, F., *Ber. Dtsch. Chem. Ges. B* **56**, 1888 (1923).

Walker, G. N., *J. Med. Chem.* **8**, 583 (1965).

Winans, C. F., and Adkins, H., *J. Am. Chem. Soc.* **54**, 306 (1932).

Winans, C. F., and Adkins, H., *J. Am. Chem. Soc.* **55**, 4167 (1933).

# Chapter 9

# Hydrogenation of Oximes

Catalytic hydrogenation of oximes may give rise to primary, secondary, or tertiary amines, alcohols derived by reductive hydrolysis, hydroxylamines or imines derived by partial reduction, or products derived by interaction with other functions. Nonetheless, the reduction can usually be well controlled despite its potential complexity.

## I. CATALYSTS

Nickel (Adkins, 1937), cobalt, palladium, platinum, rhodium, and ruthenium have been used in the hydrogenation of oximes (Breitner *et al.*, 1959). Nickel is often used in the presence of ammonia (Carmack *et al.*, 1946; King *et al.*, 1948; Shepard *et al.*, 1952), which increases the yield of primary amine, but reactions have also been carried out without ammonia (Bollinger *et al.*, 1953; Jones *et al.*, 1957). Usually high loadings of nickel are used in these reductions at elevated temperatures and pressures, and because of this runaway reactions may occur. Freifelder (1971) urges caution! Cobalt has given excellent yields of primary amines under vigorous conditions (100°C and 3000 psig), with or without ammonia present (Reeve and Christian, 1956). High-pressure reduction of 2-hydroxyimino-6-nitrohexanamide over Raney cobalt in the presence of a large excess of ammonia affords lysinamide in 80–84% yield (Fuhrmann *et al.*, 1973).

Noble metals can be used under much milder conditions. Rhodium has seldom been used (Rylander and Steele, 1965), but it has given excellent results (Freifelder *et al.*, 1962). Its use seems to be indicated when excessive secondary amine formation is a problem with other catalysts. Rhodium-on-alumina was the preferred catalyst in an improved procedure for the preparation of amino alcohols involving hydration of ethynylcarbinols, conversion to an oxime, and hydrogenation. Reduction of the oxime over platinum or palladium or with sodium in liquid ammonia or lithium aluminum hydride failed to give satisfactory yields (Newman and Lee, 1975).

$$
\underset{\underset{OH}{|}}{\overset{\overset{CH_3}{|}}{CH_3CC}}{=}CH \xrightarrow[\substack{H_2SO_4 \\ H_2O}]{HgO} \underset{OH\;O}{\overset{\overset{CH_3}{|}}{CH_3C-CCH_3}} \xrightarrow[\substack{2.\;5\%\;Rh\text{-on-}Al_2O_3 \\ H_2,\;EtOH}]{1.\;NH_2OH} \underset{OH\;NH_2}{\overset{\overset{CH_3}{|}}{CH_3C-CHCH_3}}
$$

$$80\% \qquad\qquad\qquad 94\%$$

Pappas and Gancher (1969) compared palladium- and rhodium-on-carbon for the hydrogenation of acetophenone oxime and its $O$-acetyl and $O$-benzoyl derivatives (Table I). Rhodium gives substantially less secondary amine than does palladium. Acylation is an effective means of minimizing secondary amine formation (Rosenmund and Pfannkuch, 1923, 1942). The improvement brought about by acylation may be largely the result of carboxylic acids, formed by hydrogenolysis, combining with the primary amine to prevent its conversion to secondary amine.

TABLE I[a]

$$
\underset{NOR}{\overset{}{\bigcirc\!\!-\overset{\overset{}{\|}}{C}-CH_3}} \xrightarrow{CH_3OH} \underset{NH_2}{\bigcirc\!\!-CHCH_3} + \left(\underset{}{\bigcirc\!\!-\overset{\overset{CH_3}{|}}{CH}-}\right)_2 NH
$$

| R | Primary amine (%) | | Secondary amine (%) | |
|---|---|---|---|---|
|   | Pd | Rh | Pd | Rh |
| —H | 54 | 84 | 46 | 16 |
| —COCH₃ | 83 | 100 | 17 | 0 |
| —COφ | 74 | 96 | 26 | 4 |

[a] Data of Pappas, J. J., and Gancher, E., J. Chem. Eng. Data **14**, 269 (1969). Used with permission.

Platinum catalysts are inhibited by amines, and reduction of oximes slows appreciably as the amine products build up in the system (Breitner et al., 1959). For this reason and to minimize the formation of secondary amines, hydrogenations over platinum are often carried out in acidic media (Reichert and Koch, 1935; Mousseron et al., 1947; Kreisky, 1958; Murphy, 1961; Masamune et al., 1964; Roth et al., 1964; Fryer et al., 1970). Platinum appears to be the best metal for achieving partial hydrogenation of oximes.

Ruthenium is used rarely in oxime hydrogenation. It functions best in aqueous media, but under these circumstances reductive hydrolysis is apt to be an important side reaction. In fact, the corresponding alcohol may be the major product even when no water is initially present in the system (Rylander and Steele, 1963, 1965). Ruthenium would be the preferred metal if this reductive hydrolysis reaction were needed in synthesis.

Palladium is used widely for the hydrogenation of oximes, often in acidic media (Hartung, 1928; Rosenmund et al., 1944; Duschinsky and Dolan, 1945; Turner et al., 1949; Meltzer and Lewis, 1957; Rosen and Green, 1963; Sianesi et al., 1967; Walker and Alkalay, 1967; Yamaguchi et al., 1968). The reduction proceeds under mild conditions, and the yields of product are usually high. Hydrogenation of a 3% solution of 2-formaldoximo-6-methyl-pyridine over 10% palladium-on-carbon affords 6-methyl-2-aminomethyl-pyridine in 94% yield (Fuentes and Paudler, 1975). The use of a dilute solution helps to minimize coupling reactions.

## Synergism

Some striking examples of synergism in the hydrogenation of oximes, especially oximes containing another functional group, have been reported. A mixture of palladium and platinum catalysts, such that the palladium is 30–70% of the total metal and the platinum is 70–30%, gives excellent yields of amino alcohol in the hydrogenation of α-oximinopropiophenone (Wilbert and Sosis, 1962). The rate of hydrogenation with these two catalysts together is considerably greater than that found with either alone, with the total weight of metal held constant (Karpenko and Rylander, 1958). Even small amounts, 1.5% of the total metal, of platinum or rhodium added to palladium have a beneficial effect on this reduction (Neelakantan and Hartung, 1958).

A series of metal oxide mixtures of widely varying composition (platinum–palladium, platinum–iridium, platinum-rhodium, and platinum-ruthenium) prepared by a modification of the familiar Adams' platinum oxide procedure (Adams and Shriner, 1923) were tested in the hydrogenation of cyclohexanone oxime (Table II). Most notable is the high yield of cyclohexylamine formed over rhodium oxide; inclusion of a second metal increases the yield of dicyclohexylamine. The lengthy induction periods found with some of these catalysts teach the value of being patient rather than throwing away a seemingly inactive system (Rylander et al., 1967).

**TABLE II**

**Hydrogenation of Cyclohexanone Oxime in Methanol**

| % Pt in catalyst | % Cyclohexylamine in product[a] | | | | Induction period (min)[b] | | | |
|---|---|---|---|---|---|---|---|---|
| | Pt–Rh | Pt–Ir | Pt–Ru | Pt–Pd | Pt–Rh | Pt–Ir | Pt–Ru | Pt–Pd |
| 100 | 65 | 65 | 65 | 65 | 5 | 5 | 5 | 5 |
| 95 | 76 | 54 | 61 | 47 | 5 | 5 | 10 | 5 |
| 85 | — | — | 60 | — | — | — | 10 | — |
| 75 | 61 | 58 | 52 | 54 | 5 | 5 | 15 | 5 |
| 60 | 77 | — | 48 | — | 10 | — | 15 | — |
| 50 | 52[c] | 51 | 37 | 47 | 15 | 30 | 60 | 35 |
| 40 | 69[c] | — | — | — | 40 | — | — | — |
| 25 | 59[c] | 37 | — | 40 | 90 | 110 | — | >200 |
| 5 | — | — | — | — | — | — | — | — |
| 0 | 99[c] | — | — | — | >400 | >200 | >200 | >200 |

[a] Based on cyclohexylamine plus dicyclohexylamine in the product.

[b] Approximate. Induction periods are variable, and the point of termination is subject to different interpretations.

[c] Contains small quantities of cyclohexanol.

## II.  SUBSTRATE

The products obtained in the hydrogenation of oximes are the result of several competing reactions, each of which depends in part on the structure of the substrate, including whether the oxime is syn or anti (Buchman and Milstein, 1972). Reductive coupling reactions, leading to secondary amines, are sensitive to steric effects. The amount of secondary amine decreases as the bulk of the substituents in the vicinity of the oxime increases. Large amounts of dialkylamine are obtained in the hydrogenation of acetoxime, whereas under the same conditions 3-pentanone oxime affords no dialkylamine (Rylander and Steele, 1965). Dornow and Frese (1952) noted that secondary amine formation in linear systems is limited to methyl ketones. Oximes of cyclic ketones, such as cyclopentanone and 3-methylcyclohexanone (Mousseron et al., 1947; Hückel and Thomas, 1961), are apt to form chiefly secondary amines, even in the presence of ammonia (Hückel and Kupka, 1956). Bulky substituents in the 2 position of cyclic ketones decrease secondary amine formation drastically (Masamune et al., 1964).

When oximes interact during catalytic hydrogenation, they usually do so in such a way that a new carbon–nitrogen bond is formed to ultimately afford a secondary or tertiary amine and ammonia. Ordinarily, intermediate addition compounds are not obtained, but in special cases they may be the major product. Hydrogenation of the dioxime of bicyclo[3.3.1]nonane-3,7-dione in acidic media affords 2-azaadamantan-1-amine (Gagneux and Meier, 1969; Gagneux, 1970).

NOH  NOH

12 gm                                                                      90%

In certain circumstances, interaction between oximes may take place so that a new carbon–carbon bond is formed, as in the conversion of the dioxime of 3,3,7,7-tetracarbethoxycyclooctane-1,5-dione over palladium-on-carbon to a dicyclic diamine (Cope and Kagen, 1958).

## III.  SOLVENT

The products obtained in the hydrogenation of oximes depend markedly on the solvent. Hartung (1928) was the first to realize that the mixture of primary, secondary, and tertiary amines so often obtained in neutral media could largely be avoided and good yields of primary amines obtained if the reduction were carried out in acidic media. A number of acids have been used in the hydrogenation of oximes, including acetic acid, acetic acid containing hydrogen chloride, sulfuric, phosphoric, or perchloric acid, boron trifluoride, and zinc chloride–hydrogen chloride (Rosemund et al., 1942), but they cannot necessarily be used interchangeably. For instance, either cis- (3) or trans-2-amino-1-indanol (1) can be obtained as a major product by hydrogenation of 2-hydroxyimino-1-indanone (2) under conditions that do not seem to be too dissimilar (Huebner et al., 1970). The stereochemistry assigned earlier to these compounds is reassigned by Huebner et al. (1970).

1                                          2                                          3

Acetic anhydride has been employed effectively as a solvent for the hydrogenation of oximes. The solvent acetylates amines as they are formed and thereby prevents interaction of the amine with other functional groups (Albertson et al., 1948; Cash et al., 1956; Buchanan and Sutherland, 1957).

Catalytic hydrogenation of $\beta$-acetoxy-$\alpha$-oximinopropiophenone (**5**) in acetic acid–acetic anhydride over palladium-on-carbon gave $N$-acetylnorephedrine (**6**) instead of the expected 2-acetamido-3-acetoxy-1-phenyl-1-propanol (**4**). The product presumably arose through $\alpha$-acetoamidopropiophenone (Matsumoto *et al.*, 1962).

$$\underset{\substack{| \quad | \\ \text{OH NHCOCH}_3}}{\phi\text{CHCHCH}_2\text{OCOCH}_3} \xleftarrow[\text{Ac}_2\text{O}]{\text{HOAc}} \underset{\substack{\| \; \| \\ \text{ONOH}}}{\phi\text{CCCH}_2\text{OCOCH}_3} \xrightarrow[\text{Ac}_2\text{O}]{\text{HOAc}} \underset{\substack{| \quad | \\ \text{OH NHCOCH}_3}}{\phi\text{CHCHCH}_3}$$

$$\qquad\qquad 4 \qquad\qquad\qquad\qquad 5 \qquad\qquad\qquad\qquad 6$$

## IV. PRODUCTS OF PARTIAL REDUCTION

Oximes on partial hydrogenation yield either imines by hydrogenolysis of the oxygen–nitrogen bond, or hydroxylamines by saturation of the carbon–nitrogen double bond. Imines are rarely obtained as products, inasmuch as they usually undero further reduction or reaction unless special structural features render them relatively inactive (Rapport and Kupferberg, 1973; Walker *et al.*, 1971).

It appears that hydroxylamines are best formed from oximes by reduction over platinum in acidic media (Vanon and Krajcinovic, 1928; Jones and Major, 1930; Reichert and Koch, 1935; Gilsdorf and Nord, 1952; Müller *et al.*, 1955; Major and Ohly, 1961; Rosen and Green, 1963; Benington *et al.*, 1965). Hydroxylamines can be reduced further to the amine, but this appears to be relatively slow compared to reduction of the oxime. Most of the amine formed in oxime reduction arises via an imine (Neelakantan and Hartung, 1958; Rosen and Green, 1963). The goal, then, in obtaining hydroxyamino compounds from oximes lies not so much in preventing further reduction of the hydroxyamino function, but rather in expediting its formation initially.

## V. PRODUCTS OF HYDROLYSIS

Alcohols and ketones may arise from the hydrogenation of oximes through competitive hydrolysis of intermediate imines (Breitner *et al.*, 1959; Rylander and Steele, 1963). Conditions that decrease the opportunity for hydrolysis favor increased yields of amine. The yield of cyclohexylamine from hydrogenation of cyclohexanone oxime increased from 25% in water to 82% in methanol saturated with ammonia. The beneficial effect of ammonia on yield seems to arise mostly from its competition with water in nucleophilic attack on the imine.

The yield of lysinamide (**8**) formed on hydrogenation of 2-hydroxyimino-6-nitrohexanamide (**7**) over rhodium-on-carbon in acetic acid varied appreciably with concentration of substrate in the solvent (Table III). Higher concentrations of substrate produced higher water concentrations and consequently more hydrolysis and lower yields. Pipecolinamide (**9**) is assumed to arise by attack of the terminal amino group on the carbonyl in intermediate 2-keto-6-aminohexanamide (Fuhrmann *et al.*, 1973).

**TABLE III**

$$O_2N(CH_2)_4\underset{\underset{NOH}{\|}}{C}CONH_2 \longrightarrow H_2N(CH_2)_4\underset{\underset{NH_2}{\phantom{|}}}{C}HCONH_2 \; + \;$$

(product **9**: cyclohexane ring bearing CONH$_2$ and NH)

| 7 | 8 | 9 |
|---|---|---|

| Concentration (gm 100 ml HOAc) | Yield (%) | |
|---|---|---|
| | 8 | 9 |
| 10 | 85 | 17 |
| 15 | 76 | 20 (?) |
| 20 | 65 | 35 |

Reduction of the nitro function contributes 2 moles of water to this reaction. If this water content is eliminated in the reduction by beginning with 6-amino-2-hydroxyiminohexanamide, the yield of lysinamide exceeds 95% (Fuhrmann *et al.*, 1973).

## VI.  BIFUNCTIONAL MOLECULES

The products obtained by hydrogenation of oximes may be derived by interaction of the initially formed amine, hydroxylamine, or imine with suitably disposed second functions in the same or different molecules (Ochiai and Miyamoto, 1937). Formation of five- and six-membered rings through attack of the amine or hydroxylamine on esters (Koo, 1953; Remers *et al.*, 1965), ketones (Duschinsky and Dolan, 1945), or amides (Wenkert *et al.*, 1958) is to be anticipated. When the second function is also reduced easily, the products may depend in large measure on which function is reduced more rapidly. For instance, in syntheses leading to potential folic acid antagonists, **10** was reduced over Raney nickel in ethanol under ambient conditions to afford a mixture containing 64% **11** and 10% **12**. The former product results when reduction of the nitro function precedes reduction of the oximino function, whereas **12** is formed in the reverse sequence (Elliott *et al.*, 1971).

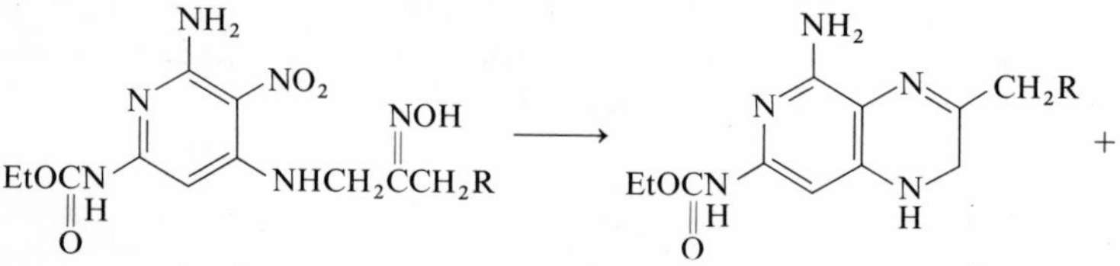

**10**

**11**

$$R = -N \overbrace{\phantom{xxxx}}^{CH_3} COOCH_3$$

(64%)

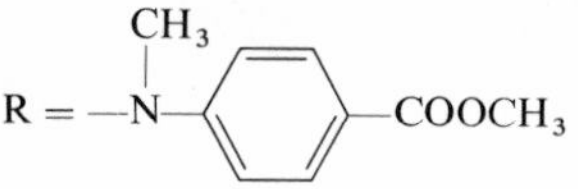

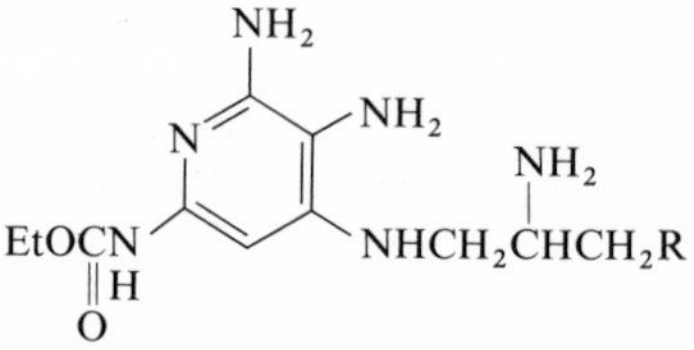

**12**

(10%)

## Oximino Ketones

Oximino ketones are obtained readily, and their reduction provides a convenient route to amino ketones and amino alcohols (Matsumoto and Hata, 1957). The reduction is often carried out in the presence of acid (Hartung and Munch, 1929) to prevent the formation of secondary and tertiary amines and dihydropyrazines that occur in neutral media (Hartung, 1931). It appears that the major product of reduction of oximino ketones in acidic solution, after absorption of 2 equivalents of hydrogen, is always the amino ketone (Murphy, 1961; Matsumoto *et al.*, 1962; Bien and Ginsburg, 1963), although its yield may be quite sensitive to small changes in the catalyst (Hartung and Chang, 1952). Excellent results have also been ob-

obtained by hydrogenation in ethanolic sodium hydroxide (Hartung and Chang, 1952). This alkaline reduction appears to be seldom used.

The products of reduction of aromatic oximino ketones depend on the aromatic moiety. When the aromatic portion is a phenol or phenyl methyl ether, the reduction stops at the amino ketone stage (Hartung, 1931; Hartung et al., 1931), but phenyl, m- or p-tolyl, or naphthyloximino ketones smoothly yield the amino alcohol (Hartung and Munch, 1929; Hartung et al., 1930, 1935).

Reduction of aliphatic oximino ketones is also structure sensitive. Hydrogenation of 4-methyl-anti-3-oximino-2-pentanone (13) over 10% palladium-on-carbon in absolute ethanol containing 3 moles of hydrogen chloride per mole of ketone gave 3-amino-4-methyl-2-pentanone hydrochloride (14) in good yield (Hoy and Hartung, 1958; Smith and Hicks, 1971), but the reduction failed when applied to 2-oximinocyclopentanone, 2-oximinocyclohexanone, anti-3-oximino-2-bornanone, and various steroidal oximino ketones (Smith and Hicks, 1971). Rhodium-on-alumina might have succeeded (Newman and Lee, 1975).

$$\underset{\textbf{13}}{(CH_3)_2CH-\overset{\overset{\displaystyle HON}{\|}}{C}-\overset{\overset{\displaystyle O}{\|}}{C}CH_3} \xrightarrow{\text{Pd-on-C}} \underset{\textbf{14}}{(CH_3)_2CH\overset{\overset{\displaystyle H_2N}{|}}{C}H-\overset{\overset{\displaystyle O}{\|}}{C}CH_3}$$

## VII.   STEREOCHEMISTRY

Hydrogenation of α-oximino ketones usually, but not always (Matsumoto and Hata, 1957), gives the amino alcohol as a single racemic modification, although two diastereoisomeric racemates might be expected (Pfau and Plattner, 1940; Weijlard et al., 1951). The high stereospecificities have been accounted for by assuming that adsorption occurs in such a way as to give rigid ringlike structure (Chang and Hartung, 1953).

Hydroxyl groups adjacent to the oximino function also tend to direct the hydrogenation so that the resulting amino function bears a cis relation to the hydroxyl (Fanta et al., 1963; Roth et al., 1964). When oximes are attached to cyclohexane or pyranose rings, there is usually a preponderance of the epimer having an axial amino group in the stable chair form (Brimacombe and How, 1963).

Addition of hydrogen to certain bridged polycyclic systems follows an exo addition rule to give endo substituents. Hydrogenation of 5-ketotetra-hydro-exo-dicyclopentadiene oxime over 5% palladium-on-carbon in absolute ethanol proceeded mainly from the exo side to afford the amine in about 80% exo-ring–endo-amine and 20% exo-ring–exo-amine (Diveley et al., 1968).

A limited study of the effect of solvent and metal on the stereochemistry of reduction of 2-, 3-, and 4-methylcyclohexanone oximes revealed little difference between ruthenium and rhodium. Palladium tended to give mainly bis(methylcyclohexyl)amine. Solvent had some effect on the stereochemistry, but it was subordinated to the influence of the oxime isomer. Over 5% rhodium-on-carbon at 100°C and 1000 psig, the percentages of trans isomer in the methylcyclohexylamine present varied in several solvents from 31–42% for 2-methylcyclohexanone oxime to 63–70% for the 3 isomer to 28% for the 2 isomer (water and ethanol only) (Rylander and Steele, 1963).

## REFERENCES

Adams, R., and Shriner, R. L., *J. Am. Chem. Soc.* **45**, 2171 (1923).

Adkins, H., "Reactions of Hydrogen," p. 92. Univ. of Wisconsin Press, Madison, 1937.

Albertson, N. F., Tullar, B. F., King, J. A., Fishburn, B. B., and Archer S., *J. Am. Chem. Soc.* **70**, 1150 (1948).

Benington, F., Morin, R. D., and Clark, L. C., Jr., *J. Med. Chem.* **8**, 100 (1965).

Bien, S., and Ginsburg, D., *J. Chem. Soc.* p. 2065 (1963).

Bollinger, F. W., Hayes, F. N., and Siegel, S., *J. Am. Chem. Soc.* **75**, 1729 (1953).

Breitner, E., Roginski, E., and Rylander, P. N., *J. Chem. Soc.* p. 2918 (1959).

Brimacombe, J. S., and How, M. J., *J. Chem. Soc.* p. 3886 (1963).

Buchanan, G. L., and Sutherland, J. K., *J. Chem. Soc.* p. 2334 (1957).

Buchman, O., and Milstein, D., *Tetrahedron Lett.* p. 4099 (1972).

Carmack, M., Bullitt, O. H., Jr., Handrick, G. R., Kissinger, L. W., and Von, I., *J. Am. Chem. Soc.* **68**, 1222 (1946).

Cash, W. D., Semeniuk, F. T., and Hartung, W. H., *J. Org. Chem.* **21**, 999 (1956).

Chang, Y. T., and Hartung, W. H., *J. Am. Chem. Soc.* **75**, 89 (1953).

Cope, A. C., and Kagen, F., *J. Am. Chem. Soc.* **80**, 5499 (1958).

Diveley, W. R., Buntin, G. A., and Lohr, A. D., *J. Org. Chem.* **33**, 616 (1968).

Dornow, A., and Frese, A., *Arch. Pharm.* (*Weinheim, Ger.*) **285**, 463 (1952).

Duschinsky, R., and Dolan, L. S., *J. Am. Chem. Soc.* **67**, 2079 (1945).

Elliott, R. D., Temple, C., Jr., Frye, J. L., and Montgomery, J. A., *J. Org. Chem.* **36**, 2818 (1971).

Fanta, P. E., Pandya, L. J., Groskopf, W. R., and Su, H. J., *J. Org. Chem.* **28**, 413 (1963).

Freifelder, M., "Practical Catalytic Hydrogenation," p. 262. Wiley, New York, 1971.

Freifelder, M., Smart, W. D., and Stone, G. R., *J. Org. Chem.* **27**, 2209 (1962).

Fryer, R. I., Earley, J. V., Evans, E., Schneider, J., and Sternbach, L. H., *J. Org. Chem.* **35**, 2455 (1970).

Fuentes, O., and Paudler, W. W., *J. Org. Chem.* **40**, 1210 (1975).

Fuhrmann, R., Pisanchyn, J., and Koff, F., *Ann. N.Y. Acad. Sci.* **214**, 243 (1973).

Gagneux, A. R., U.S. Patent 3,493,577, Feb. 3, 1970.

Gagneux, A. R., and Meier, R., *Tetrahedron Lett.* p. 1365 (1969).

Gilsdorf, R. T., and Nord, F. F., *J. Am. Chem. Soc.* **74**, 1837 (1952).

Hartung, W. H., *J. Am. Chem. Soc.* **50**, 3370 (1928).

Hartung, W. H., *J. Am. Chem. Soc.* **53**, 2248 (1931).

Hartung, W. H., and Chang, Y. T., *J. Am. Chem. Soc.* **74**, 5927 (1952).

Hartung, W. H., and Munch, J. C., *J. Am. Chem. Soc.* **51**, 2262 (1929).

Hartung, W. H., Munch, J. C., Deckert, W. A., and Crossley, F., *J. Am. Chem. Soc.* **52**, 3317 (1930).

Hartung, W. H., Munch, J. C., Miller, E., and Crossley, F., *J. Am. Chem. Soc.* **53**, 4149 (1931).

Hartung, W. H., Munch, J. C., and Crossley F. S., *J. Am. Chem. Soc.* **57**, 1091 (1935).

Hoy, K. L., and Hartung, W. H., *J. Org. Chem.* **23**, 967 (1958).

Huebner, C. F., Donoghue, E. M., Novak, C. J., Dorfman, L., and Wenkert, E., *J. Org. Chem.* **35**, 1149 (1970).

Hückel, W., and Kupka, R., *Chem. Ber.* **89**, 1694 (1956).

Hückel, W., and Thomas, K. D., *Justus Liebigs Ann. Chem.* **645**, 177 (1961).

Jones, L. W., and Major, R. T., *J. Am. Chem. Soc.* **52**, 669 (1930).

Jones, R., Jr., Price, C. C., and Sen, A. K., *J. Org. Chem.* **22**, 783 (1957).

Karpenko, I., and Rylander, P. N., unpublished observations, Engelhard Ind. Res. Lab., Newark, New Jersey, 1958.

King, F. E., Henshall, T., and Whitehead, R. L. St. D., *J. Chem. Soc.* p. 1373 (1948).

Koo, J., *J. Am. Chem. Soc.* **75**, 723 (1953).

Kreisky, S., *Monatsh. Chem.* **89**, 685 (1958).

Levin, N., Graham, B. E., and Kolloff, H. G., *J. Org. Chem.* **9**, 380 (1944).

Major, R. T., and Ohly, K. W., *J. Med. Chem.* **4**, 51 (1961).

Masamune, T., Ohne, M., Koshi, M., Ohuchi, S., and Iwardare, T., *J. Org. Chem.* **29**, 1419 (1964).

Matsumoto, T., and Hata, *J. Am. Chem. Soc.* **79**, 5506 (1957).

Matsumoto, T., Nishida, T., and Shirahama, H., *J. Org. Chem.* **27**, 79 (1962).

Meltzer, R. I., and Lewis, A. D., *J. Org. Chem.* **22**, 612 (1957).

Mousseron, T., Froger, M., Granger, P., and Winternitz, F., *Bull. Soc. Chim. Fr.* p. 843 (1947).

Müller, E., Fries, D., and Metzger, H., *Chem. Ber.* **88**, 1891 (1955).

Murphy, J. G., *J. Org. Chem.* **26**, 3104 (1961).

Neelakantan, L., and Hartung, W. H., *J. Org. Chem.* **23**, 964 (1958).

Newman, M. S., and Lee, V., *J. Org. Chem.* **40**, 381 (1975).

Ochiai, E., and Miyamoto, Y., *Yakugaku Zasshi* **57**, 583 (1937).

Pappas, J. J., and Gancher, E., *J. Chem. Eng. Data* **14**, 269 (1969).

Pfau, A. S., and Plattner, P. A., *Helv. Chim. Acta* **23**, 768 (1940).

Rapport, R. L., and Kupferberg, H. J., *J. Med. Chem.* **16**, 599 (1973).

Reeve, W., and Christian, J., *J. Am. Chem. Soc.* **78**, 860 (1956).

Reichert, B., and Koch, W., *Arch. Pharm. (Weinheims Ger.)* **273**, 265 (1935).

Remers, W. A., Roth, R. H., and Weiss, M., *J. Org. Chem.* **30**, 2910 (1965).

Rosen, W. E., and Green, M. J., *J. Org. Chem.* **28**, 2797 (1963).

Rosenmund, K. W., and Pfannkuch, E., *Ber. Dtsch. Chem. Ges. B* **56**, 2258 (1923).

Rosenmund, K. W., and Pfannkuch, E., *Ber. Dtsch. Chem. Ges. B* **75**, 1859 (1942).

Rosenmund, K. W., Karg, E., and Marcus, F. K., *Ber. Disch. Chem. Ges. B* **75**, 1859 (1942).

Roth, W., Pigman, W., and Danishefsky, I., *Tetrahedron* **20**, 1675 (1964).

Rylander, P. N., and Steele, D. R., *Engelhard Ind, Tech. Bull.* **4**, 20 (1963).

Rylander, P. N., and Steele, D. R., *Engelhard Ind., Tech. Bull.* **5**, 113 (1965).

Rylander, P. N., Hasbrouck, L., Hindin, S. G., Iverson, R., Karpenko, I., and Pond, G., *Engelhard Ind., Tech. Bull.* **8**, 99 (1967).

Shepard, E. R., Noth, J. F., Porter, H. D., and Simmans, C. K., *J. Am. Chem. Soc.* **74**, 4611 (1952).

Sianesi, E., Re, D. A., Magistretti, M. J., and Setnikar, I., *J. Med. Chem.* **10**, 1144 (1967).

Smith, H. E., and Hicks, A. A., *J. Org. Chem.* **36**, 3659 (1971).

Turner, R. A., Huebner, C. F., and Scholz, C. R., *J. Am. Chem. Soc.* **71**, 2801 (1949).

Vanon, G., and Krajcinovic, M., *Bull. Soc. Chim. Fr.* **43**, 231 (1928).

Walker, G. N., and Alkalay, D., *J. Org. Chem.* **32**, 2213 (1967).

Walker, G. N., Alkalay, D., Engle, A. R., and Kempton, R. J., *J. Org. Chem.* **36**, 466 (1971).

Weijlard, J., Pfister, K., III, Swanezy, E. F., Robinson, C. A., and Tishler, M., *J. Am. Chem. Soc.* **73**, 1216 (1951).

Wenkert, E., Bernstein, B. S., and Udelhofen, J. H., *J. Am. Chem. Soc.* **80**, 4899 (1958).

Wilbert, G., and Sosis, P., U.S. Patent 3,028,429, Apr. 3, 1962.

Yamaguchi, S., Ito, S., Suzuki, I., and Inque, N., *Bull. Chem. Soc. Jpn.* **41**, 2073 (1968).

# Chapter 10

# Reductive Alkylation

Primary or secondary amines or ammonia may interact with aldehydes or ketones in the presence of hydrogen and a hydrogenation catalyst to produce a new amine. Two discrete steps are required for alkylation. First is the condensation of the carbonyl compound with the amine to give an addition product or, by elimination of water, an imine. Second, the addition product or imine is hydrogenated. Occasionally, the intermediate imine is isolated (Emerson, 1948), but usually the two steps are achieved in a single operation. Some workers allow the amine and carbonyl compound to stand together for some time before the hydrogenation is begun (Archer *et al.*, 1957), but this procedure is not always necessary or desirable (Heyl *et al.*, 1952).

Reductive alkylations can be carried out with substrates that are not carbonyl compounds or amines but that can be transformed into these compounds during the course of reduction. Functions reducible to amines include azo (British Patent 771,063, Mar. 27, 1957), hydrazo (Emerson *et al.*, 1941), nitroso, nitro (Emerson and Uraneck, 1941; Bowman and Stroud, 1950; Kaplan and Conroy, 1963; Kilbourne *et al.*, 1965), oximino, and hydroxyl-amines. Phenols (Dankert and Permoda, 1951), acetals, or ketals (Johnson and Crosby, 1962) may provide precursors for carbonyl compounds.

## I. SUBSTRATES

Primary or secondary amines formed in a reductive alkylation are themselves suitable substrates for the reaction and may be further alkylated. Increased steric hindrance in the intermediate products works toward limiting the reaction (Skita and Keil, 1928; Skita *et al.*, 1933), whereas electronic effects may promote further reaction. For instance, reductive alkylation of glycine over palladium with various straight-chain aliphatic aldehydes gave only *N,N*-dialkyl derivatives; the monoalkyl derivative was not obtained in any case, for the monoalkyl intermediates were more reactive in competition than the starting material. However, considerable amounts of *N*-monoalkyl derivatives are formed when the reaction is carried out with branched-chain aldehydes such as isobutyraldehyde and isovaleraldehyde (Ikutani, 1969). Cyclic ketones tend to produce more secondary amine on reductive alkylation with ammonia than do linear ketones of comparable carbon number (Hückel and Kupka, 1956). Reductive alkylation may occur with unusual ease if cyclization occurs as a consequence of the reaction (Pachter and Suld, 1960).

Selective reductive alkylations can be achieved as a consequence of differing degrees of steric hindrance around competing carbonyl functions. Selective attack of the 3-oxo group was the basis for preparing isomeric dimethylamino derivatives of 3,17-dioxo(5α)androstane (**1**). Alkylation of **1** with dimethylamine over palladium-on-carbon gave 3β-dimethylamino-17-oxo(5α)androstane (**3**); the sterically hindered 17-carbonyl did not react to any appreciable extent. Treatment of **1** with hydroxylamine followed by reduction of the dioxime over platinum oxide gave a 3α-amino-17-hydroxy-amino intermediate, which on sodium–alcohol reduction afforded 3α,17β-diamino(5α)androstane (**2**). The dimethylamino derivative **4** was formed by reductive alkylation with formaldehyde and palladium-on-carbon. The 3β,17β-diamino and 3β,17β-dimethylamino compounds are obtained by reduction of the dioxime with sodium in amyl alcohol followed by reductive alkylation (Glaser and Gabbay, 1970).

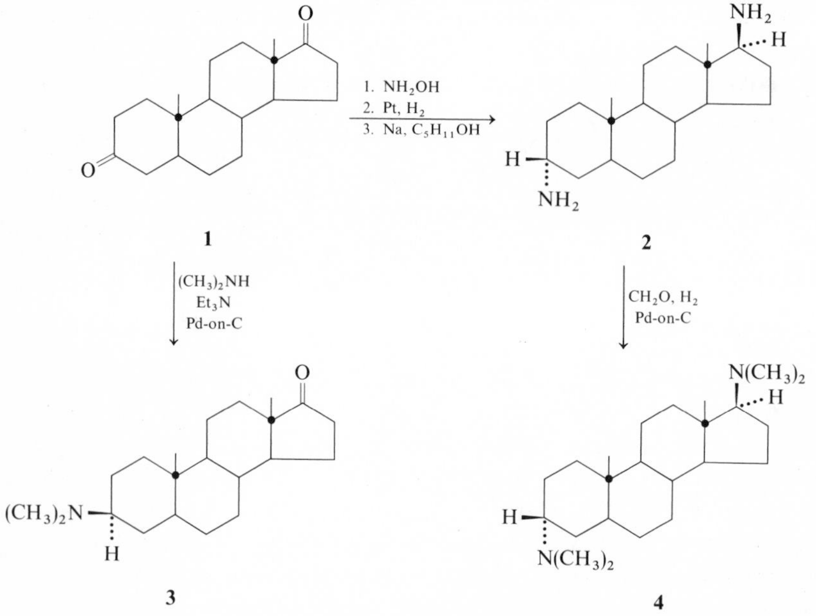

1. NH$_2$OH
2. Pt, H$_2$
3. Na, C$_5$H$_{11}$OH

**1**

(CH$_3$)$_2$NH
Et$_3$N
Pd-on-C

**2**

CH$_2$O, H$_2$
Pd-on-C

**3**

**4**

Hydrazine is a suitable amine component for reductive alkylations. Hydrazine with diketones provides an entry into 1,2-diaza cyclic systems. Treatment of 2-(1,1-dimethylacetonyl)cyclopentanone with hydrazine hydrate followed by reduction over platinum oxide in acetic acid affords, at 50 psig and 25°C, 4,5,5-trimethyl-2,3-diazabicyclo[4.3.0]nonane in 38% yield. Apparently no effort was made to limit hydrogen absorption (Hennion and Quinn, 1970).

Reductive alkylation of acid hydrazides has been used as a means of preparing mono- and disubstituted hydrazides. When the reduction is carried out stepwise, the substituents may be different (Malz et al., 1976a). Methyl- and dimethylhydrazines, which are rocket fuel ingredients, are prepared by reductive alkylation of acethydrazide with formaldehyde over palladium followed by hydrolysis (Malz et al., 1977; Malz and Greenfield, 1979).

## A.   Imines

Imines are likely intermediates in reductive alkylation reactions, but they are not usually isolated (Emerson, 1948). Isolation may in fact be inferior to direct reductive alkylation (Layer, 1963). With certain combinations of amine and carbonyl compound, interaction to form an addition compound or imine does not occur readily, and reductive alkylation of such systems fails. The condensation reaction can itself be catalyzed, and a variety of condensing catalysts and reagents have been used for this purpose. The list includes basic materials such as triethylamine or guanidine (Henry and Finnegan, 1954), acids such as zinc chloride, hydrochloric acid, or acetic acid (Skita and Stühmer, 1955), acidic carbon, and drying agents such as calcium oxide or anhydrous magnesium sulfate (Heinzelman and Aspergren, 1953). Acids also serve the function of neutralizing the inhibiting effects of the saturated amines formed on imine hydrogenation (Maxted and Walker, 1948; Maxted and Biggs, 1957; Breitner et al., 1959; Freifelder, 1961).

Small amounts of acid may be effective, as illustrated by a convenient route to 1-alkyl-3,3-dimethylazetidines through reductive alkylation of 2,2-

dimethyl-3-hydroxypropanal, formation of a suitable leaving group, and ring closure. The aldehyde is obtained readily by base-catalyzed condensation of formaldehyde with isobutyraldehyde. A mixture of 125 gm benzylamine, 6 drops concentrated hydrochloric acid, 550 ml absolute ethanol, and 0.25 gm platinum oxide at 3 atm affords 3-benzylamino-2,2-dimethyl-1-propanol in 91% yield (Anderson and Wills, 1968).

## B.   Amine and Carbonyl Precursors

Many compounds that are not themselves suitable components for reductive alkylation can nonetheless be used successfully if they can first be transformed into an amine or carbonyl. The technique may have the advantage of saving a separate reduction step. For instance, the use of the nitro function as an amine precursor is illustrated by the conversion of *m*-nitrobenzaldehyde to *m*-dimethylaminobenzaldehyde.

This reaction failed when applied to 6-nitrosalicylaldehyde (**5**), but through the use of 4-ethoxy-5-nitro-1,3-benzodioxane (**6**) and subsequent hydrolysis of **7** it was possible to obtain **8** (Ando and Emoto, 1974). Preparation of tertiary aromatic amines is sometimes difficult because of facile nuclear condensation, but excellent results have been obtained nonetheless with careful attention to detail (Pearson and Bruton, 1951).

Pyridines can serve as aliphatic amine precursors in reductive alkylation. The catalyst must be chosen so that the pyridine ring is reduced in preference to the carbonyl function. Palladium-on-carbon proved to be satisfactory for this task in the reduction of **9** to **10** (Stefanović *et al.*, 1970). The C-5 double bond survived these vigorous conditions. Reduction of a similar system over platinum oxide gave a mixture of pyridyl alcohol and cyclized product, whereas over palladium-on-carbon, under mild conditions, only the double bond was reduced (Sam *et al.*, 1964).

Acetals, ketals (Johnson and Crosby, 1962), and phenols (Kilbrourne *et al.*, 1965) have been used as precursors of aldehydes and ketones in reductive alkylation reactions. Palladium is probably the most useful catalyst in reactions involving phenols due to its great effectiveness in converting phenols to cyclohexanones and its low activity for reduction of aliphatic ketones. The acetal group may protect an aldehyde during reductive alkylation with another carbonyl function. For example, aminoacetaldehyde dialkyl acetal has been used effectively in reductive alkylations to gain entry into a variety of tetrahydroisoquinolines (Bobbitt *et al.*, 1965; Bobbitt and Moore, 1968; Coomes *et al.*, 1973; Gensler *et al.*, 1968; Bobbitt and Shibuya, 1970).

## II.  CATALYSTS

Nickel, rhodium, ruthenium, palladium, and platinum catalysts have been used successfully in reductive alkylations. By far the most frequently used catalyst is platinum (Emerson, 1948), but it is not clear whether this preference is due to a demonstrated superiority or to habit. Palladium-on-carbon is used commercially in a series of reductive alkylations, but as the molecular weight of the ketone increases platinum-on-carbon becomes suitable. In the laboratory, platinum oxide is often used. Some workers prereduce this catalyst to avoid lengthy induction periods (Cope and Hancock, 1942; Iles and Worrall, 1961; Manske and Johnson, 1929), but this procedure is not always necessary (Freifelder, 1963).

### Sulfided Platinum Metal Catalysts

Excellent results in the reductive alkylation of anilines (Dovell and Greenfield, 1965) have been obtained with platinum metal sulfide catalysts (manufactured by Engelhard Industries, Newark, New Jersey). These catalysts, which are usually on a support such as carbon, are used in diverse industrial applications. In general, they behave like base metal sulfide catalysts but are much more active. They are resistant to poisoning and are unique in that they can be used for reductive alkylation of haloanilines and halonitro compounds without dehalogenation (Dovell and Greenfield, 1967).

$$Cl \longleftarrow \text{—} NO_2 + CH_3COCH_3 \xrightarrow[\substack{1300 \text{ psig} \\ 180^\circ C}]{2.5 \text{ gm } 5\% \text{ RhSx-on-C}} Cl \longleftarrow \text{—} NHCH(CH_3)_2$$

$$31.5 \text{ gm} \qquad\qquad\qquad\qquad\qquad\qquad\qquad\qquad\qquad\qquad\qquad 34 \text{ gm}$$

### III.  STEREOCHEMISTRY

Reductive alkylation with optically active substrates may afford compounds containing new optically active centers. The reaction has been of considerable interest, especially as a means of preparing optically active amino acids by transamination (Hiskey and Northrup, 1961, 1965). The magnitude of the induced asymmetry may depend on the structure of the substrate, on the catalyst (Hiskey and Northrup, 1961), on the solvent (Harada and Matsumoto, 1968), and on the temperature (Harada and Yoshida, 1970, 1973; Harada et al., 1973). Alanine, derived from pyruvic acid or benzyl pyruvate and various optically active alkylamines, had optical purities of 60–80% in less polar solvents and 30–50% in more polar solvents

(Hara and Matsumoto, 1968). These differences were thought to be due to changing conformations of the adsorbed substrate.

$$
\begin{array}{c}
CH_3 \\
| \\
C{=}O + \phi\overset{*}{C}HNH_2 \\
| \\
COOH \quad R
\end{array}
\longrightarrow
\begin{array}{c}
CH_3 \\
| \\
CHNH\overset{*}{C}H\phi \\
| \\
R \\
COOH
\end{array}
\longrightarrow
\begin{array}{c}
CH_3 \\
| \\
{}^{*}CHNH_2 + \phi CH_2R \\
| \\
COOH
\end{array}
$$

Transamination of oxaloacetic acid with $S(-)$-$\alpha$-ethylbenzylamine failed to give any aspartic acid but gave instead optically active alanine through decarboxylation and debenzylation (Matsumoto and Harada, 1968).

$$
\begin{array}{c}
COOH \\
| \\
CH_2 \\
| \\
C{=}O \\
| \\
COOH
\end{array}
+ NH_2\overset{*}{C}H\phi \underset{\underset{25^\circ C}{EtOH}}{\overset{Pd(OH)_2\text{-on-C}}{\longrightarrow}}
\begin{array}{c}
COO^- \\
| \\
CH_2 \\
| \\
C{=}NCH\phi \\
\quad | \\
\quad CH_3 \\
COO^-
\end{array}
\longrightarrow
\begin{array}{c}
CH_3 \\
| \\
{}^{*}CHNH_2 + CO_2 \\
| \\
COOH
\end{array}
$$

A convenient synthesis of optically active alkyl- and methoxy-substituted amphetamines is through reductive alkylation of phenylacetones with chiral $\alpha$-methylbenzylamine followed by hydrogenolysis of the resulting N-($\alpha$-phenethyl)phenylisopropylamines. Enantiomeric purities range from 96 to 99%. The reductive alkylation may be carried out in one (Weinges and Graab, 1970) or two stages (Nichols *et al.*, 1973). In the latter, which is preferred, the intermediate imines were preformed by azeotropic water removal in refluxing benzene. The imines were reduced directly without isolation.

One can usually predict which isomer will predominate in a reductive alkylation if one assumes that hydrogen is adsorbed on and added from the catalyst, that the catalyst approaches the substrate from its least hindered side (Prelog, 1956), and that the intermediate species undergoing reduction is the imine and not the hydroxyamino addition compound. Some complications may arise if the imine intermediate can undergo tautomerization. Identical mixtures of amines were obtained in reductive alkylation either of pivaldehyde with *trans*-4-methylcyclohexylamine or of 4-methylcyclohexanone with neopentylamine over 5% palladium-on-carbon.

The authors surmised that these results obtain because of an equilibrium among tautomeric imines (Goldberg and Lam, 1969).

Optical activity is retained in the reduction or reductive alkylation of optically pure 3-formylpinane (Himmele and Siegel, 1976).

Reductive alkylation of sterically bulky *d*-camphor with methylamine in methanol over platinum oxide at 900 psig and 120°C affords *N*-methylisobornylamine in 92% yield. Formation of this product suggests cis addition of hydrogen from the endo side of the intermediate imine. Under the same conditions the reaction was about half complete over palladium-on-carbon and had scarcely begun over Raney nickel (Kiyooka and Suzuki, 1974).

## REFERENCES

Anderson, A. G., Jr., and Wills, M. T., *J. Org. Chem.* **33**, 2123 (1968).
Ando, M., and Emoto, S., *Bull. Chem. Soc. Jpn.* **47**, 501 (1974).
Archer, S., Lewis, T. R., Unser, M. J., Hoppe, J. O., and Lape, H., *J. Am. Chem. Soc.* **79**, 5783 (1957).
Bobbitt, J. M., and Moore, T. E., *J. Org. Chem.* **33**, 2958 (1968).
Bobbitt, J. M., and Shibuya, S., *J. Org. Chem.* **35**, 1181 (1970).
Bobbitt, J. M., Kiely, J. M., Khanna, K. L., and Ebermann, R., *J. Org. Chem.* **30**, 2247 (1966).
Bowman, R. E., and Stroud, H. H., *J. Chem. Soc.* p. 1342 (1950).
Breitner, E., Roginski, E., and Rylander, P. N., *J. Chem. Soc.* p. 2918 (1959).
Coomes, R. M., Falck, J. R., Williams, D. K., and Stermitz, F. R., *J. Org. Chem.* **38**, 3701 (1973).
Cope, A. C., and Hancock, E. M., *J. Am. Chem. Soc.* **64**, 1503 (1942).
Dankert, L. J., and Permoda, D. A., U.S. Patent 2,571,016, Oct. 9, 1951.
Dovell, F. S., and Greenfield, H., *J. Am. Chem. Soc.* **87**, 2767 (1965).
Dovell, F. S., and Greenfield, H., U.S. Patent 3,350,450, Oct. 31, 1967.
Emerson, W. S., *Org. React.* **4**, 174 (1948).
Emerson, W. S., and Uraneck, C. A., *J. Am. Chem. Soc.* **63**, 749 (1941).
Emerson, W. S., Reed, S. K., and Merner, R. R., *J. Am. Chem. Soc.* **63**, 972 (1941).
Freifelder, M., *J. Org. Chem.* **26**, 1835 (1961).
Freifelder, M., *J. Med. Chem.* **6**, 813 (1963).
Gensler, W. J., Shamasundar, K. T., and Marburg, S., *J. Org. Chem.* **33**, 2861 (1968).
Glaser, R., and Gabbay, E. J., *J. Org. Chem.* **35**, 2907 (1970).
Goldberg, S. I., and Lam, F. L., *J. Am. Chem. Soc.* **91**, 5113 (1969).
Harada, K., and Matsumoto, K., *J. Org. Chem.* **33**, 4467 (1968).
Harada, K., and Yoshida, T., *Chem. Commun.* p. 1071 (1970).
Harada, K., and Yoshida, T., *J. Org. Chem.* **37**, 4366 (1973).
Harada, K., Iwasaki, T., and Okawara, T., *Bull. Chem. Soc. Jpn.* **46**, 1901 (1973).
Heinzelman, R. V., and Aspergren, B. D., *J. Am. Chem. Soc.* **75**, 3409 (1953).
Hennion, G. F., and Quinn, F. X., *J. Org. Chem.* **35**, 3054 (1970).
Henry, R. A., and Finnegan, W. G., *J. Am. Chem. Soc.* **76**, 926 (1954).
Heyl, D., Luz, E., Harris, S. A., and Folkers, K., *J. Am. Chem. Soc.* **74**, 414 (1952).
Himmele, W., and Siegel, H., *Tetrahedron Lett.* p. 911 (1976).
Hiskey, R. G., and Northrop, R. C., *J. Am. Chem. Soc.* **83**, 4798 (1961).
Hiskey, R. G., and Northrop, R. C., *J. Am. Chem. Soc.* **87**, 1753 (1965).
Hückel, W., and Kupka, R., *Chem. Ber.* **89**, 1694 (1955).
Ikutani, Y., *Bull. Chem. Soc. Jpn.* **42**, 2330 (1969).

Iles, R. W., and Worrall, W. S., *J. Org. Chem.* **26**, 5233 (1961).

Johnson, H. E., and Crosby, D. G., *J. Org. Chem.* **27**, 2205 (1962).

Kaplan, F., and Conroy, H., *J. Org. Chem.* **28**, 1593 (1963).

Kilbrourne, H. W., VanVerth, J. E., and Wilder, G. R., U.S. Patent 3,219,703, Nov. 23, 1965.

Kiyooka, S., and Suzuki, K., *Bull. Chem. Soc. Jpn.* **47**, 2081 (1974).

Layer, R. W., *Chem. Rev.* **63**, 489 (1963).

Malz, R. E., Jr., Amidon, R. W., and Greenfield, H., U.S. Patent 3,965,174, June 22, 1976a.

Malz, R. E., Jr., and Greenfield, H., *in* "Catalysis in Organic Syntheses, 1978." (W. S. Jones, ed.), Academic Press, New York, in press.

Maltz, R. E., Jr., Amidon, R. W., and Greenfield, H., U.S. Patent 4,045,484, Aug. 30, 1977.

Manske, R. H. F., and Johnson, T. B., *J. Am. Chem. Soc.* **51**, 580 (1929).

Matsumoto, K., and Harada, K., *J. Org. Chem.* **33**, 4526 (1968).

Maxted, E. B., and Biggs, M. S., *J. Chem. Soc.* p. 3844 (1957).

Maxted, E. B., and Walker, A. G., *J. Chem. Soc.* p. 1093 (1948).

Nichols, D. E., Barfknecht, C. F., Rusterholz, D. B., Benington, F., and Morin, R. D., *J. Med. Chem.* **16**, 480 (1973).

Pachter, I. J., and Suld, G., *J. Org. Chem.* **25**, 1680 (1960).

Pearson, D. E., and Bruton, J. D., *J. Am. Chem. Soc.* **73**, 864 (1951).

Prelog, V., *Bull. Soc. Chim. Fr.* p. 987 (1956).

Sam, J., England, J. D., and Alwani, D. W., *J. Med. Chem.* **7**, 732 (1964).

Skita, A., and Keil, F., *Ber. Dtsch. Chem. Ges. B* **61**, 1452 (1928).

Skita, A., and Stühmer, W., Ger. Patent 932,677, Sept. 5, 1955.

Skita, A., Keil, F., and Baesler, E., *Ber. Dtsch. Chem. Ges. B* **66**, 858 (1933).

Stefanović, M., Mićović, I. V., Jeremić, D., and Miljković, D., *Tetrahedron* **26**, 2609 (1970).

Weinges, F., and Graab, G., *Chem.-Ztg., Chem. Appar.* **94**, 728 (1970).

# Chapter 11

# Hydrogenation of
# Carbocyclic Aromatics

Carbocyclic aromatics are readily hydrogenated to the fully saturated derivative. Reduction is considered to proceed stepwise, but ordinarily the partially hydrogenated intermediates do not accumulate to any appreciable extent unless special structural features are present (Folkers and Johnson, 1933; Cram and Allinger, 1955; Rylander and Rakoncza, 1964; Siegel and Garti, 1977). The major problems connected with aromatic hydrogenations are those of controlling hydrogenolysis or reduction of other functions, selectivity in polycyclic aromatics, and stereochemistry. Hydroxyl and amino substituents on the aromatics introduce special problems; reduction of these compounds is discussed in Sections II and III.

## I. CARBOCYCLIC AROMATICS

### A. Catalysts

Palladium, platinum, rhodium, and ruthenium catalysts are often used for the hydrogenation of carbocyclic aromatic compounds. Iridium is rarely used, osmium is very sluggish, and nickel, unless exceptionally active, requires vigorous conditions. Greenfield (1973) established the relative activity order for hydrogenation of benzene and isobutylbenzene as rhodium > ruthenium ≫ platinum ≫ palladium ≫ nickel > cobalt, in agreement with other workers.

Platinum is often used as platinum oxide, a catalyst sensitive to both intrinsic and extrinsic trace impurities (Burger and Mosettig, 1936; Fieser and Hershberg, 1937, 1938). Sodium salts contained in the catalyst can be potent inhibitors (Adams and Marshall, 1928; Baker and Schuetz, 1947; Keenan *et al.*, 1954), whereas small amounts of strong acids frequently function as effective promoters, perhaps through interaction with sodium components (Brown *et al.*, 1936; Shriner and Witte, 1941; Phillips and Mentha, 1956). In some cases, weak acids, such as acetic acid, are more effective than strong acids (Ferber and Brückner, 1939). Platinum-on-alumina is used industrially for the hydrogenation of benzene to high-purity cyclohexane (Teter, 1959).

Palladium hydrogenates aromatic rings very slowly under mild conditions, and it is frequently recommended for reductions in which ring saturation is to be avoided. Nonetheless, palladium finds important commercial use in both batch and continuous hydrogenations carried out at elevated temperatures and pressures ranging from 3000 to 5000 psig. Palladium-on-carbon is a preferred industrial catalyst for the hydrogenation of benzoic acid to cyclohexanecarboxylic acid, a caprolactam intermediate. The saturated product is used as a solvent in this hydrogenation, which gives almost quantitative yields (Taverna and Chita, 1970).

Although most palladium-catalyzed ring reductions are carried out at elevated pressure, ring saturation has also been achieved at atmospheric pressure. The products obtained in the reduction of **2** depend on the temperatures. At room temperature only debenzylation and dehydrohalogenation occur (**1**), whereas at 90°C the carbocyclic ring is saturated as well, affording **3** (Grethe *et al.*, 1968).

Rhodium and ruthenium are very useful for the hydrogenation of aromatic systems, especially when hydrogenolysis is to be avoided. Rhodium is often used effectively under mild conditions (Egli and Eugster, 1975), whereas ruthenium is usually used at elevated temperatures and pressures. Water has a powerful promoting effect on ruthenium hydrogenation, an effect not shared to the same extent by other noble metal catalysts (Table I) (Rylander *et al.*, 1963).

## B.   Effect of Structure

Linear free-energy relationships (Hammett, 1970) involving both steric and electronic parameters (Mochida and Yoneda, 1968) have been used

TABLE I

**Effect of Water on Hydrogenation of Benzoic Acid**[a]

| H$_2$O (gm) | Benzoic acid reduced/min (gm) | |
| --- | --- | --- |
| | 500 mg 5% Ru-on-C | 1500 mg 5% Ru-on-C |
| 0 | 0.1 | 0.4 |
| 15 | 0.6 | 2.0 |
| 30 | 1.2 | 3.0 |
| 45 | 1.5 | — |
| 60 | 1.6 | — |

[a] Each experiment was carried out with 30 gm benzoic acid, 70 gm hexahydrobenzoic acid, and water as noted at 130°C and 2225 psig.

successfully in heterogeneous catalysis (Kraus, 1967). However, interpreting the experimental data is more difficult than in the case of homogeneously catalyzed reactions, since the substituent itself gives electronic and/or steric interactions with the heterogeneous catalyst. Kieboom (1976), interpreting the data of Yoshida (1974) on the hydrogenation of monoalkylbenzenes over ruthenium, concluded that the reaction rate constant is almost completely determined by steric interactions, whereas the adsorption equilibrium constant is influenced by electronic as well as steric effects. The lack of dependence of the reaction rate constant on electronic effects requires that atomic hydrogen, rather than hydride ion, be the reducing agent (Yoshida, 1974). An increase in the size of a substituent decreases the reaction rate but increases the strength of adsorption (Kieboom, 1976).

Predicting the effect of substitution on the rate of hydrogenation of aromatic is particularly difficult, for the relative rates depend on the catalyst as well as the test conditions (Smith, 1957, 1967). A useful but fallible generalization applicable to alkylbenzenes is that the rates of hydrogenation decrease as the number of substituents increases, and the rates increase as the symmetry of the substitution pattern of the ring increases, even when the higher symmetry leads to a greater degree of substitution (Trahanovsky and Bohlen, 1972).

Highly substituted, strained aromatic systems tend to undergo hydrogenation readily, even over palladium under mild conditions (Rapoport and Pasky, 1956; Rapoport and Smolinsky, 1960). The release of peri strain in the transition state is assumed to be the driving force for the unusual preferential saturation of the highly hindered ring in 1,4-di-*tert*-butylnaphthalene (Franck and Yanagi, 1968).

## C.  Olefin Intermediates

Olefin intermediates rarely accumulate in the hydrogenation of aromatics because they are usually hydrogenated much more readily than the starting material. Nonetheless, olefins have been detected in widely varying amounts during the hydrogenation of benzene (Hartog et al., 1965), 2-tert-butyl-benzoic acid (van Bekkum, 1965), 1,2-di-tert-butylbenzene (van de Graaf et al., 1968), biphenyl (Rylander and Vaflor, 1974), and other compounds (Siegel et al., 1963a,b; Siegel, 1966). In special cases, the accumulation of olefin is substantial. In the hydrogenation of 1,3,5-tri-tert-butylbenzene the intermediate olefin, cis-1,3,5-tri-tert-butylcyclohexene, accumulates to the extent of 65, 48, and 12% of the total mixture over rhodium, platinum, and palladium, respectively (van Bekkum et al., 1969). This ordering of metals for the accumulation of cyclohexene intermediates seems general. It is attributed to the stronger adsorption of the olefin relative to the aromatic over palladium and to weaker olefin adsorption relative to the aromatic over rhodium (van Bekkum et al., 1969).

Siegel and Garti (1977) found the maximal percentage of 1,4-di-tert-butylcyclohexene formed in hydrogenation of 1,4-di-tert-butylbenzene over 5% rhodium-on-alumina to vary with pressure, over the range 0.34–150 atm, from about 4 to 36%; surprisingly, the maximal percentage occurs at an intermediate pressure, 1.35 atm. The maximal rate of reduction also occurs at an intermediate pressure; above 7.7 atm the rate falls steadily with increasing pressures up to 150 atm, the highest reported. The authors suggested that rhodium may contain at least two types of active site, one catalyzes the hydrogenation of arenes and the other may catalyze the hydrogenation of both arenes and alkenes. Treatment of the catalyst with hydrogen may convert one form to another. Rhodium catalysts that had been treated with hydrogen at 140°C lost activity steadily over a period of 9 days when sealed in a bottle containing air.

Partial ring saturation is of limited synthetic utility per se, but the intervention of olefin intermediates has an important influence on the stereochemistry of the final products as well as the selectivity of reduction, especially hydrogenolysis. Intermediate olefins can undergo alkylation with the aromatic if the hydrogenation is carried out in the presence of an acidic catalyst. Phenylcyclohexane can be obtained in this way in high yield by hydrogenation of benzene when hydrogenation activity and acidity are properly balanced (Slaugh and Leonard, 1969).

Carbon–carbon bonds may also form readily during the hydrogenation of aromatic rings held in propinquity. Hydrogenation of [2.2]metacyclo-phane (5) does not give the expected dodecahydro[2.2]metacyclophane (4) but rather a product of intraannular cyclization, the perhydropyrene 6, one of 14 possible configurational isomers (Langer and Lehner, 1973a,b,c).

4                                    5                                        6

## D.  Selective Hydrogenation of Polycylic Compounds

Many synthetic problems require the selective hydrogenation of only one aromatic ring in a molecule containing several. High selectivity is often achieved rather easily as a consequence of the substrate structure (Smith *et al.*, 1949; Zaugg *et al.*, 1958; Baltzly *et al.*, 1961; Freifelder, 1964). In other cases, selectivity varies markedly with catalyst and reaction conditions. The point is illustrated by the data of Table II on the selective hydrogenation of biphenyl to phenylcyclohexane (Rylander and Steele, 1968; Rylander and Vaflor, 1974). Cyclohexenylcyclohexane and dicyclohexyl were the only other products detected in the hydrogenation of either biphenyl or phenyl-cyclohexane.

Of the metals tested, palladium clearly makes the most selective catalyst for the synthesis of phenylcyclohexane. Solvent has a pronounced effect on the maximal yield of phenylcyclohexane, the yield going from 49% in ethanol to 78% in tetrahydrofuran when rhodium-on-carbon is used. Temperature

**TABLE II**

Effect of Catalyst, Solvent, Temperature, and Support on Selectivity in Biphenyl Hydrogenation[a]

| Catalyst | Solvent | Temperature (°C) | Maximal % in product | |
|---|---|---|---|---|
| | | | Phenyl-cyclohexane | Cyclohexenyl-cyclohexane |
| PdO$_2$ | Methanol | 100 | 94 | 0 |
| 5% Pd-on-Al$_2$O$_3$ | Methanol | 100 | 92 | 0 |
| 5% Pd-on-C | Methanol | 100 | 89 | 0 |
| 5% Pt-on-C | Methanol | 100 | 66 | 0 |
| 5% Rh-on-C | Methanol | 100 | 62 | 0 |
| 5% Rh-on-Al$_2$O$_3$ | Methanol | 100 | 75 | 3.9 |
| Rh$_2$O$_3$ | Methanol | 100 | 62 | 4.6 |
| 5% Ru-on-C | Methanol | 100 | 62 | 0 |
| 5% Rh-on-C | Ethanol | 50 | 49 | 0.3 |
| 5% Rh-on-C | Methanol | 50 | 50 | 0 |
| 5% Rh-on-C | tert-Butanol | 50 | 57 | 0.5 |
| 5% Rh-on-C | Cyclohexane | 50 | 59 | 0.4 |
| 5% Rh-on-C | Ethyl acetate | 50 | 66 | 1.0 |
| 5% Rh-on-C | Tetrahydrofuran | 50 | 78 | 2.6 |
| 5% Rh-on-C | Methanol | 25 | 45 | 0 |
| 5% Rh-on-C | Methanol | 50 | 50 | 0 |
| 5% Rh-on-C | Methanol | 75 | 57 | 0 |
| 5% Rh-on-C | Methanol | 100 | 62 | 0 |
| 5% Rh-on-C | Methanol | 125 | 73 | 0 |

[a] Pressure, 500 psig.

is also an important variable; high selectivity is favored by elevated temperatures. Selectivity varies somewhat with support; alumina gives a higher selectivity than carbon. The maximal yield or phenylcyclohexane is only slightly changed by pressure in the range 1–80 atm (data not shown).

Cyclohexenylcyclohexane was found in more than trace amounts only over rhodium, in agreement with the findings of others that rhodium tends more than palladium or platinum to release olefin from the catalyst surface (van Bekkum *et al.*, 1969).

### Fused Rings

Selectivity in partial saturation of fused ring systems depends importantly on both the substrate structure and the catalyst. The descending order of selectivity for conversion of naphthalene to tetralin for noble metals is Pd > Pt > Rh > Ir > Ru. The unique ability of palladium to stop at the tetralin stage is in keeping with its ability to selectively hydrogenate olefins in the presence of aromatics and supports the idea that naphthalene may be adsorbed as if it were a cyclic diolefin fused to an aromatic system (Weitkamp, 1966). The hydrogenation of anthracene provides another illustration of the differences among metals (Table III). Products may arise by direct hydrogenation or by isomerization and disproportionation of intermediates. Palladium, more than platinum or rhodium, tends toward selective hydrogenation of a terminal ring (Cowen and Eisenbraun, 1977).

**TABLE III**[a]

| Catalyst | Reaction time (hr) | % In product | | |
|---|---|---|---|---|
| | | A | B | C |
| 5% Pd-on-C | 5 | 59 | 32 | 7 |
| 5% Pt-on-C | 7 | 79 | 6 | 10 |
| 5% Rh-on-$Al_2O_3$ | 9 | 85 | 3 | 7 |

[a] Data of Cowen and Eisenbraun (1977). Used with permission.

Raney copper and copper chromite are selective catalysts for hydrogenation of 9-aminoalkylanthracenes to the 9,10-dihydro derivatives. The temperature threshold for regiospecific addition of hydrogen is 125°C, and a minimum of 20% by weight of catalyst is required (Blackburn and Mlynarski, 1973).

Fu and Harvey (1977), after examining the hydrogenation of various polycyclic aromatic hydrocarbons, concluded that palladium regiospecifically reduces the electron-rich K-region bond, the region of minimal bond delocalization energy, but platinum attacks elsewhere. For instance, hydrogenation of benz[a]anthracene over 10% palladium-on-carbon at 20 psig and 25°C in ethyl acetate affords 5,6-dihydrobenzanthracene in 97% yield, whereas over platinum catalysts 8,9,10,11-tetrahydrobenzathracene is obtained in 95% yield.

Interesting differences among catalysts were uncovered by Mylroie and Stenberg (1973) in studies on the 9,10-disubstituted, 9,10-bridged anthracenes. Only ruthenium-on-carbon and palladium-on-carbon were found to be suitable for complete ring saturation without hydrogenolysis of the substitutents. Hydrogenation 9,10-hydroxymethyltriptycene (**8**) over ruthenium-on-carbon or palladium-on-carbon afforded high yields of isomeric octadecahydro-9,10-bis(hydroxymethyl)triptycenes (**9**). Ruthenium gave a product (mp 232°–234°C) that was believed to be either the cis,syn, cis,trans or the cis,syn,cis,anti,cis,syn isomer, whereas palladium gave a product (mp 310°–315°C) that was believed to be the trans,anti,trans,anti, trans isomer. Unless small amounts of methanesulfonic acid were added to the ruthenium-catalyzed reduction, the product contained substantial amounts of olefinic material. Hydrogenation over ruthenium dioxide caused extensive hydrogenolysis, affording the deoxy compound **7** in 90% yield. Hydrogenolysis over rhenium heptoxide–tetrahydropyran complex was more severe, and the main product was 1,2,3,4,5,6,7,8-octahydro-9,10-dimethylanthracene (**10**).

Further differences between ruthenium and palladium are revealed by the hydrogenation of anthracene bridged by a 9,10-norbornyl or 9,10-ethano (12) group. Ruthenium-on-carbon reduced only one ring (11) whereas palladium-on-carbon reduced both (13) (Mylroie and Stenberg, 1973).

| 11 | 12 | 13 |

Palladium-on-carbon was effective in converting 9-chloromethyltriptycene to the octadecahydro-9-chloromethyltriptycene if the solvent was 1 $N$ ethanol–hydrogen chloride; without hydrogen chloride present the halogen was lost during ring saturation.

As substitution on an aromatic ring increases, hydrogenation becomes more difficult; nonetheless, highly substituted rings can be reduced. Perhydrotriphenylene (15) can be obtained by catalytic hydrogenation of aromatic compounds having the same skeleton if attention is paid to the catalyst and reaction conditions. Dodecahydrotriphenylene (14) which can be derived from cyclohexanone, is a particularly good starting material. The best catalyst is palladium-on-carbon; Raney nickel is much less effective. The solvent is important; high yields are obtained in saturated hydrocarbons, whereas the reduction does not proceed in methanol or ethanol. The chief isomer formed is 15, which was shown to be derived from successive epimerizations of other stereoisomers. Functional substituents (16) are apt to be lost during the hydrogenation because of the vigorous reaction conditions required (Farina and Audisio, 1970).

| 14 | 15 | 16 |

## E.   Benzyl Compounds

Benzyl compounds carrying oxygen or nitrogen functions undergo hydrogenolysis readily over palladium, usually with little or no ring reduction

Hartung and Simonoff, 1953). This course of reduction is discussed fully in Chapter 15. Hydrogenation in the reverse sense, ring saturation without hydrogenolysis, is best accomplished over ruthenium (Arnold, 1951; Freifelder and Stone, 1958; Freifelder et al., 1964) or rhodium (Stocker, 1962, 1964; Wu et al., 1962; Galantay, 1963; Kalm, 1964) and to a lesser extent over platinum (Levine and Sedlecky, 1959; Nishimura, 1959; Grogan et al., 1964; Ichinohe and Ito, 1964). Small amounts of acetic acid are sometimes useful in promoting ring saturation, but larger amounts tend to favor hydrogenolysis (Nishimura, 1959; Stocker, 1962).

The hydrogenation of methylphenylcarbinol over various metal hydroxides illustrates the importance of the catalyst in determining the course of reduction (Table IV). Ruthenium proved to be much more effective than rhodium for the production of methylcyclohexylcarbinol; complete hydrogenolysis occurred over palladium (Taya et al., 1968). However, other workers (Stocker, 1962) using rhodium-on-alumina under mild conditions obtained much higher yields of the saturated carbinol.

**TABLE IV**

| Catalyst | % In product | | |
|----------|:---:|:---:|:---:|
|          | A  | B  | C |
| Palladium hydroxide  | 0   | ····100··· | |
| Rhodium hydroxide    | 53  | 47  | 0 |
| Ruthenium hydroxide  | 100 | 0   | 0 |

In general, ring saturation can be carried out under mild conditions over rhodium, whereas to obtain satisfactory rates with ruthenium more vigorous conditions are required. Hydrogenation of 7-carbomethoxyl-2,3-dihydro-4(1$H$)-quinolone proceeds smoothly over 5% ruthenium-on-carbon to give a 7-carbomethoxy-1-aza-4-decalol containing less than 4% of hydrogenolyzed benzyl oxygen (Hirsch and Schwartzkopf, 1974).

$$CH_3OC\text{-...}O\text{-quinolinone} \longrightarrow CH_3OC\text{-...}OH\text{-decahydroquinolinol}$$

## F. Stereochemistry

The major stereoisomer resulting from the hydrogenation of aromatics under mild conditions is usually the all-cis isomer (Burwell, 1957, 1969; McDonald and Reitz, 1970; Allinger et al., 1971; Balasubrahmanyam and Balasubramanian, 1973; Cargill et al., 1973) as if all the hydrogen had been added to one side of the molecule (Linstead et al., 1942a,b). Under vigorous conditions the fully hydrogenated products may be isomerized, and the resulting mixtures will not reflect the course of the original saturation (Farina and Audisio, 1970). Some trans isomers are usually produced even under the mildest conditions. Their formation is frequently accounted for by total or partial desorption and readsorption in a new orientation of some partially hydrogenated intermediate (Hartog and Zwietering, 1963; Siegel et al., 1962, 1963). Hydrogen is usually considered to add to the aromatic from the side adsorbed on the catalyst, but there is some evidence that topside addition of hydrogen may occur as well (Farina et al., 1976).

Some of the parameters influencing the stereochemistry of aromatic hydrogenation are illustrated by the hydrogenation of xylenes (Siegel and Dunkel, 1957; Schuetz and Caswell, 1962; Siegel et al., 1962, 1963; Hartog and Zweitering, 1963; Siegel and Ku, 1965). Practical interest in this reaction is limited, but it serves as a model by which to judge the effect that reaction variables might have on the stereochemistry of alkyl aromatic hydrogenation generally.

Rylander and Steele (1962a), working mainly with rhodium and ruthenium, examined the effect of metal, metal concentration, catalyst concentration, catalyst support, temperature, pressure, agitation, and solvent on the stereochemistry of xylene hydrogenation. They concluded that temperature was the most effective single way to influence stereochemistry aside from choice of metal. The trans content of dimethylcyclohexanes arising from p-xylene hydrogenation over rhodium increased from 14 to 44% over the range $15°–160°C$. Metal concentration had some influence on the results, with the percentage of more stable isomer for all xylenes increasing with metal concentration. Support had a minor influence on stereochemical results; the direction of change brought about by the support depended on the metal used, which made generalizations difficult to formulate. Pressure and solvent usually have relatively minor influences on stereochemistry. Pressure dependency is sufficient, however, to make it a powerful tool for the elucidation of mechanisms of saturation (Siegel, 1966; Siegel and Garti, 1977).

Good estimations of the stereochemical composition of cycloparaffins derived from hydrogenation of xylenes and other aromatics (Mahmoud and Greenlee, 1962) were obtained when it was assumed that the composition would be the same as that derived from a mixture of cycloalkenes obtained by random cis addition of hydrogen to the aromatic ring (Siegel, 1966). However, other workers in explaining very high cis hydrogenation of *o*-xylene over osmium and iridium assumed either selective formation of 1,2-dimethycyclohexene as the olefinic intermediate or only very slight desorption of 2,3-dimethylcyclohexene from the catalyst surface (Nishimura *et al.*, 1970).

The assumption that the major isomer is cis or cis,cis can be applied with reasonable assurance to most nonphenolic di or trisubstituted aromatics, especially if one isomer occurs in far greater amounts than the other. However, when the isomers are formed in nearly equal amounts, the greater amount is not safely assumed to consist of the cis isomer. A case in point is the hydrogenation of the disodium salt of 1,1,3-trimethyl-5-carboxy-3-(*p*-carboxyphenyl)indan (**17**) over 5% ruthenium-on-carbon in water. The product is 47% *cis*- and 53% *trans*-4'-[3-(1,3,3-trimethyl-5-carboxy-indanoyl)]cyclohexanecarboxylic acid (**18**) derived by exclusive hydrogenation of the disubstituted ring (Steitz, 1970).

## II. ANILINES

Anilines are readily hydrogenated to the saturated amine. The reduction may be accompanied by several side reactions including hydrogenolysis of the carbon–nitrogen bond, reductive hydrolysis, and reductive coupling. These major features of aniline hydrogenation are readily accounted for by a generally accepted mechanism that involves partially hydrogenated aromatic rings (Greenfield, 1964).

The olefinic intermediates can then undergo saturation with formation of cyclohexylamines or hydrogenolysis of an allylic or vinylic amino species, affording ammonia and a cycloalkane.

Isomerization of an olefinic intermediate can give an imine, hydrolysis of which leads to cyclohexanones or cyclohexanols.

Interaction of the imine with a cyclohexylamine gives an addition product, which can undergo hydrogenolysis or elimination of ammonia followed by hydrogenation to afford dicyclohexylamines.

The multiplicity of steps required for dicyclohexylamine formation precludes an *a priori* ordering of catalysts, but experimentally the sequence for increasing coupling, Ru < Rh ≪ Pd ≳ Pt, is the same as the order for increasing tendency toward hydrogenolysis (Rylander *et al.*, 1973).

## A. Catalysts

A variety of catalysts have been used successfully for the hydrogenation of anilines, including supported and unsupported cobalt, nickel, palladium, platinum, rhodium, ruthenium, and iridium (Rylander and Steele, 1962b). Base metals require high temperatures and pressures (Winans, 1940), whereas some noble metals can be used under ambient conditions, albeit inefficiently. In recent years, rhodium, at lower pressures, and ruthenium, at higher pressures, have become preferred catalysts in laboratory and industrial practice, for both have high activity and little tendency toward hydrogenolysis and reductive coupling.

Several groups of workers have examined the use of mixed-metal catalysts, either as fused oxide alloys (Nishimura and Taguchi, 1963b; Rylander *et al.*, 1967a,b) or coprecipitated on supports (Ikedate and Suzuki, 1971) or as physical mixtures of separate catalysts (Rylander *et al.*, 1973). Efforts to correlate performance with electronic structure in alloyed catalysts have not been very successful (Ikedate and Suzuki, 1971), which is not surprising inasmuch as significant changes in both activity and selectivity can be achieved by merely mixing two separate catalysts (Rylander et al., 1973).

An interesting difference among catalysts has been noted with regard to the composition of coupled products. Palladium-on-carbon catalysts, especially when prepared from palladium nitrate, give high yields of N-phenylcyclohexylamine in hydrogenations of aniline in acetic acid. This amine is also formed over platinum, but to a lesser extent, whereas none is obtained over rhodium (Ikedate et al., 1968; Ikedate and Suzuki, 1969).

## B.  Supports

The most efficient use of a metal is made if it is supported; in addition, the support has some influence on the products of reduction, although it is usually secondary to the larger influence of the metal. The percentage of dicyclohexylamine formed in the hydrogenation of aniline without solvent over supported rhodium decreased in the order carbon (30%), barium carbonate (20%), alumina (18%), barium sulfate (14%), calcium carbonate (9%). Of some interest is the finding that the amount of coupling over rhodium-on-alumina can be increased substantially by the addition of carbon to the alumina catalyst. Carbon is often used in catalytic hydrogenation as a sop for poisons. It is usually considered to have no other influence on the reaction, but evidently this assumption is sometimes incorrect (Rylander et al., 1973). Coupled products increase with increasing reaction temperature (Greenfield, 1973).

## C.  Additives

Hydrogenation of anilines is at times markedly influenced by the presence of various additives, such as sodium carbonate (Kalina and Pasek, 1969), alkali and alkaline earth hydroxides, ammonia, and amines. Ammonia is sometimes added to suppress diamine formation, but it does so at the expense of rate. If aqueous ammonia is used, the rate is better, but oxygenated compounds derived by hydrolysis are formed (Maxted, 1951; Greenfield, 1964).

Lithium hydroxide is an especially effective promoter in the hydrogenation of anilines over ruthenium hydroxide catalysts and in some instances rhodium catalysts. The additive both accelerates the rate of reduction and suppresses dicyclohexylamine formation (Nishimura et al., 1971b). The increase in rate is probably due to elimination of the inhibiting effect of ammonia through preventing or weakening its adsorption on the catalyst. Its action in minimizing dicyclohexylamine formation is probably similarly related to decreasing the adsorption of either cyclohexylamine or intermediate addition products (Nishimura *et al.*, 1966). An effective level of lithium hydroxide hydrate is 30–100 mg per 65 mg ruthenium hydroxide catalyst; with 5% ruthenium-on-carbon effective levels of lithium hydroxide hydrate are between 120 and 360 mg per 360 mg of catalyst. The upper limits of the promoter are not necessarily the maximum that can be used; they are merely the maximum reported. Lithium hydroxide was found to be more effective than sodium, potassium, barium, and calcium hydroxides.

The inhibiting effects of ammonia and amines, such as methylamine and dimethylamine, formed by hydrogenolysis can also be reduced by the technique of venting and repressuring the reactor (Bauer, 1968). The technique has general applicability and is often used successfully to force sluggish reactions to completion.

## D.  Solvents

Solvents may have a marked influence on both the rate of reduction and product composition. Isopropanol and *tert*-butanol are excellent solvents for the reduction of anilines to cyclohexylamines; the lower alcohols methanol and ethanol give substantially more coupled products. In the reduction of aniline over ruthenium oxide, 11.9 and 6.9% dicyclohexylamine was formed in methanol and ethanol, respectively, whereas in isopropanol or *tert*-butanol the amount of coupled product was only 0.3%. The rate of reduction is also much slower in the lower alcohols, probably due to inhibition by ammonia formed in the coupling reaction (Nishimura *et al.*, 1966). The lower alcohols also tend to produce substantially more hydrogenolysis in sensitive compounds, such as alkoxyanilines (Nishimura *et al.*, 1968) and *N,N*-dimethyl-*p*-phenylenediamine (Nishimura *et al.*, 1971b), than do isopropanol and *tert*-butanol. Solvents of low dielectric constant, such as ethers or alkanes, are useful in minimizing hydrogenolysis.

Aqueous solvents, particularly acidic aqueous solvents, promote reductive hydrolysis (Kuhn and Haas, 1958; Rylander and Steele, 1963a). Reductions of anilines over ruthenium-on-alumina in aqueous ethanol are several times faster than in anhydrous ethanol (Freidlin *et al.*, 1976). Carboxylic acid solvents permit good rates of reduction by neutralizing the inhibiting effect of basic nitrogen on catalytic activity, but they tend to increase the amount of coupled products (Nishimura and Taguchi, 1963b).

## E.  Synthetic Applications

Reduction of an aromatic amine to the saturated cyclic amine provides an excellent route to this type of compound, and many such reductions have been carried out successfully, especially over rhodium and ruthenium (Barkdoll *et al.*, 1953; Freifelder and Stone, 1962; Freifelder *et al.*, 1965; Nishimura *et al.*, 1966; Freidlin *et al.*, 1973). Diamines, such as bis(4-aminophenyl)methane, can be reduced smoothly to the corresponding cyclohexyl compounds despite the inherent possibility of the catalysts being deactivated by extended coupling reactions. High yields of bis(4-aminocyclohexyl)methane were obtained by hydrogenation over ruthenium dioxide in dioxane or in anhydrous ammonia or without solvent at 110°C and 2000–3000 psig (Barkdoll *et al.*, 1953). This hydrogenation has also been carried out quantitatively over 5% rhodium-on-alumina at 117°C and 100 psig. The catalyst is regenerated easily by two treatments with 1 part concentrated ammonia and 3 parts water at 60°–75°C, followed by drying at 90°C (Farrissey and Frulla, 1970), or by washing with glacial or aqueous acetic acid (British Patent 1,222,037 Feb. 10, 1971). The hydrogenation can also be carried out over cobalt oxide–sodium carbonate-on-pumice (French Patent 1,568,539 May 23, 1969) or over 25% cobalt–2% manganese-on-pumice (Grosskinsky and Merkel, 1972), but pressures of the order of 250 atm are required.

Phenylenediamines are best reduced to the corresponding cyclohexyldiamines over ruthenium-on-alumina in ethanol. The 1,3-diamine was obtained in 91% yield, and the 1,4-diamine in 88% yield. The percentage of cis isomer varied between only 70 and 84% in a number of solvents and over a range of temperatures and pressures (Litvin *et al.*, 1973).

Anilines containing carbonyl functions may give rise to bi- and tricyclic compounds through interaction of these functions once ring rigidity has been lost by hydrogenation (Augustine and Pierson, 1969b). Witte and Boekelheide (1972), as part of a total synthesis of ibogamine (Snieckus *et al.*, 1972), prepared the isoquinuclidone portion by selective hydrogenation of a trisubstituted aniline (**19**) over ruthenium oxide in isopropanol at 150°C and 2000 psig. Oxidation of **20** with dipyridine–chromium(VI) oxide afforded **21** in 54% overall yield. The formation of a single isomer in this type of cyclization is favored by lower reaction temperatures (Augustine and Vag, 1975).

19                                    20                                    21

A similar reduction of the substituted aniline **22** over ruthenium-on-carbon in ethanol at 160°C and 2000 psig afforded an epimeric alcohol mixture of the tricyclic compound **23**. The authors (Augustine and Pierson, 1969a) suggest that formation of an epimeric alcohol mixture indicates that reduction of the ketone preceded cyclization; cyclization before carbonyl reduction should give a single isomer. These examples emphasize the generality that ruthenium and rhodium are preferred catalysts for ring saturation without hydrogenolysis of an aromatic ketone.

|   22   |   23   |
|--------|--------|
|        | (81%)  |

Attempted cyclization of **24** by hydrogenation over ruthenium failed to give the isoquinuclidone skeleton but gave instead **25** as the major product. The authors (Witte and Boekelheide, 1972) concluded that spontaneous ring closure does not occur with amides but requires a free, basic amino group.

|   24   |   25   |
|--------|--------|
|        | (42%)  |

An unusual reductive cyclization was reported by Augustine and Vag (1975). Hydrogenation of **26** did not yield the expected cyclized lactam, but rather **27** and **28**, compounds in which the amino function, surprisingly, was lost. The loss of nitrogen was contrary to previous reports on the competitive loss of aromatic nitrogen and oxygen functions (Freifelder and Stone, 1962; Nishimura and Yoshino, 1969). Lability of an amino function was also observed in the hydrogenation of 3,4-diaminobenzoic acid, which produced a mixture of bicyclic lactams lacking an amino substitutent (Augustine and Vag, 1975).

**26**                                    **27**            **28**
                                          (15%)           (65%)

Hydrogenation of anilines in dilute aqueous acids gives products derived by hydrolysis (Baer and Kienzle, 1969). Hydrolytic cleavage is facilitated by substitution on the nitrogen atom. The reduction is believed to go through an imine type of intermediate (Kuhn and Haas, 1958).

Reductive hydrolysis of anilines carrying chiral substituents gives chiral cyclohexanones of up to 30% optical purity (Kuhn et al., 1968).

## III.   PHENOLS AND PHENYL ETHERS

Phenols and phenyl ethers are hydrogenated readily. The reduction can proceed in several ways to afford products derived by full or partial hydrogenation of the aromatic ring and with or without loss of the oxygen function. With attention to choice of catalyst and conditions, good yields of the various possible products usually can be obtained.

## A.   Hydrogenolysis of Phenols

Catalytic hydrogenolysis of phenols without ring reduction does not occur cleanly, but the reaction acquires synthetic utility by prior conversion of the phenol to a suitable derivative, such as ethers (**30**), made by interaction of the phenol with 2-chlorobenzoxazole (**29**) or 1-phenyl-5-chlorotetrazole (Musliner and Gates, 1966; Sakai *et al.*, 1969; Barfknecht *et al.*, 1970).

The course of reduction is solvent dependent; the yield of dimethylindane (**31**) dropped to 40% when acetic acid was the solvent. Urethanes derived from phenyl isocyanate can be used similarly (Eisenbraun *et al.*, 1973; Weaver *et al.*, 1973). Phenolic hydroxyl can also be easily displaced by conversion to 2,4,6-tris(aryloxy)-*s*-triazines and hydrogenolysis over palladium. The rate of hydrogenolysis is not influenced by the electronic character of the substituents, but it is sensitive to steric factors. The reaction is catalyzed by amines and by adsorbents for cyanuric acid, such as carbon, alumina, or silica. Amines, such as triethylamine, are thought to function by solubilizing cyanuric acid and aiding its transport from the catalyst to other adsorbents. Yields are generally excellent (van Muijlwilk *et al.*, 1974).

## B.   Conversion of Phenols to Cyclohexanones

Phenols can be converted directly to the corresponding cyclohexanones in excellent yield either in liquid or in vapor-phase hydrogenation (British Patent 890,095 Feb. 8, 1962).

Palladium is the preferred catalyst for simple phenols. In a comparison of metals, all isomeric cresols were converted to the corresponding methyl-cyclohexanones in yields ranging from 52 to 77% over rhodium or ruthenium, whereas quantitative yields were obtained over palladium (Matsumoto *et al.*, 1969). The reason for the exceptional effectiveness of palladium is still obscure. An attractive explanation based on random or nearly random reduction of the ring to afford dihydro and tetrahydro derivatives followed by isomerization of the resulting double bond has been ruled out on the grounds that hydrogenolysis products, which would result from intermediate allyl-type alcohols, have not been found (Takagi *et al.*, 1970b). Alternatively, it has been suggested that in cresols the keto rather than the enol form of the cresol is reduced or that the cresol is reduced to give only particular dihydro intermediates.

A variety of palladium catalysts have been used with success in the partial hydrogenation of phenols. These include palladium-on-carbon (Kaye and Matthews, 1963; Little and Cheena, 1971), palladium chloride-on-carbon (Hart and Cassis, 1951), palladium hydroxide-on-barium sulfate (Kuhn and Haas, 1958), and palladium-on-calcium carbonate (Oparina *et al.*, 1973). Many of the catalyst variations feature adjustments of alkalinity (Duggan *et al.*, 1963; Smeykal *et al.*, 1969; Areshidz *et al.*, 1971; Thelen *et al.*, 1972). Selectivity in hydrogenation of phenol over palladium-on-alumina is said to depend on the thickness of the active layer of palladium (Lyubarskii *et al.*, 1971, 1973). Surface-coated palladium on large-pore carriers has been used to minimize overhydrogenation (Baltz *et al.*, 1971). The large number of different palladium catalysts that have been used in the industrially impor-tant hydrogenation of phenols need not cause confusion as to choice of catalyst; a palladium-on-carbon or palladium-on-alumina powder will prob-ably give quite satisfactory results for most synthetic applications.

Other catalysts may prove to be better than palladium in the partial hydrogenation of dihydric aromatics. The yield of dihydroresorcinol from the hydrogenation of resorcinol in sodium hydroxide over 5% rhodium-on-carbon is 87% (Smith and Stump, 1961), whereas over palladium-on-carbon the yield is only 50–60% (Esch and Schaeffer, 1960). The course of this reduction is very solvent dependent; no trace of dihydroresorcinol was found in rhodium-catalyzed reductions in ethanol (Rylander and Himelstein, 1964). High yields of dihydroresorcinols can also be obtained over Raney nickel in alkaline solution at high pressures (50–100 atm) (Thompson, 1947; Grob and Kieger, 1965; Teuber *et al.*, 1966). Pyrogallol, on partial reduction over nickel, affords dihydropyrogallol (Pecherer *et al.*, 1948). 3,5-Dihydroxy-phenylacetic acid was reduced to the corresponding dione in 77% yield over rhodium-on-alumina in base. No effort was made to limit hydrogen uptake (Mokotoff and Cavestri, 1974).

21.5 gm

Wenkert *et al.* (1968), after examining the hydrogenation of various pyridines, concluded that any aromatic nucleus capable of unmasking a stable vinylogous amide unit during reduction is prone to stoppage at an intermediate stage. Conversion of the aminophenol **32** over Raney nickel to the vinylogous amide **33** illustrates this point (Valenta *et al.*, 1964).

## C. Ring Saturation with Hydrogenolysis

Achieving ring reduction with accompanying hydrogenolysis of a hydroxy or alkoxy substituent depends very much on the substrate structure, catalyst, and conditions. Extensive hydrogenolysis may occur especially over platinum (Edson, 1934; Ruggli *et al.*, 1941; Tomita and Uyeo, 1942; Gauthier, 1945; Woodward and Doering, 1945; Price *et al.*, 1947; Johnson *et al.*, 1956) and, limited data suggest, over iridium (Rylander and Steele, 1965a,b; Takagi, 1970; Takagi *et al.*, 1970a; Nishimura *et al.*, 1971a). Either of these metals would appear to offer the best chance of achieving ring reduction with accompanying hydrogenolysis, and conversely they are best avoided when hydrogenolysis is not desired. Hydrogenolysis, as well as the rate of ring reduction, is enhanced by acids, such as acetic, perchloric, sulfuric, and hydrochloric acids (Heckel and Adams, 1925; Karrer and Kehl, 1930; Ferber and Brückner, 1939; Linstead *et al.*, 1942b; Billman and Buehler. 1953; Rylander and Steele, 1965a,b).

## D. Ring Saturation without Hydrogenolysis

Reduction of phenols and phenyl ethers to the corresponding cyclo-hexanols and cyclohexyl ethers can usually be accomplished readily. Compounds vary widely in their susceptibility to hydrogenolysis, and the optimal catalyst may depend on the substrate structure. In general, preferred catalysts are ruthenium and rhodium and to a lesser extent palladium, but exceptions exist (Section H). Hydrogenolysis during ring reduction is decreased with increased pressure and decreased temperature and by the use of solvents with low dielectric constant (Nishimura and Yoshino, 1969).

## 1. Palladium

Before the advent of ruthenium and rhodium as catalysts for hydrogenation, palladium was the preferred catalyst for reduction of phenols and phenyl ethers without hydrogenolysis. The use of palladium, with rare exceptions (Schrecker and Hartwell, 1953), requires elevated temperatures and pressures for satisfactory rates; in addition, the use of elevated pressure may diminish hydrogenolysis markedly (Levin and Pendergrass, 1947). Some examples of the successful use of palladium are the hydrogenation of di(4-hydroxyphenyl)methane, 2% Pd-on-SrCo$_3$, 165°C, 1600 psig (Novello and Christy, 1951); of 4-tert-butylphenol, 5% Pd-on-BaCO$_3$, 175°C, 1800 psig (Sommerville and Theimer, 1960); and of pyrogallol, 2% Pd-on-SrCO$_3$, 80°C, 2400 psig (Christian et al., 1951).

Palladium-on-carbon in methanol is especially useful for the selective hydrogenation of p-phenylphenol to p-cyclohexylphenol, 90% yield at 50°C and 500 psig, after 3 moles of hydrogen are absorbed. Continued hydrogenation results in 84% yield of 4-dicyclohexyl methyl ether (cis–trans ratio 1.2). In ethanol the ethyl ether is similarly formed, but in isopropanol the final product is 4-cyclohexylcyclohexanol with no trace of ether (Rylander and Vaflor, 1974). Ether formation in the hydrogenation of phenols in methanol or ethanol is expected, with the yield of ether depending markedly on the catalyst (Nishimura et al., 1967). Hydrogenation of p-phenylphenol over platinum, rhodium, ruthenium, or iridium-on-carbon, in sharp contrast to palladium, gives complex mixtures of products after 3 moles of hydrogen are absorbed (Rylander and Vaflor, 1974).

Surprisingly, palladium reduced one ring of o,o'-biphenol preferentially to afford o-(2-hydroxycyclohexyl)phenol in 76% yield. Hydrogenation over rhodium does not lead to an accumulation of this intermediate but instead affords perhydrodibenzofuran in 85% yield, presumably through cyclization and hydrogenolysis of o-(2-hydroxycyclohexyl)cyclohexanone. Perhydrodibenzofuran can be obtained quantitatively by hydrogenation of dibenzofuran over rhodium (Rylander and Steele, 1967).

## 2.  Rhodium

Supported rhodium, rhodium black, rhodium oxide, and rhodium hydroxide are useful catalysts for the hydrogenation of phenols and phenyl ethers. Rhodium catalysts saturate these compounds, usually with little hydrogenolysis under ambient conditions, but better efficiency of the catalyst is obtained if more vigorous conditions are used. These catalysts are of special advantage when used with sensitive compounds such as dihydroxy- and dialkoxybenzenes, phloroglucinol, pyrogallol (Gilman and Cohn, 1957; Smith and Thompson, 1957; Smith and Stump, 1961), methoxyphenols, guaiacol (Eliel and Brett, 1963), gallic acid (Burgstahler and Bithos, 1960) and 2,6-dimethylhydroquinone (Stolow et al., 1972), all of which undergo appreciable hydrogenolysis when reduced with catalysts such as platinum oxide. Extensive hydrogenolysis occurs in the hydrogenation of ethyl p-hydroxyphenylacetate over platinum catalysts, but, over 5% rhodium-on-alumina in ethanol at 3000 psig and 40°C, ethyl 4-hydroxycyclohexylacetate is obtained in 83% yield (Whitehead et al., 1961). Essentially quantitative yields of hydroxycyclohexane-4-carboxylic acid are obtained in the hydrogenation of p-hydroxybenzoic acid over 5% rhodium-on-alumina at 1700 psig and 25°C in acetic acid (Giudici and Bruice, 1970). Other catalysts give yields as low as 25%. Benzofurans are reduced to the perhydro derivatives over rhodium-on-carbon (Rylander and Steele, 1967) or rhodium-on-alumina (Cantor and Tarbell, 1964) without hydrogenolysis. Hydrogenation of 2,2-dimethylbenzofuran over 5% rhodium-on-alumina in acetic acid at 60 psig affords stereospecifically cis-2,2-dimethyloctahydrobenzofuran in 94% yield (Meyers and Baburao, 1964).

Rhodium hydroxide gives excellent results in ring saturation of the hydrogenolysis-sensitive compounds hydroquinone dimethyl ether and phenyl ether (Takagi et al., 1965). Rhodium black is also very effective in minimizing hydrogenolysis during ring saturation of dimethoxybenzenes, more so than all other noble metals (Takagi *et al.*, 1970b). Over all catalysts, hydrogenolysis of the meta isomer is the most extensive. This has been attributed to stabilization, by the meta electron-releasing group, of the carbonium ion type of intermediate formed on the catalyst when allyl type of ethers undergo hydrogenolysis.

The same explanation was advanced to account for the meta isomer undergoing the most hydrogenolysis in methoxyanilines (Nishimura *et al.*, 1968). In support of this suggestion is the observation that the meta isomer of dimethoxybenzenes or methoxyanilines does not undergo the most extensive hydrogenolysis when the reductions are carried out in the presence of 1 mole of acetic acid.

Alkoxyanilines undergo hydrogenolysis readily with most catalysts, but, over rhodium oxide prepared by fusion of rhodium chloride with lithium nitrate, alkoxycyclohexylamines are obtained in excellent yield at 80°C and 1500 psig. Hydrogenolysis amounts to only 3–6%, and coupling reactions to form diamines are almost completely suppressed. Rhodium oxides prepared by fusion with sodium nitrate or potassium nitrate are not nearly as effective (Nishimura *et al.*, 1968).

### 3.   Ruthenium

Ruthenium oxide, ruthenium hydroxide, ruthenium-on-carbon, and ruthenium-on-alumina have given excellent results in ring saturation of phenols and phenyl ether at elevated temperatures and pressures (Johnwon *et al.*, 1956; Walton *et al.*, 1956; Rodig and Ellis, 1961; Ireland and Schiess, 1963; Takagi *et al.*, 1965; Karakhanov *et al.*, 1969; Wharton *et al.*, 1972). Hydrogenolysis is usually kept to low levels, even with compounds in which it occurs extensively over other catalysts. Apparently, over ruthenium, hydrogenolysis of benzyl hydroxyl occurs more readily than cleavage of the phenyl–oxygen bond (Frank, 1949; Freifelder *et al.*, 1965); rhodium may be more appropriate for the hydrogenation of compounds containing both functions if hydrogenolysis of the benzylic oxygen over ruthenium proves to be excessive.

Ruthenium hydroxide is an effective catalyst for the hydrogenation of phenyl ether and 1,4-dimethoxybenzene (Takagi *et al.*, 1965). Addition of water to these systems markedly increases the rate, as has been observed with other compounds and with other ruthenium catalysts (Rylander *et al.*,

1963), but the yield of expected products is diminished due to hydrolysis of intermediates (Takagi *et al.*, 1965). Quantitative yields of methylcyclohexanols are obtained in the reduction of cresols over ruthenium hydroxide without solvent at 80°C and 1500 psig (Takagi, 1970).

### 4.  Nickel

Satisfactory results have been reported in the complete reduction of phenols to the saturated derivatives over nickel. Hydrogenations are carried out frequently in the presence of added alkali, but, even if not added deliberately, alkali is usually present in the catalyst. Alkali suppresses formation of atomic hydrogen and favors formation of hydride ion; in the hydrogenation of 2-naphthol over nickel this condition favors reduction of the phenolic ring (Krause, 1972). High yields of cyclohexanols are obtained in the reduction of alkylphenols over Raney nickel with (Ungnade and Nightingale, 1944) or without (Banks *et al.*, 1956) added alkali. Phenoxide ion has a marked accelerating effect on the rate, whereas water has a depressing effect. The acceleration in rate is not observed when aqueous alkali is added (Ungnade and Nightingale, 1944). Fair yields (ca. 80%) of ethyl 2-, 3-, and 4-hydroxycylohexanecarboxylates are obtained from the corresponding phenols over Raney nickel at 4000 psig and 200°C (Ungnade and Morriss, 1948).

### 5.  Binary Catalysts

Platinum–rhodium oxide catalysts prepared by fusion of a mixture of platinum and rhodium salts with sodium nitrate display marked synergistic effects, possess an unusual resistance to poisoning, and frequently afford a product composition different from that obtained with either rhodium oxide or platinum oxide alone (Rylander *et al.*, 1967a,b). These catalysts, which were pioneered by Nishimura (1960), have been recommended for use when hydrogenolysis of carbon–oxygen bonds is to be avoided (Nishimura and Taguchi, 1963a). Catalytic characteristics depend on composition, as illustrated by the data of Table V (Rylander and Kilroy, 1968) on the hydrogenation of diphenyl ether, a compound that undergoes facile hydrogenolysis.

The maximal yield of dicyclohexyl ether occurs with 70% Rh–30% Pt, the composition usually recommended by Nishimura and co-workers for minimal hydrogenolysis. A much higher yield of dicyclohexyl ether is obtained with this catalyst at elevated pressures (Nishimura and Taguchi, 1963a). Maximal yield is not obtained with the same catalyst composition as maximal activity. Maximal activity is obtained with about 40% Pt–60% Rh. This activity, part of which is attributable to higher surface area,

TABLE V

**Hydrogenation of Diphenyl Ether over Platinum–Rhodium Oxide**[a]

| Platinum–rhodium oxide | Yield of dicyclohexyl ether (%) |
|---|---|
| 100% Pt | 0 |
| 95% Pt–5% Rh | 6 |
| 75% Pt–25% Rh | 25 |
| 50% Pt–50% Rh | 42 |
| 30% Pt–70% Rh | 48 |
| 5% Pt–95% Rh | 25 |
| 100% Rh | 20 |

[a] Conditions, ambient; solvent, cyclohexane.

is four to five times greater than that shown by either rhodium oxide or platinum oxide.

## E.  Solvents

Solvent, as well as catalyst, may have an important influence on selectivity in phenol hydrogenation, as illustrated by the reduction of 2-naphthol (**34**), to 1,2,3,4-tetrahydro-2-naphthol (**35**) and 5,6,7,8-tetrahydro-2-naphthol (**36**) (Kajitani et al., 1975). The catalysts examined were Urushibara (Taira, 1961) and Raney (Adkins and Billica, 1948) nickel and cobalt.

Table VI illustrates the effect of solvent and catalyst on selectivity and their interrelation. In general, hydrogenation of the phenolic ring occurs preferentially over Raney nickel, and that of the benzene ring over Urushibara nickel B (UNi B.) Selectivity may change markedly with solvent, but some catalysts are much more sensitive than others to the effect of solvent (Kajitana et al., 1975).

Earlier workers (Stork, 1947; Adkins and Kusek, 1948) noted that addition of amines promotes selective hydrogenation of the benzene ring in 2-naphthol. The effect is complex, however, and the change in selectivity on promotion by amines depends on both the amine structure and the catalyst, as illustrated by the data of Table VII (Kajitani et al., 1975).

**TABLE VI**

Selectivity in Hydrogenation of 2-Naphthol in Various Solvents[a]

| | Selectivity over various catalysts[b] | | | | |
|---|---|---|---|---|---|
| Solvent | UNi B | UNi A | UCo B | RaNi | RaCo |
| Methanol | 22 | 28 | 56 | 65 | — |
| Ethanol | 34 | 35 | 55 | 67 | 52 |
| Isopropanol | 47 | 44 | 50 | 81 | — |
| Ether | 45 | 62 | 55 | 81 | 54 |
| Butyl ether | 54 | 65 | — | — | — |

[a] Data of Kajitani *et al.* (1975). Used with permission.
[b] Selectivity is equal to

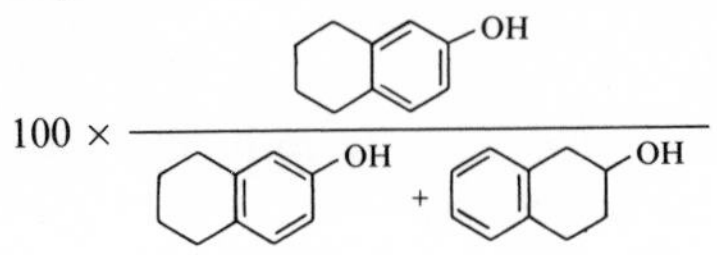

**TABLE VII**

Selectivity in Hydrogenation of 2-Naphthol Promoted by Various Amines[a,b]

| | Selectivity over various catalysts[c] | | | |
|---|---|---|---|---|
| Amine | UNi B | UCo B | RaNi | RaCo |
| Triethylamine | 33 | 57 | 33 | 55 |
| N-Ethylmorpholine | 40 | 59 | 46 | 52 |
| Diethylamine | 41 | 62 | 34 | 51 |
| Isopropylamine | 54 | 63 | 57 | 59 |
| n-Butylamine | 70 | 65 | 64 | 62 |

[a] Data of Kajitiani *et al.* (1975). Used with permission.
[b] Solvent, ethanol.
[c] Selectivity is equal to

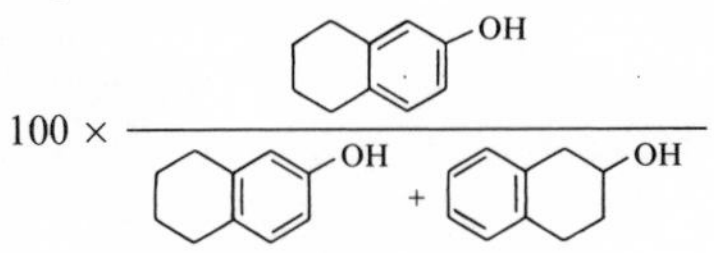

**TABLE VIII**

**Hydrogenation of Diphenyl Ether**

| Catalyst | Solvent | Pressure (atm) | Yield of dicyclohexyl ether (%) |
|---|---|---|---|
| 5% Rh-on-C | Ethanol | 68 | 50 |
| 5% Rh-on-C | Ethanol | 1 | 21 |
| 5% Pt-on-C | Ethanol | 68 | 40 |
| 5% Pt-on-C | Ethanol | 1 | 14 |
| 70% Rh–30% Pt oxide | Acetic acid | 100 | 71[a] |
| 70% Rh–30% Pt oxide | Cyclohexane | 1 | 48 |
| 5% Pd-on-C | Ethanol | 68 | 90 |

[a] Reported by Nishimura and Taguchi (1963a); see text.

The extent of hydrogenolysis is also influenced by the solvent, generally increasing with increasing polarity. In the hydrogenation of *p*-methoxy-aniline over ruthenium hydroxide the percentage of hydrogenolysis fell with solvent in the order methanol (35%), ethanol (30%), propanol (26%), butanol (22%), isopropanol (16%), isopropyl ether (16%), methylcyclohexane (13%), tetrahydrofuran (11%), and *tert*-butanol (8%). Interestingly, *tert*-butanol gives the least hydrogenolysis, less than the less polar, nonalcoholic solvents isopropyl ether and methylcyclohexane (Nishimura and Yoshino, 1969).

## F. Pressure

In general, hydrogenolysis is diminished by elevated pressure. The effect is illustrated by the data of Table VIII on the hydrogenation of diphenyl ether (Rylander and Kilroy, 1968), a compound that undergoes hydro-genolysis easily. The 71% yield reported by Nishimura and Taguchi (1963a) (see Table VIII) would have undoubtedly been higher in a nonacidic solvent. Palladium at elevated pressure is especially effective for conversion of diphenyl ether to dicyclohexyl ether.

## G. Stereochemistry

Various authors have assumed that hydrogenation of simple aromatics, such as xylenes (Siegel *et al.*, 1962), cresols (Takagi *et al.*, 1967, 1968), and alkyl phenyl ethers (Nishimura *et al.*, 1972), proceeds by formation of dihydro intermediates, the probability of formation being the same for each irrespective of steric considerations. The assumption is supported by kinetic and stereochemical measurements and leads to conclusions that are often in agreement with experimental results. However, exceptions exist (Takagi *et al.*, 1970a), notably in the palladium-catalyzed hydrogenation of hydroxy aromatics.

Cyclohexanols are formed from phenols by two independent pathways represented as follows:

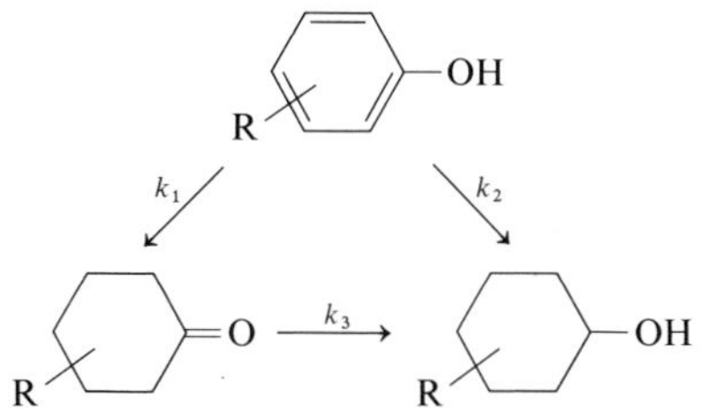

The stereochemical results depend on how much of the reaction proceeds through the cyclohexanone ($k_1$) and how much goes directly to the alcohol ($k_2$) as well as on the extent to which the adsorption–desorption equilibrium of the ketone and other dihydro and tetrahydro intermediates is established. The stereochemistry of hydrogenation of phenols, phenyl ethers, and phenyl acetates varies with solvent, temperature, and catalyst as well as substrate structure (Rylander and Steele, 1963b; Zymalkowsky and Strippel, 1964; Schweers, 1969; Takagi, 1967; Ruzicka *et al.*, 1970). Table IX gives experimental results on the stereochemical consequences of hydrogenation of

TABLE IX

**Stereochemistry of Cresol and Methylcyclohexanone Hydrogenation**[a,b]

| Isomer | Catalyst | % Cis isomer in methylcyclohexanol | |
| --- | --- | --- | --- |
| | | From cresol | From methylcyclohexanone |
| Ortho | Ru | 60 | 58 |
| | Pd | 54 | 54 |
| | Rh | 78 | 64 |
| Meta | Ru | 56.5 | 51 |
| | Pd | 62 | 62 |
| | Rh | 64 | 49 |
| Para | Ru | 59.5 | 53 |
| | Pd | 46 | 45 |
| | Rh | 73 | 70 |

[a] Data of Takagi (1970). Used with permission.
[b] Conditions: Ru and Rh hydroxides, 5% palladium-on-carbon, 80°–120°C, 100 atm.

isomeric cresols and the corresponding methylcyclohexanones (Takagi *et al.*, 1970a).

The data are indicative of the magnitude of the change that can be expected from variations in catalysts. In this series, rhodium is the preferred catalyst for maximizing the percentage of cis isomer. The differences in results from the hydrogenation of cresols and the corresponding methylcyclohexanones make it possible to calculate the percentage of ketone that is intermediate in cresol hydrogenation. In another study, the generalization was made that the percentage of the most stable epimeric methylcyclohexanol derived from cresol hydrogenation increased with catalyst in the order rhodium-on-carbon, ruthenium-on-carbon, and platinum oxide (Rylander and Steele, 1963b).

Temperature has some effect on stereochemistry. Over rhodium-on-carbon in the range 30°–100°C, the cis isomer content of the methylcyclohexanol from *p*-cresol decreased from 61 to 49%, and that from *m*-cresol decreased from 79 to 62% (Rylander and Steele, 1963b). Similar temperature effects were reported by Takagi (1970). Over a wide range, pressure has relatively little effect on stereochemistry (Takagi, 1970).

Solvent changes the stereochemistry of reduction of phenols, but the solvent effect is not independent of the catalyst used. The data of Table X comparing hydrogenation of *p*-cresol over rhodium and ruthenium hydroxides reveal that reductions over rhodium are much more solvent

**TABLE X**

**Hydrogenation of *p*-Cresol over Rhodium and Ruthenium Hydroxides[a]**

|  | Cis–trans ratio in methylcyclohexanol | |
|---|---|---|
| Solvent | Rhodium[b] | Ruthenium |
| *n*-Hexane | 3.1 | 1.53 |
| Cyclohexane | 2.7 | 1.47 |
| Ethyl acetate | 2.7 | 1.50 |
| Tetrahydrofuran | 2.3 | — |
| Cyclohexanol | 2.1 | 1.53 |
| *tert*-Butanol | 2.7 | 1.40 |
| Isopropanol | 2.2 | 1.48 |
| Ethanol | 2.1 | 1.56 |
| Methanol | 1.7 | 1.60 |
| Water | 1.1 | 1.45 |

[a] Data of Takagi (1970). Used with permission.
[b] Approximate values for Rh; number abstracted from graph.

sensitive than are those over ruthenium. Because the data are insufficient, it is not possible to generalize beyond noting that solvent can cause very significant changes in composition. Over rhodium the cis–trans ratio tends to decrease with increasing dielectric constant of the solvent (Takagi, 1970).

## H.   Some Mechanistic Considerations

A careful examination of the hydrogenation of ethyl p-tolyl ether revealed some interesting orderings of noble metal catalysts with respect to the formation of various products (Nishimura et al., 1972). The major products of reduction are ethyl 4-methylcyclohexyl ether and methylcyclohexane together with smaller amounts of 4-methylcyclohexanone, its diethyl acetal, and 4-methylcyclohexanol. The effectiveness of catalysts in the formation

$$H_3C-\langle\bigcirc\rangle-OC_2H_5 \xrightarrow[H_2]{C_2H_5OH} H_3C-\langle\bigcirc\rangle-OC_2H_5 + H_3C-\langle\bigcirc\rangle +$$

$$H_3C-\langle\bigcirc\rangle\genfrac{}{}{0pt}{}{OC_2H_5}{OC_2H_5} + H_3C-\langle\bigcirc\rangle-OH$$

of hydrogenolysis products increased in the order Pd < Ru ≪ Rh < Ir < Pt. Clearly, palladium and ruthenium are preferred catalysts when hydrogenolysis is to be avoided. The reason for the unexpectedly small amount of hydrogenolysis occurring over palladium has been stated to be the lack of formation of allyl and homoallyl ether intermediates and to the formation mainly of the enol ether intermediate (Nishimura et al., 1972). An initial dihydro intermediate is assumed which undergoes selective hydrogenation to the enol ether, a result expected from analogous examples (Bell et al., 1967). The latter compound is known to undergo hydrogenolysis to only a slight extent over palladium (Nishimura et al., 1971a). The fact that only a small amount of hydrogenolysis occurs over ruthenium is attributed to its poor activity toward hydrogenolysis of intermediate enol, allyl, and homoallyl ethers. The amount of enol ether formed over ruthenium is low relative to that formed over palladium. Hydrogenolysis products predominate in reductions over rhodium. At the disappearance of the starting material, the product contains 6% acetal, 18% methylcyclohexanol, 26% ethyl methylcyclohexyl ether, and 48% methylcyclohexane. Methylcyclohexanol can be derived by reduction of methylcyclohexanone. This ketone, which appears as an intermediate in 7% yield, was assumed to arise from hydrolysis of the acetal, despite careful drying of the system. Alternatively, some of the alcohol might be derived by direct hydrogenolysis of the ethyl carbon–oxygen bond. This type of hydrogenolysis over rhodium has been observed in related systems (Rylander and Himelstein,

1967). Extensive hydrogenolysis over iridium and platinum has been assumed to be caused by formation of allyl and enol ether intermediates, both of which are susceptible to hydrogenolysis over these metals. In addition, platinum affords toluene directly by hydrogenolysis of the aromatic–oxygen bond.

The relative tendencies of ethyl p-tolyl ether, and presumably other aromatics as well, to undergo hydrogenolysis over various metals seem to be related to three major features of the catalyst: (1) its tendency to form certain specific dihydro and tetrahydro intermediates, (2) its activity toward hydrogenolysis of these intermediates, and (3) its activity toward hydrogenolysis of aromatic ethers before ring reduction. These characteristics are summarized in Table XI.

TABLE XI

Characteristics of Noble Metals Relating to Hydrogenolysis of Ethyl p-Tolyl Ether

| Catalyst | Characteristic |
|---|---|
| Palladium | Little tendency to form allyl and homoallyl ethers; low activity for hydrogenolysis of enol ethers (related to high isomerization activity) |
| Ruthenium | Low activity for hydrogenolysis of enol, allyl, and homoallyl ethers (related to inability to ionize adsorbed hydrogen) |
| Rhodium | Moderate activity for hydrogenolysis of enol and allyl ethers; tendency to form enol ethers; tendency to cause alkyl–oxygen bond cleavage in alkyl aromatic ethers |
| Platinum | Tendency to form and cleave enol and allyl ethers; tendency to cleave aromatic–oxygen bonds |
| Iridium | Tendency to form and cleave enol and allyl ethers |

An interesting link relating the activity of catalysts for hydrogenolysis in enol ethers with isomerization activity and activity toward acetal formation was suggested by Nishimura et al. (1971a). Activity toward acetal formation relative to hydrogenation increases with metal in the order Os < Ru < Ir ≪ Rh < Pd < Pt. Since acetal formation is acid catalyzed, this order is assumed to be related to the ability of these metals to ionize adsorbed hydrogen. Isomerization activity for the metals is related to the affinity between the initial half-hydrogenated state and the metal. The weaker the affinity the greater the tendency toward reversal and possible isomerization and the less the tendency toward elimination.

$$H_3C-\langle\rangle-OC_2H_5 \underset{\longleftarrow}{\overset{+MH}{\longrightarrow}} H_3C-\langle\rangle\begin{matrix}OC_2H_5\\M\end{matrix} \overset{-HOC_2H_5}{\longrightarrow} H_3C-\langle\rangle-M$$

Elimination is also acid catalyzed and should be favored by those metals effective in acetal formation. The ratio of acetal formation activity to isomerization activity produced the following sequence for hydrogenolysis: Pd $\simeq$ Ru $\ll$ Os $<$ Rh $<$ Ir $\ll$ Pt; this is in excellent accord with experimental findings.

## REFERENCES

Adams, R., and Marshall, J. R., *J. Am. Chem. Soc.* **50**, 1970 (1928).

Adkins, H., and Billica, H. R., *J. Am. Chem. Soc.* **70**, 695 (1948).

Adkins, H., and Kusek, G., *J. Am. Chem. Soc.* **70**, 412 (1948).

Allinger, N. L., Gorden, B. J., Tyminski, I. J., and Wuesthoff, M. T., *J. Org. Chem.* **36**, 739 (1971).

Areshidz, K. I., Sikharulidze, N. G., and Dzhaoshvil, O. A., Br. Patent 1,257,607, Dec. 22, 1971; *Chem. Abstr.* **76**, 45816f (1972).

Arnold, H. W., U.S. Patent 2,555,912, June 5, 1951.

Augustine, R. L., and Pierson, W. G., *J. Org. Chem.* **34**, 1070 (1969a).

Augustine, R. L., and Pierson, W. G., *J. Org. Chem.* **34**, 2235 (1969b).

Augustine, R. L., and Vag, L. A., *J. Org. Chem.* **40**, 1074 (1975).

Baer, H. H., and Kienzle, F., *J. Org. Chem.* **34**, 3848 (1969).

Baker, R. H., and Schuetz, R. D., *J. Am. Chem. Soc.* **69**, 1250 (1947).

Balasubrahmanyam, S. N., and Balasubramanian, M., *Tetrahedron* **29**, 683 (1973).

Baltz, H., Blume, H., Lunau, J., Meye, H., Oberender, H., Schaefer, H., and Timm, D., Ger. Patent 2,059,938, July 8, 1971; *Chem. Abstr.* **75**, 632422 (1971).

Baltzly, R., Mehta, N. B., Russell, P. B., Brooks, R. E., Grivsky, E. M., and Steinberg, A. M., *J. Org. Chem.* **26**, 3669 (1961).

Banks, C. V., Hooker, D. T., and Richards, J. J., *J. Org. Chem.* **21**, 747 (1956).

Barfknecht, C. F., Smith, R. V., and Reif, V. V., *Can. J. Chem.* **48**, 2138 (1970).

Barkdoll, A. E., England, D. C., Gray, H. W., Kirk, W., Jr., and Whitman, G. M., *J. Am. Chem. Soc.* **75**, 1156 (1953).

Bauer, C. R., U.S. Patent 3,376,341, Apr. 2, 1968.

Bell, J. M., Garrett, R., Jones, V. A., and Kubler, D. G., *J. Org. Chem.* **32**, 1307 (1967).

Billman, J. H., and Buehler, J. A., *Proc. Indiana Acad. Sci.* **63**, 120 (1953).

Billman, J. H., and Buehler, J. A., *Proc. Indiana Acad. Sci.* **63**, 120 (1953).

Blackburn, D. W., and Mlynarski, J. J., *Ann. N.Y. Acad. Sci.* **214**, 158 (1973).

Brown, J. H., Durand, H. W., and Marvel, C. S., *J. Am. Chem. Soc.* **58**, 1594 (1936).

Burger, A., and Mosettig, E., *J. Am. Chem. Soc.* **58**, 1857 (1963).

Burgstahler A. W., and Bithos, Z. J., *J. Am. Chem. Soc.* **82**, 5466 (1960).

Burwell, R. L., Jr., *Chem. Rev.* **57**, 895 (1957).

Burwell, R. L., Jr., *Acc. Chem. Res.* **2**, 289 (1969).

Cantor, S. E., and Tarbell, D. S., *J. Am. Chem. Soc.* **86**, 2902 (1964).

Cargill, R. L., Foster, A. M., Good, J. J., and Davis, F. K., *J. Org. Chem.* **38**, 3829 (1973).

Christian, W. R., Gogek, C. J., and Purves, C. B., *Can. J. Chem.* **29**, 911 (1951).

Cowen, K. D., and Eisenbraun, E. J., Ph.D. Thesis of K. D. Cowen, Oklahoma State Univ., Stillwater, 1977.

Cram, D. J., and Allinger, N. L., *J. Am. Chem. Soc.* **77**, 6289 (1955).

Duggan, R. J., Murray, E. J., and Winstrom, L. O., U.S. Patent 3,076,810, Feb. 5, 1963.

Edson, N. L., *J. Soc. Chem. Ind., London* **53**, 138 (1934).

Egli, R., and Eugster, C. H., *Helv. Chim. Acta* **58**, 2321 (1975).

Eisenbraun, E. J., Burnham, J. W., Weaver, J. D., and Webb, T. E., *Ann. N.Y. Acad. Sci.* **214**, 204 (1973).

Eliel, E. L., and Brett, T. J., *J. Org. Chem.* **28**, 1923 (1963).

Esch, B., and Schaeffer, H. J., *J. Am. Pharm. Assoc. Sci. Ed.* **49**, 786 (1960).

Farina, M., and Audisio, G., *Tetrahedron* **26**, 1827 (1970).

Farina, M., Morandi, C., Mantica, E., and Botta, D., *J. Chem. Soc., Chem. Commun.* p. 816 (1976).

Farrissey, W. J., Jr., and Frulla, F. F., Ger. Patent 1,948,566, Apr. 2, 1970; *Chem. Abstr.* **72**, 132163b (1970).

Ferber, E., and Brückner, H., *Ber. Dtsch. Chem. Ges. B* **72**, 995 (1939).

Fieser, L. F., and Hershberg, E. B., *J. Am. Chem. Soc.* **59**, 2502 (1937).

Fieser, L. F., and Hershberg, E. B., *J. Am. Chem. Soc.* **60**, 940 (1938).

Folkers, K., and Johnson, T. B., *J. Am. Chem. Soc.* **55**, 1140 (1933).

Franck, R. W., and Yanagi, K., *J. Org. Chem.* **33**, 811 (1968).

Frank, C. E., U.S. Patent 2,478,261, Aug. 9, 1949.

Freidlin, L. K, Litvin, E. F., Oparina, G. K., Gurskii, R. N., Istratova, R. V., and Videneeva, L. V., *Zh. Org. Khim.* **9**, 959 (1973); *Chem. Abstr.* **79**, 42018y (1973).

Freidlin, L. K. Litvin, E. F., Yakubenok, V. V., and Vaisman, I. L., *Izv. Akad. Nauk SSSR, Ser. Khim.* No. 5 p. 976 (1976).

Freifelder, M., *J. Org. Chem.* **29**, 979 (1964).

Freifelder, M., and Stone, G. R., *J. Am. Chem. Soc.* **80**, 5270 (1958).

Freifelder, M., and Stone, G. R., *J. Org. Chem.* **27**, 3568 (1962).

Freifelder, M., Anderson, T., Ng, Y. H., and Papendick, F., *J. Pharm. Sci.* **53**, 967 (1964).

Freifelder, M., Ng, Y. H., and Helgren, P. F., *J. Org. Chem.* **30**, 2485 (1965).

Fu, P. P., and Harvey, R. G., *Tetrahedron Lett.* p. 415 (1977).

Galantay, E., *Tetrahedron* **19**, 319 (1963).

Gauthier, B., *Ann. Chim. (Paris)* **20**, 581 (1945).

Gilman, G., and Cohn, G., *Adv. Catal.* **9**, 733 (1957).

Giudici, T. A., and Bruice, T. C., *J. Org. Chem.* **35**, 2386 (1970).

Greenfield, H., *J. Org. Chem.* **29**, 3082 (1964).

Greenfield, H., *Ann. N.Y. Acad. Sci.* **214**, 233 (1973).

Grethe, G., Lee, H. L., Uskokovic, M., and Brossi, A., *J. Org. Chem.* **33**, 494 (1968).

Grob, C. A., and Kieger, H. R., *Helv. Chim. Acta* **48**, 799 (1965).

Grogan, C. H., Geschickter, C. F., and Rice, L. M., *J. Med. Chem.* **7**, 78 (1964).

Grosskinsky, O. A., and Merkel, A., Ger. Patent 2,039,818, Feb. 17, 1972; *Chem. Abstr.* **76**, 140045n (1972).

Hammett, L. P., "Physical Organic Chemistry," 2nd Ed. McGraw-Hill, New York, 1970.

Hart, H., and Cassis, F. A. Jr., *J. Am. Chem. Soc.* **73**, 3179 (1951).

Hartog, F., and Zwietering, P., *J. Catal.* **2**, 79 (1963).

Hartog, F., Tebben, J. H., and Weterings, C. A. M., *Proc. Int. Congr. Catal. 3rd, Amsterdam 1964* **2**, 1210 (1965).

Hartung, W. H., and Simonoff, R., *Org. React.* **7**, 263 (1953).

Heckel, H., and Adams, R., *J. Am. Chem. Soc.* **47**, 1712 (1925).

Hirsch, J. A., and Schwartzkopf, G., *J. Org. Chem.* **39**, 2044 (1974).

Ichinohe, Y., and Ito, H., *Bull. Chem. Soc. Jpn.* **37**, 887 (1964).

Ikedate, K., and Suzuki, S., *Nippon Kagaku Zasshi* **90**(1), 91 (1969); *Chem. Abstr.* 96287t (1969).

Ikedate, K., and Suzuki, S., *Bull. Chem. Soc. Jpn.* **44**, 325 (1971).

Ikedate, K., Suzuki, T., and Suzuki, S., *Nippon Kagaku Zasshi* **89**(3), 304 (1968); *Chem. Abstr.* **69**, 51335v (1968).

Ireland, R. E., and Schiess, P. W., *J. Org. Chem.* **28**, 6 (1963).

Johnson, W. S., Rogier, E. R., and Ackerman, J., *J. Am. Chem. Soc.* **78**, 6322 (1956).

Kajitani, M., Watanabe, Y., Iimura, Y., and Sugimori, A., *Bull. Chem. Soc. Jpn.* **48**, 2848 (1975).

Kalina, M., and Pasek, J., *Kinet. Katal.* **10**(3), 574 (1969); *Chem. Abstr.* **71**, 54119 (1969).

Kalm, M. J., *J. Med. Chem.* **7**, 427 (1964).

Karakhanov, E. A., Saginova, L. G., Karakhanov, R. A., and Viktorova, E. A., *Vestn. Mosk. Univ., Khim.* **24**, 133 (1969); *Chem. Abstr.* **71**, 90529 (1969).

Karrer, P., and Kehl, W., *Helv. Chim. Acta* **13**, 50 (1930).

Kaye, I. A., and Matthews, R. S., *J. Org. Chem.* **28**, 325 (1963).

Keenan, C. W., Giesemann, B. W., Smith, H. A., *J. Am. Chem. Soc.* **76**, 229 (1954).

Kieboom, A. P. G., *Bull. Chem. Soc. Jpn.* **49**, 331 (1976).

Kraus, M., *Adv. Catal.* **17**, 75 (1967).

Krause, A., *Rev. Roum. Chim.* **17**(3), 539 (1972).

Kuhn, R., and Haas, H. J., *Justus Liebigs Ann. Chem.* **611**, 57 (1958).

Kuhn, R., Driesen, H. E., and Haas, H. J., *Justus Liebigs Ann. Chem.* **718**, 78 (1968).

Langer, E., and Lehner, H., *Tetrahedron Lett.* p. 1143 (1973a).

Langer, E., and Lehner, H., *Monatsh. Chem.* **104**(5), 1154 (1973b).

Langer, E., and Lehner, H., *Monatsh. Chem.* **104**(6), 1484 (1973c).

Levin, R. H., and Pendergrass, J. H., *J. Am. Chem. Soc.* **69**, 2436 (1947).

Levine, M., and Sedlecky, R., *J. Org. Chem.* **24**, 115 (1959).

Linstead, R. P., Doering, W. E., Davis, S. B., Levine, P., and Whetstone, R. R., *J. Am. Chem. Soc.* **64**, 1985 (1942a).

Linstead, R. P., Whetstone, R. R., and Levine, P., *J. Am. Chem. Soc* **64**, 2014 (1942b).

Little, E. D., and Cheema, Z. K., Ger. Patent 2, 125,846, Dec. 16, 1971; *Chem. Abstr.* **76**, 72113f (1972).

Litvin, E. F., Freidlin, L., K, Oparina, G. K., Kheifets, V. I., Yakubenok, V. V., Pivonenkova, L. P., and Bychkova, M. K., *Izv. Akad. Nauk SSSR, Ser. Khim.* p. 854 (1973); *Chem. Abstr.* **79**, 31603j (1973).

Lyubarskii, G. D., Strelets, M. M., and Sololski, V. A., Ger. Patent 2,025,726, Dec. 9, 1971; *Chem. Abstr.* **76**, 99211y (1972).

Lyubarskii, G. D., Buyanova, N. E., Ratner, I. D., and Strelets, M. M., *Kinet. Katal.* **14**(4), 1020 (1973).

McDonald, R. N., and Reitz, R. R., *J. Org. Chem.* **35**, 2666 (1970).

Mahmoud, B. H., and Greenlee, K. W., *J. Org. Chem.* **27**, 2369 (1926).

Matsumoto, M., Suzuki, T., and Suzuki, S., *Kogyo Kagaku Zasshi* **72**, 881 (1969).

Maxted, E. B., *Adv. Catal.* **3**, 129 (1951).

Meyers, A. I., and Baburao, K., *J. Heterocycl. Chem.* **1**, 203 (1964).

Mochida, I., and Yoneda, Y., *J. Catal.* **11**, 183 (1968).

Mokotoff, M., and Cavestri, R. C., *J. Org. Chem.* **39**, 409 (1974).

Musliner, W. J., and Gates, J. W., Jr., *J. Am. Chem. Soc.* **88**, 4271 (1966).

Mylroie, V. L., and Stenberg, J. F., *Ann. N.Y. Acad. Sci.* **214**, 255 (1973).

Nishimura, S., *Bull. Chem. Soc. Jpn.* **32**, 1155 (1959).

Nishimura, S., *Bull. Chem. Soc. Jpn.* **33**, 566 (1960).

Nishimura, S., and Taguchi, H., *Bull. Chem. Soc. Jpn.* **36**, 353 (1963a).

Nishimura, S., and Taguchi, H., *Bull. Chem. Soc. Jpn.* **36**, 873 (1963b).

Nishimura, S., and Yoshino, H, *Bull. Chem. Soc. Jpn.* **42**, 499 (1969).

Nishimura, S., Shu, T., Hara, T., and Takagi, Y., *Bull. Chem. Soc. Jpn.* **39**, 329 (1966).

Nishimura, S., Itaya, T., and Shiota, M., *Chem. Commun.* p. 442 (1967).

Nishimura, S., Uchino, H., and Yoshino, H., *Bull. Chem. Soc. Jpn.* **41**, 2194 (1968).
Nishimura, S., Mochizuki, F., and Kobayakawa, S., *Bull. Chem. Soc. Jpn.* **43**, 1919 (1970).
Nishimura, S., Katagiri, M., Watanabe, T., and Uramoto, M., *Bull. Chem. Soc. Jpn.* **44**, 166 (1971a).
Nishimura, S., Kono, Y., Otsuki, Y., and Fukaya, Y., *Bull. Chem. Soc. Jpn.* **44**, 240 (1971b).
Nishimura, S., Uramoto, M., and Watanabe, T., *Bull. Chem. Soc. Jpn.* **45**, 216 (1972).
Novello, F. C., and Christy, M. E., *J. Am. Chem. Soc.* **73**, 1267 (1951).
Oparina, G. K., Chernyshova, M. P., Kheifets, V. I., and Gluzman, S. S., *Zh. Vses. Khim. Ova.* **18**(3), 346 (1973); *Chem. Abstr.* **79**, 77778y (1973).
Pecherer, B., Jampolsky, L. M., and Wuest, H. M., *J. Am. Chem. Soc.* **70**, 2587 (1948).
Phillips, A. P., and Mentha, J., *J. Am. Chem. Soc.* **78**, 140 (1956).
Price, C. C., Enos, H. I., Jr., and Kaplan, W., *J. Am. Chem. Soc.* **69**, 2261 (1947).
Rapoport, H., and Pasky, J. Z., *J. Am. Chem. Soc.* **78**, 3788 (1956).
Rapoport, H., and Smolinsky, G., *J. Am. Chem. Soc.* **82**, 1171 (1960).
Rodig, O. R., and Ellis, L. C., *J. Org. Chem.* **26**, 2197 (1961).
Ruggli, P., Leupin, O., and Businger, A., *Helv. Chim. Acta* **24**, 339 (1941).
Ruzicka, V., Vrbsky, I., and Medonos, V., *Sb. Vys. Sk. Chem.-Technol. Praze, Org. Chem. Technol.* No. 15, p. 21 (1970); *Chem. Abstr.* **76**, 13999b (1972).
Rylander, P. N., and Himelstein, N., *Engelhard Ind., Tech. Bull.* **5**, 43 (1964).
Rylander, P. N., and Himelstein, N., *Engelhard Ind., Tech. Bull.* **8**, 53 (1967).
Rylander, P. N., and Kilroy, M., *Engelhard Ind., Tech. Bull.* **9**, 14 (1968).
Rylander, P. N., and Rakoncza, N. F., U.S. Patent 3,163,679, Dec. 22, 1964.
Rylander, P. N., and Steele, D. R., *Engelhard Ind., Tech. Bull.* **3**, 91 (1962a).
Rylander, P. N., and Steele, D. R., *Engelhard Ind., Tech. Bull.* **3**, 19 (1962b).
Rylander, P. N., and Steele, D. R., *Engelhard Ind., Tech. Bull.* **4**, 20 (1963a).
Rylander, P. N., and Steele, D. R., *Engelhard Ind., Tech. Bull.* **3**, 125 (1963b).
Rylander, P. N., and Steele, D. R., *Engelhard Ind., Tech. Bull.* **5**, 113 (1965a).
Rylander, P. N., and Steele, D. R., *Engelhard Ind., Tech. Bull.* **6**, 41 (1965b).
Rylander, P. N., and Steele, D. R., *Engelhard Ind., Tech. Bull.* **7**, 153 (1967).
Rylander, P. N., and Steele, D. R., U.S. Patent 3,387,048, June, 4, 1968.
Rylander, P. N., and Vaflor, X., *Am. Chem. Soc., Northeast Reg. Meet. 6th, Burlington, 1974.*
Rylander, P. N., Rakoncza, N., Steele, D. R., and Bollinger, M., *Engelhard Ind., Tech. Bull.* **4**, 95 (1963).
Rylander, P. N., Hasbrouck, L., Hindin, S. G., Karpenko, I., Pond, G., and Starrick, S., *Engelhard Ind., Tech. Bull.* **8**, 25 (1967a).
Rylander, P. N., Hasbrouck, L., Hindin, S. G., Iverson, R., Karpenko, I., and Pond, G., *Engelhard Ind., Tech. Bull.* **8**, 93 (1967b).
Rylander, P. N., Hasbrouck, L., and Karpenko, I., *Ann. N.Y. Acad. Sci.* **214**, 100 (1973).
Sakai, S., Kubo, A., Hamamoto, T., Wakabayashi, M., Takahashi, K., Ohtani, Y., and Haginiva J., *Tetrahedron Lett.* p. 1489 (1969).
Schrecker, A. W., and Hartwell, J. L., *J. Am. Chem. Soc.* **75**, 5917 (1953).
Schuetz, R. D., and Caswell, L. R., *J. Org. Chem.* **27**, 486 (1962).
Schweers, W. Holzforschung **23**(4), 120 (1969).
Shriner, R. L., and Witte, M., *J. Am. Chem. Soc.* **62**, 2134 (1941).
Siegel, S., *Adv. Catal.* **16**, 123 (1966).
Siegel, S., and Dunkel, M., *Adv. Catal.* **9**, 15 (1957).
Siegel, S., and Garti, N., *in* "Catalysis in Organic Syntheses, 1977" (G. V. Smith, ed.), pp. 9–23. Academic Press, New York, 1977.
Siegel, S., and Ku, V., *Proc. Int. Congr. Catal. 3rd, Amsterdam, 1964* 1199–1207 (1965).
Siegel, S., and Smith, G. V., *J. Am. Chem. Soc.* **82**, 6082, 6087 (1960).

Siegel, S., Smith, G. V., Dmuchovsky, B., Dubbell, D., and Halpren, W., *J. Am. Chem. Soc.* **84**, 3136 (1962).
Siegel, S., Ku, V., and Halpren, W., *J. Catal.* **2**, 348 (1963).
Slaugh, L. H., and Leonard, J. A., *J. Catal.* **13**, 385 (1969).
Smeykal, K., Naumann, H. J., Schaefer, H., Veit, J., Becker, K., and Block, A., Ger. Patent 1,298,098, June 26, 1969; *Chem. Abstr.* **71**, 60843c (1969).
Smith, H. A., *in* "Catalysis" (P. H. Emmett, ed.), Vol. 5, pp. 175– 256. Reinhold, New York, 1957.
Smith, H. A., *Ann. N.Y. Acad. Sci.* **145**, 72 (1967).
Smith, H. A., and Stump, B. L., *J. Am. Chem. Soc.* **83**, 2739 (1961).
Smith, H. A., and Thompson, R. G., *Adv. Catal.* **9**, 727 (1957).
Smith, H. A., Alderman, D. M., Jr., Shacklett, C. D., and Welch, C. M., *J. Am. Chem. Soc.* **71**, 3772 (1949).
Snieckus, V. A., Onouchi, T., and Boekelheide, V., *J. Org. Chem.* **37**, 2845 (1972).
Sommerville, W. T., and Theimer, E. T., U.S. Patent 2,927,127, Mar. 1, 1960.
Steitz, A., Jr., *Org. Chem.* **35**, 854 (1970).
Stocker, J. H., *J. Org. Chem.* **27**, 2288 (1962).
Stocker, J. H., *J. Org. Chem.* **27**, 2288 (1962).
Stocker, J. H., *J. Org. Chem.* **29**, 3593 (1964).
Stolow, R. D., Giants, T. W., Krikorian, R. R., Litchman, M. A., and Wiley, D. C., *J. Org. Chem.* **37**, 2894 (1972).
Stork, G., *J. Am. Chem. Soc.* **69**, 576 (1947).
Taira, S., *Bull. Chem. Soc. Jpn.* **34**, 1294 (1961).
Takagi, Y., *J. Catal.* **8**, 100 (1967).
Takagi, Y., *Sci. Pap. Inst. Phys. Chem. Res.* (*Jpn.*) **64**, 39 (1970).
Takagi, Y., Naito, T., and Nishimura, S., *Bull. Chem. Soc. Jpn.* **38**, 2119 (1965).
Takagi, Y., Nishimura, S., Taya, K., and Hirota, K., *J. Catal.* **8**, 100 (1967).
Takagi, Y., Nishimura, S., and Hirota, K., *J. Catal.* **12**, 214 (1968).
Takagi, Y., Nishimura, S., and Hirota, K., *Bull. Chem. Soc. Jpn.* **43**, 1846 (1970a).
Takagi, Y., Ishii, S., and Nishimura, S., *Bull. Chem. Soc. Jpn.* **43**, 917 (1970b).
Taverna, M., and Chita, M., *Hydrocarbon Process. Nov.*, p. 137 (1970).
Taya, K., Hiramoto, M., and Hirota, K., *Sci. Pap. Inst. Phys. Chem. Res.* (*Jpn*) **62**(4), 145 (1968).
Teter, J. W., U.S. Patent 2,898,387, Aug. 4, 1959.
Teuber, H. J., Cornelius, D., and Wölche, U., *Justus Liebigs Ann. Chem.* **696**, 116 (1966).
Thelen, H., Halcour, K., Schwerdtel, W., and Swodnek, W., Ger. Patent 2,045,882, Mar. 23, 1972; *Chem. Abstr.* **76**, 153236a (1972).
Thompson, R. B., *Org. Synth.* **27**, 21 (1947).
Tomita, M., and Uyeo, S., *Nippon Kagaku Zasshi* **63**, 1189 (1942).
Trahanovsky, W. S., and Bohlen, D. H., *J. Org. Chem.* **37**, 2192 (1972).
Ungnade, H. E., and Morriss, F. V., *J. Am. Chem. Soc.* **70**, 1898 (1948).
Ungnade, H. E., and Nightingale, D. E., *J. Am. Chem. Soc.* **66**, 1218 (1944).
Valenta, Z., Deslongchamps, P., Ellison, R., Wiesner, K., *J. Am. Chem. Soc.* **86**, 2533 (1964).
van Bekkum, H., *Proc. Int. Congr. Catal., 3rd, Amsterdam, 1964* **2**, 1208 (1965).
van Bekkum, H., Buurmans, H. M. A., van Minnen-Pathuis, G., and Wepster, B. M., *Rec. Trav. Chim. Pays-Bas* **88**, 779 (1969).
van de Graaf, B., van Bekkum, H., and Wepster, B. M., *Rec. Trav. Chim. Pays-Bas* **87**, 777(1968).
van Muijlwijk, A. W., Kieboom, A. P. G., and van Bekkum, H., *Rec. Trav. Chim. Pays-Bas* **93**, 204 (1974).

Walton, E., Wilson, A. N., Haven, A. C., Jr., Hoffman, C. H., Johnston, E. L., Newhall, W. F.,
    Robinson, F. M., and Holly, F. W., *J. Am. Chem. Soc.* **78**, 4760 (1956).
Weaver, J. D., Eisenbraun, E. J., and Harris, L. E., *Chem. Ind. (London)* p. 187 (1973).
Weitkamp, A. W., *J. Catal.* **6**, 431 (1966).
Wenkert, E., Dave, K. G., Haglid, F., Lewis, R. G., Oishi, T., Stevens, R. V., and Terashima,
    M., *J. Org. Chem.* **33**, 747 (1968).
Wharton, P. S., Sundin, C. E., Johnson, D. W., and Kluender, H. C., *J. Org. Chem.* **37**, 34 (1972).
Whitehead, C. W., Traverso, J. J., Marshall, F. J., and Morrison, D. E., *J. Org. Chem.* **26**,
    2809 (1961).
Winans, C. F., *Ind. Eng. Chem.* **32**, 1215 (1940).
Witte, J., and Boekelheide, V., *J. Org. Chem.* **37**, 2849 (1972).
Woodward, R. B., and Doering, W. E., *J. Am. Chem. Soc.* **67**, 860 (1945).
Wu, Y. H., Gould, W. A., Lobeck, W. G., Jr., Roth, H. R., and Feldkamp, R. F., *J. Med. Chem.*
    **5**, 752 (1962).
Yoshida, T., *Bull. Chem. Soc. Jpn.* **47**, 2061 (1974).
Zaugg, H. E., Michaels, R. J., Glenn, H. J., Swett, L. R., Freifelder, M., Stone, G. R., and
    Weston, A. W., *J. Am. Chem. Soc.* **80**, 2763 (1958).
Zymalkowsky, F., and Strippel, G., *Arch. Pharm. (Weinheim, Ger.)* **297**, 727 (1964).

# Chapter 12

# Hydrogenation of
# Heterocyclic Compounds

This chapter is confined to the more common unsaturated heterocyclic compounds the reduction of which ordinarily leaves the ring intact. Compounds that undergo facile ring cleavage, such as isoxazoles, are discussed in Chapter 14.

## I. PYRIDINES

A great variety of pyridines have been hydrogenated. Acidic media are frequently used to prevent inhibition of the catalyst by the basic nitrogen atom of the products (Maxted and Walker, 1948; Freifelder, 1963a). Acid also changes the species actually undergoing hydrogenation from pyridine to pyridinium ions. To account for the more facile reduction of pyridinium ions, it has been suggested that they are adsorbed flat on the catalyst, whereas the free base may be adsorbed edgewise (Skomoroski and Schriesheim, 1961). Acid may also affect other portions of the molecule. Hydrogenolysis of the butane ring in **1** occurs readily when reduction is carried out in excess hydrochloric acid, but, in solutions containing just slightly more than 2 equivalents of acid, ring opening does not occur and **2** is obtained in 88% yield. The same reduction can be achieved in neutral media over rhodium, but the yield is less (Heitmeier *et al.*, 1971).

| 1 | 2 |

## A.  Catalysts

Palladium, platinum (Aeberli *et al.*, 1969), rhodium, ruthenium, and nickel have been used for the hydrogenation of pyridines. Nickel requires high temperatures and pressures; when nickel is used in lower alcohols, *N*-alkylation, and in certain compounds dealkylation, become important side reactions (Freifelder, 1963b). *N*-Alkylation is the major reaction in the reduction of pyridines in alcohols over rhenium sulfide (Mistryukov *et al.*, 1971); the yields are often very high. The relative effectiveness of noble metals depends on the reaction conditions and solvent employed (Rylander and Steele, 1962). Rhodium is the most effective catalyst under mild conditions; under more vigorous conditions (1000 psig and 100°C) rhodium is still the most effective, but the differences among catalysts are not as striking. Under these conditions ruthenium is excellent, especially in water, and in alcohols *N*-alkylation is avoided. Palladium in acetic acid is also a good catalyst at temperatures above 70°C (Walker, 1962; Walker *et al.*, 1965; Steck *et al.*, 1963). In the hydrogenation of 2,2′ -dipyridyl to piperidyl-pyridine, palladium is the most selective of the noble metals (Rylander and Steele, 1965). Over palladium-on-carbon an attempted reduction of pyrido [2, 1-*a*] isoindolium bromide failed, but the hydrogenation proceeded smoothly over platinum oxide in ethanol containing hydrobromic acid to afford the dodecahydro derivative (Bradsher and Voigt, 1971).

## B.  Quaternary Compounds

Quaternary pyridinium salts are readily reduced over platinum metal catalysts, offering a convenient way of obtaining *N*-substituted piperidines (Hamilton and Adams, 1928; Grigor'eva *et al.*, 1957; Hayes *et al.*, 1956). Besides changing the species undergoing hydrogenation, quaternization diminishes catalyst inhibition by the nitrogen atom. Quaternary pyridinium salts can be reduced more easily than hydro halide salts, which in turn can be reduced more easily than the free pyridine base. This sequence offers some possibility for controlling selectivity in complex compounds (Howton and Golding, 1950).

## C.  Pyridinecarboxylic Acids

Pyridinecarboxylic acids are reduced readily, but hydrogenation of 3-pyridinecarboxylic acids is apt to be accompanied by extensive decarboxylation (Sorm, 1948; Freifelder, 1962). Decarboxylation can be prevented by hydrogenation in the presence of 1 equivalent of base, such as sodium bicarbonate, sodium hydroxide (Raasch, 1962), or ammonia (Freifelder and

Stone, 1961). Excellent yields of piperidinecarboxylic acids have been obtained by hydrogenation over ruthenium dioxide (95°C, 1500 psig) (Freifelder and Stone, 1961), 5% rhodium-on-alumina (25°C, 60 psig) (Freifelder, 1963b), platinum oxide (McElvain and Adams, 1923; Freifelder, 1962), and palladium-on-carbon (60°C, 50 psig) (Freifelder, 1963a).

## D. Aminopyridines

2-Aminopyridines and 4-aminopyridines, like the corresponding hydroxy-pyridines, can exist in tautomeric form and they all behave similarly, whereas 3-aminopyridine resembles aniline (Nienburg, 1973). No particular dfficul-ties are to be expected in the reduction of 3-aminopyridines. Quantitative yields of 3-aminopiperidine were obtained by hydrogenation of-3-amino-pyridine over platinum oxide in hydrocholoric acid (Nienburg, 1937). Rhodium and ruthenium would also be expected to be effective. 4-Amino-pyridine is reduced with some difficulty over platinum catalysts (Orthner, 1927), even at 80 atm (Yakhontov *et al.*, 1958). However, the tertiary amino compound *N*-(4-pyridyl)morpholine was reduced readily in water over ruthenium dioxide at elevated temperatures and pressure (Freifelder and Stone, 1961).

2-Aminopyridines and *N*-mono- and *N,N*-disubstituted aminopyridines, like 2-hydroxypyridines, tend to adsorb only 2 moles of hydrogen and resist further reduction (Birkofer, 1942; Freifelder *et al.*, 1964) unless hydro-genolysis of the amino function occurs (Graves, 1924; Robison *et al.*, 1957). Difficulty in reducing the tetrahydro derivative further is expected since the resulting function is an amidine, a structure known to be reduced with diffi-culty and one susceptible to hydrogenolysis.

## E. Hydroxypyridines

In the hydrogenation of hydroxypyridines, as in the hydrogenation of phe-nols, some hydrogenolysis of the hydroxyl function might be expected (Biel *et al.*, 1952; Dart and Henbest, 1959). Nonetheless, satisfactory to excellent yields of 3- and 4-hydroxypiperidines have been obtained by reduction of hydroxypyridines over ruthenium at elevated temperatures and pressure (Hall, 1958), over platinum oxide in acetic acid (Marion and Cockburn, 1949), and as the hydrochloride salt in ethanol (Koelsch and Carney, 1950). Rhodium-on-carbon or rhodium-on-alumina (Büchi and Wüest, 1971; Möhrle and Weber, 1971) appears to be especially effective (Biel and Aiman, 1967; Biel and Blicke, 1962; Shapiro *et al.*, 1962), giving excellent yields of hydroxypiperidines in cases in which platinum oxide caused extensive hydrogenolysis (Shapiro *et al.*, 1959).

Hydrogenation of 2-hydroxypyridines (2-pyridones) stops spontaneously at the 2-piperidone stage; the amide structure is resistant to hydrogenation under mild conditions.

## F.   Pyridylalkylamines

The basic nitrogen atom in pyridylalkylamines tends to inhibit ring saturation by nickel catalysts, and nickel has rarely proved useful (Britton and Horsley, 1958). Excellent results have been obtained in neutral media over ruthenium (100°C, 1000 psig), which is the catalyst of choice if pressure equipment is available (Freifelder and Stone, 1961). Good results were also obtained over platinum oxide in acidic media (Freifelder, 1963a; Minor et al., 1962). 3-Aminomethylpyridine was reductively alkylated with acetone using platinum and saturated over ruthenium oxide. This product was then cyclized to the first bicyclic urea with a bridgehead nitrogen atom (Hall and Johnson, 1972).

Reduction of the complex pyridine **3** over platinum oxide in ethanol and hydrogen chloride affords the hexahydroindolizin-3-one **4**,

but hydrogenation of the malonate **5** gives ethyl octahydro-3-oxoindolizine-2-carboxylate (**6**) with loss of aniline (Sprake and Watson, 1976).

5                                                    6

## G.   Partial Ring Reduction

2-Amino- and 2-Hydroxypyridines usually undergo only partial reductions, as noted in preceding sections. In addition, many examples of partial
hydrogenation of pyridine rings containing 3-acyl, aldehydo, keto, cyano,
and other functions have been reported (Lyle and Mallett, 1967; Freifelder,
1964; Wenkert and Wickberg, 1965). The substituent usually survives unchanged in the 2-piperidine products. Wenkert *et al.* (1968) noted that hydrogenation of any aromatic nucleus capable of unmasking a stable vinylogous
amide unit may be prone to stoppage sooner than the hexahydro stage.
Partial hydrogenation may be favored by limiting hydrogen absorption, by
limited catalysts, and perhaps by catalyst deactivators (Supniewski and
Serafinowna, 1963; Lyle and Lyle, 1954; Lyle *et al.*, 1955).

Hydrogenation of diethyl 3,5-pyridinedicarboxylate (7) over palladium-
on-carbon in ethanol affords the dihydro derivative 8 after absorption of
1 mole of hydrogen; uninterrupted hydrogenation affords the tetrahydro-
pyridine 9 in high yield. The author (Eisner, 1969) believed that 9 was formed
by disproportionation of 8 to 7 and 9 rather than by direct hydrogenation.

7                                                    8

9

The reluctance of the N—C=C—C=O grouping to undergo hydrogenation is further shown in the partial hydrogenation of the azepine 10 to the
dihydro derivative 11, (van Bergen and Kellogg, 1971).

**10**                                                **11**

Partial hydrogenation of $N$-alkylpyridinium salts in the presence of 1 equivalent of alkali over nickel or palladium hydroxide-on-barium sulfate provides a convenient synthesis of $\omega$-alkylaminovaleraldehydes, a class of compounds otherwise obtained with difficulty. The reduction is best carried out at 0°–5°C to prevent destruction of the product by the alkaline solution and overhydrogenation to the alkylpiperidine (Schöpf et al., 1957).

## H.   Selective Hydrogenation of Pyridines

It is an accepted generality that pyridines bearing phenyl substituents, either away from or attached to the ring, are hydrogenated preferentially in the pyridine ring, barring adverse steric effects (Freifelder, 1971). For example, reduction of 4-benzylpyridine gives almost entirely 4-benzyl-piperidine (Freifelder et al., 1962; Overhoff and Wibaut, 1931). Recently it was found that a reverse selectivity obtains if reduction of pyridine compounds, including fused ring compounds, is carried out in strong acid (as opposed to dilute acid). Sulfuric acid and 12 $N$ hydrochloric acid are suitable, but trifluoroacetic acid is preferred because the rate of hydrogenation is much higher (Vierhapper and Eliel, 1975). This interesting discovery is worthy of further elaboration.

96%

The balance of selectivity is sometimes tilted easily by small changes in steric or electronic effects. 1-(2-Pyridyl)-2-propanone (**12**) is reduced selectively at the pyridyl ring to afford 2-piperidylpropanone (**13a**) (Wibaut and Kloppenburg, 1946; Wibaut et al., 1944), but 1-(3-substituted 2-pyridyl)-2-propanones (**14**) are not, except under special conditions (**13b**). Selective reduction of the pyridine ring was achieved over rhodium-on-alumina in water containing 1 equivalent of hydrobromic acid at 50 psig and 50°–70°C. The system is specific; hydrogenation of the free base in methanol or in acetic acid over rhodium, or of the hydrobromide salt in methanol, or of the free base in acetic acid over platinum oxide, or of the free base in acetic acid over palladium failed to give selective reduction. In addition, selectivity could be achieved only if the substrate was carefully purified. The products are the cis isomers (Barringer *et al.*, 1973).

**12,** R = H
**14,** R = OCH$_3$, OC$_2$H$_5$, OCH(CH$_3$)$_2$, CH$_3$

**13a,** R = H
**13b,** R = OCH$_3$, OC$_2$H$_5$, OCH(CH$_3$)$_2$, CH$_3$

In general, nitro functions are reduced more readily than aromatic systems, but exceptions exist. One exception is preferential saturation of the aromatic ring in 1-(α-carboethoxy-β-indolyl-2-nitrobutane), a result attributed to steric hindrance at the nitro function (Young and Snyder, 1961). Another exception is the selective conversion of 2-α-pyridyl-5-nitro-5-methyl-1,3-dioxane to 2-piperidyl-5-nitro-5-*cis*-methyl-1,3-dioxane over platinum oxide in acetic acid. In isopropanol, platinum-on-carbon preferentially and partially reduces the nitro group to 2-α-pyridyl-5-hydroxyamino-5-*cis*-methyl-1,3-dioxane, whereas Raney nickel in isopropanol affords 2-α-pyridyl-5-amino-5-*cis*-methyl-1,3-dioxane (Aeberli and Houlihan, 1967). Some of these selectivity differences are due to solvent; acetic acid would be expected to facilitate ring reduction.

Suitably placed pyridine rings can undergo alkylation reactions through their intermediate products. Reduction of the quaternized pyridine **15** over 10% palladium-on-carbon in absolute ethanol at 30 psig gave predominantly the expected piperidyl compound **16** (79%) accompanied by a tetracyclic compound (**17**). Formation of **17** was rationalized as alkylation of the indole 3 position by an iminium intermediate (Sundberg *et al.*, 1972).

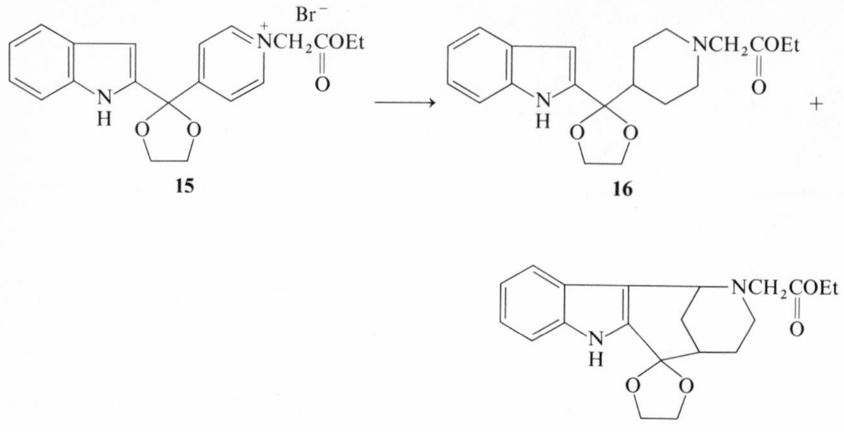

15

16

+

17

A synthesis of ormosanine (**20**) involved stereoselective hydrogenation of the pyridine ring in **18** over platinum in acetic acid. The predominance of one isomer (**19**; 20:1) was attributed to hydrogenation of a hydrogen-bonded pyridine ring being attacked from the less hindered side. Support for this suggestion comes from the finding that selectivity is completely lost if the lactam carbonyl is reduced before the pyridine ring (Liu *et al.*, 1970).

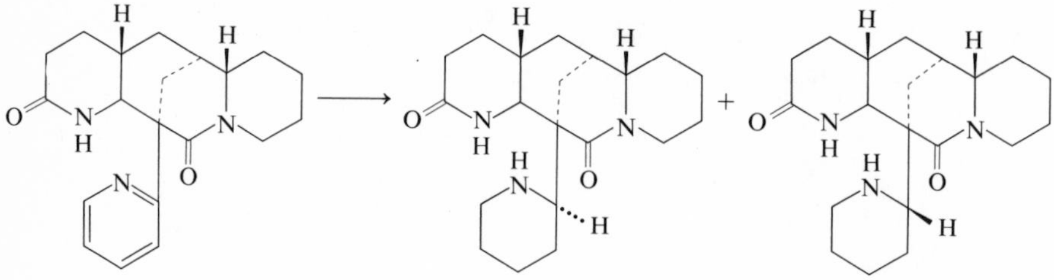

18                                    19                                    20

## II.  PYRROLES

Pyrroles are hydrogenated with more difficulty than carbocyclic aromatics, and in compounds containing both types of rings reduction probably proceeds nonselectively or with preference for the carbocyclic ring if the rings are not fused (Craig and Hixon, 1930). The reverse selectivity in the reduction below is probably best attributed to trisubstitution in the carbocyclic ring (Dolby *et al.*, 1972).

29 gm                                     75% after distillation

Pyrroles are frequently hydrogenated in acidic medium, such as ethanolic hydrogen chloride or acetic acid (deJong and Wibaut, 1930; Craig and Hixon, 1930; Atkinson *et al*., 1964), although acidic medium is not necessary. Acid facilitates reduction probably by neutralizing the inhibiting effects of the amine products. A careful purification of the pyrrole may be necessary to ensure satisfactory reductions (Andrews and McElvain, 1929). Reductions may proceed stepwise through intermediate pyrrolines (Wibaut and Proost, 1933; Fuhlhage and VanderWerf, 1958; Yamamoto, 1956).

**Catalysts**

Platinum oxide has been frequently used with success for the hydrogenation of pyrroles (Atkinson *et al*., 1964; Cantor and VanderWerf, 1958; Dann and Dimmling, 1953; Kray and Reinecke, 1967; Putokhin, 1930; Rapoport *et al*., 1954). For low-pressure hydrogenations, rhodium-on-carbon or rhodium-on-alumina are currently the catalysts of choice (Adams *et al*., 1960; Fuhlhage and VanderWerf, 1958; Overberger *et al*., 1955; Patterson *et al*., 1962; Schweizer and Light, 1966). Quantitative yields of the pyrrolidine **22** are obtained on reduction of **21** over 5% rhodium-on-carbon in acetic acid (Weinstein and Craig, 1976). Means of selectively reducing either the keto group or the pyrrole ring have proved to be elusive.

21                                        22

Pyrroles can be reduced over nickel catalysts, but high temperatures and pressures are required, and the yields may not be good. Excellent results were reported for the reduction of pyrroles without solvent over ruthenium dioxide at 100°–200°C and 1500 psig (Freifelder, 1971). Palladium is also a satisfactory catalyst for pyrrole hydrogenation. An interesting contrast between palladium and platinum was reported in the hydrogenation of 3-(2-pyrryl)pyridine. Over palladium-on-carbon in acetic acid the pyrrole ring is selectively reduced (Späth and Kuffner, 1935), whereas over platinum oxide the pyridine ring is reduced (Ochiai *et al.*, 1936). Isomerization accompanied reduction and hydrogenolysis of the pyrrole ring and 9-amino substituent in 9-(*N*,*N*-disubstituted amino)-9*H*-pyrrolo[1,2-*a*]indoles with either the hydrochloride salt or free base. The product is the more stable indole derivative (Raines *et al.*, 1971).

Facile hydrogenation of 2,3-dihydro-1*H*-pyrrolizin-1-one (**23**) occurs over 5% rhodium-on-carbon to afford 1-hydroxypyrrolidizine (**24**). Meinwald and Ottenheym (1971) assigned this product a trans configuration, whereas earlier workers believed the hydroxyl and C-8 bridgehead hydrogen atom to be cis (Adams *et al.*, 1960).

## III.  INDOLES

Indoles are reduced with relative difficulty (Robinson, 1969). Either ring or both rings may be hydrogenated, and mixtures often result. Adkins and Coonradt (1941) obtained 2,3-dihydroindole by reduction of indole over copper chromite. Indoles can also be reduced under milder conditions over Raney nickel. *N*-Acetylindoles are reduced to the 2,3-dihydro derivatives more selectively than unsubstituted indoles, whereas *N*-alkyl compounds are reduced with more difficulty, and octahydroindole derivatives result (Toth and Gerecs, 1971). In acidic media, the benzene ring of indole is reduced in preference to the pyrrole ring (Karrer and Waser, 1949; Janot *et al.*, 1952; Young and Snyder, 1961).

High yields of indolines can be obtained by reduction of indoles over platinum oxide under ambient conditions in solutions of ethanol and fluoroboric acid (42%, w/w). It is important that either the indole to be reduced be stable in strongly acidic solutions, or the rate of hydrogenation be rapid relative to the rate of polymerization. The method is applicable to indole itself (Smith and Utley, 1965). Indoline is obtained in 80% yield by hydrogenation of indole over palladium hydroxide-on-barium sulfate in acetic acid–hydrochloric acid at 60°C. Hydrogenation of indoline over this catalyst in glacial acetic acid affords octahydroindole in 80% yield, but in 1 $N$ hydrochloric acid 3,3a,4,5,6,7-hexahydro-2$H$-indole is obtained in 64% yield (Butula and Kuhn, 1968).

Hydrogenation of indolizines to the 5,6,7,8-tetrahydro compounds was not achieved over palladium or platinum catalysts, but the reduction proceeded smoothly at room temperature over massive amounts of Raney nickel in alcohol (Walter and Margolis, 1967).

## IV.  QUINOLINES AND ISOQUINOLINES

Quinolines and isoquinolines are usually reduced preferentially in the hetero ring regardless of catalyst, but under special conditions this preference can be reversed. As with pyridines, reduction of the free base of these compounds proceeds less readily than reduction of acid salts or quaternary derivatives. Alkyl substituents in the nitrogen ring increase the relative ease of saturation of the carbocyclic ring and vice versa (von Braun and Lemke, 1930; von Braun et al., 1922, 1923, 1924). However, 2-phenylquinoline affords first the 1,2,3,4-tetrahydro derivative, then the 2-cyclohexyl compound, and finally the decahydro derivative (Oldham and Johns, 1939). Hydrogenation of 1,1′-biisoquinoline in acetic acid over rhodium-on-carbon results in saturation of the carbocyclic rings with formation of 5,5′,6,6′,7,7′,8,8′-octahydro-1,1′-biisoquinoline as the sole crystalline product (Nielsen, 1970). Surprisingly, 3-hydroxyquinoline-8-carboxylic acid is hydrogenated to the 5,6,7,8-tetrahydro derivative (Ochiai et al., 1960).

Hydrogenation to the decahydro derivatives occurs more readily in quinolines than in isoquinolines (Witkop, 1948; Rapala et al., 1957;

Freifelder, 1963a), but in neither type is total reduction particularly easy. Total reduction is facilitated if the inhibiting effects of the nitrogen atom are diminished through salt (Hückel and Stepf, 1927; Overhoff and Wibaut, 1931), or zwitterion formation (Rapala et al., 1957), or acetylation (Woodward and Doering, 1945).

## A.  Catalysts

Palladium (Cavallito and Haskell, 1944; Freifelder, 1964; Wenkert and Wickberg, 1965; Grethe et al., 1968), platinum (Hamilton and Adams, 1928; Wieland et al., 1928), rhodium, ruthenium (Freifelder, 1963a), and nickel (Sugimoto and Kugita, 1958) have been used effectively in the hydrogenation of quinolines. In general, palladium seems to be the preferred metal for selective hydrogenation of the hetero ring, whereas platinum, and ruthenium are probably more suitable for complete saturation. When more than one task is to be accomplished, it may be expeditious to use two catalysts together. By the use of a mixture of platinum oxide and palladium-on-calcium carbonate, 5-bromo-8-nitro-2-methylisoquinolinium p-tosylate is converted in a single step to 8-amino-2-methyl-1,2,3,4-tetrahydroisoquinoline (Mathison and Morgan, 1974).

By the use of potassium borohydride as a source of hydrogen in the presence of palladium-on-carbon, selective hydrogenation of the hetero ring in isoquinolines was achieved without reduction of an aromatic nitro substituent (Neumeyer et al., 1968).

## B.  Reductive Alkylation

Suitably structured quinolines or isoquinolines can undergo inter- or intramolecular reductive alkylation at either carbon or nitrogen. For instance, isoquinolines afford 4-substituted tetrahydroisoquinolines (Grewe et al., 1964).

Hydrogenation of 8-butyrylamino-6-methoxyquinoline (**25**) in acetic acid over platinum oxide affords 8-methoxy-2-(*n*-propyl)-5,6-dihydro-4-imidazo[*i,j*]quinoline (**26**) in 98% yield (Carroll *et al.*, 1976).

**25**                                      **26**

## C.  Reverse Selectivity

Under most conditions, including the use of dilute acid solvents, quinolines and isoquinolines are reduced preferentially, if not exclusively, in the hetero ring (Freifelder, 1963a) (Table I). However, in strong hydrochloric acid the preference can be altered sharply (Ginos, 1975).

**TABLE I[a]**

|  | Product composition (%) | |
|---|---|---|
| Solvent | A | B |
| MeOH | 87 | 13 |
| MeOH, 1 N in dry HCl | 30 | 70 |
| MeOH, 4 N in dry HCl | 13 | 87 |
| Conc HCl[b] | — | 95 |

[a] Data of Ginos (1975). Used with permission.
[b] Vierhapper and Eliel (1974).

Similar but less pronounced changes were found with quinoline hydrochloride as a function of solvent. Sulfuric acid or trifluoroacetic acid can be used as well as hydrochloric acid. The rates tend to be appreciably higher in the fluoro acid (Vierhapper and Eliel, 1974).

## V. IMIDAZOLES AND IMIDAZOLONES

The imidazole ring is reduced with difficulty. Imidazoles carrying carbo-cyclic aromatic substituents are reduced preferentially at the carbocyclic ring (Hartmann and Panizzon, 1938; Hoffman, 1953; Schubert et al., 1962). Reduction of 4-(benzylmethylaminomethyl)imidazole in methanol con-taining hydrogen chloride over 5% palladium-on-carbon split off only the benzyl group and not the 4-methyleneimidazole, which can be considered a benzyl type of radical (Turner et al., 1949). Benzimidazoline was reduced smoothly over platinum oxide in acetic anhydride to diacetylbenzimidazoline in 86% yield (Bauer, 1961).

Imidazolones, like imidazoles, are reduced with relative difficulty. Plati-num oxide, palladium-on-carbon (Duschinsky et al., 1947), and, under more vigorous conditions, nickel (McKennis and DuVigneaud, 1946) have been used successfully. Hydrogenation of 4-methyl-5-benzoylimidazolone-2 pro-ceeds sequentially over platinum oxide, first removing the benzyl oxygen, then saturating the phenyl ring, and finally reducing the imidazolone nucleus. The deoxy compound is best obtained over palladium-on-carbon. Preferential reduction of the ketonic functions is not limited to benzyl ketones; acyl groups are reduced as well (Duschinsky and Dolan, 1945).

## VI. PYRAZINES

Pyrazines undergo hydrogenation to either fully or partially reduced derivatives (Behun and Levine, 1961; Felder et al., 1960; Kipping, 1929, 1932; Magers and Berends, 1959; Mertes and Patel, 1966; Munk and Schultz, 1952; Rogers and Becker, 1966; Taylor and McKillop, 1965). Increased substitution tends to arrest the reduction at a partially reduced stage. Mild conditions, especially over palladium or platinum, are often sufficient, although for sterically hindered rings more vigorous conditions may be needed. Freifelder (1971) suggested that ruthenium may be the catalyst of choice if vigorous conditions are required. As with pyridines, the use of acid solvent helps to neutralize the inhibiting effect of the basic nitrogen.

Early workers thought that partial hydrogenation of the ring afforded the antiaromatic 1,4-dihydropyrazines, but these have been shown to be in fact 1,2-dihydro derivatives. When R = ethyl (27) the reduction stops spontaneously at the 1,2-dihydro stage, but when R = methyl (28) the initial 1,2-dihydro derivative is reduced further to tetra- and hexahydro compounds (Williams et al., 1972). 1,4-Dialkyl-2,6-diphenyl-1,4-dihydro-pyrazines have been prepared and hydrogenated. Hydrogen absorption stops after 1 mole of hydrogen is absorbed. (Lown et al., 1974).

$$\text{ROOC} \diagdown \text{N} \diagup \text{COOR} \qquad \text{ROOC} \diagdown \overset{\text{H H}}{\text{N}} \diagup \text{COOR}$$
$$\text{ROOC} \diagup \text{N} \diagdown \text{COOR} \qquad \text{ROOC} \diagup \text{N} \diagdown \text{COOR}$$

**27**, R = $C_2H_5$
**28**, R = $CH_3$

Benzopyrazines (quinoxalines) are reduced readily, with the nitrogen ring being preferentially saturated. Good yields of 1,2,3,4-tetrahydroquinoxalines were obtained by reduction over platinum oxide (Cavagnol and Wiselogle, 1947) or palladium-on-carbon (Munk and Schultz, 1952) under mild conditions. Low-pressure hydrogenations over massive amounts of nickel have been reported (Broadbent et al., 1960). Under vigorous conditions hydrogenation of benzopyrazines may continue on to the decahydroquinoxalines. Freifelder (1971) suggested that ruthenium may be the catalyst of choice for complete saturation.

## VII.  PYRIMIDINES

Hydrogenation of pyrimidines can be achieved easily, but the reaction may be complex, with hydrogenolysis, hydrolysis, or alcoholysis occurring concomitantly. Palladium (Bauer, 1961; Brown and Evans, 1962; Evans, 1964; Smith and Christiensen, 1955; Wempen et al., 1965), platinum (Aft and Christiensen, 1962; Whitlock et al., 1963), and rhodium (Cline et al., 1959; Cohn and Doherty, 1956; Hanze, 1967; Fox and Van Praag, 1960) have been effective in the hydrogenation of pyrimidines. Acidic solutions are commonly, but not necessarily, employed. Reduction usually stops at the tetrahydro stage (Henze and Winthrop, 1957). Reduction beyond this stage usually occurs with hydrogenolysis of the ring, as might be expected from the *gem*-diamine structure.

Hydroxypyrimidines exist primarily in the pyrimidone structure (Brown et al., 1955; Short and Thompson, 1952), and catalytic hydrogenation of these compounds usually ceases with the carbonyl function intact, even under vigorous conditions (Batt et al., 1954). Hydrolysis may occur during reduction of pyrimidones in aqueous media (Kny and Witkop, 1959).

## VIII.  FURANS

Hydrogenation of furans may afford products of ring addition or of ring cleavage (Zelenkova and Totonomarev, 1951). The ratio of hydrogenolysis to hydrogenation depends on the substrate, catalyst, and reaction environment (Smith, 1957; Shuikin and Vasilevskaya, 1964). The generality has been made that regardless of catalyst an increase in temperature will favor hydrogenolysis (Shuikin et al., 1962; Chouikine and Belsky, 1961). One might guess that an increase in pressure will favor hydrogenation relative to hydrogenolysis. The presence of acid favors hydrogenolysis (Smith and Fuzek, 1949; Londergan et al., 1953).

## A. Catalysts

Palladium (Starr and Hixon, 1934; McCarthy and Kahl, 1956; Huffman and Browder, 1964; Lew, 1965; Brust et al., 1966; Moffatt, 1966; Darling and Wills, 1967; Kaufman et al., 1970), rhodium (Bissell and Finger, 1959; Paquette et al., 1967), and ruthenium (Ponomarev and Chegolya, 1962; Ponomarev et al., 1966; Rylander and Steele, 1967; Webb and Borcherdt, 1951) have given high yields of tetrahydrofurans from furans. Ruthenium requires elevated pressure (> 1000 psig) for satisfactory rates. Both ruthenium dioxide and ruthenium-on-carbon have maintained activity for long periods of use (Ponomarev and Chegolya, 1962). Nickel (Paul and Hilly, 1939; Boekelheide and Morrison, 1958; Mitsui et al., 1960; Mitsui and Saito, 1960; Tarbell and Weaver, 1941) also gives mainly ring-saturated derivatives, but it is much less active than ruthenium. Supported platinum or platinum oxide tends to cause hydrogenolysis of the carbon–oxygen bond (Smith and Fuzek, 1949; Inamura, 1970). The Nishimura catalyst (70% rhodium–30% platinum), which is often used when hydrogenolysis is to be minimized, gives mainly ring-saturated furans (Nishimura, 1960; Nishimura and Taguchi, 1963).

Ambient hydrogenation of 2,5-diferrocenylfuran over 5% rhodium-on-carbon in ethanol afforded quantitative yields of cis-diferrocenyltetrahydrofuran, whereas reduction of this compound over platinum or nickel gave mainly the hydrogenolysis product 1,4-diferrocenylbutane (Yamakawa and Moroe, 1968).

## B.  Solvents

Pronounced solvent effects have been found in the hydrogenation of furans. Hydrogenation of 3,4-dimethyl-2,5-bis(3,4-dimethoxyphenyl)furan (29) over palladium oxide gives a mixture of nordihydroguaiaretic acid tetramethyl ether (30) and isogalbulin (31) in ratios that depend markedly on the solvent (Table II). Good yields of 30 can also be obtained by the use of palladium chloride as catalyst in tetrahydrofuran with sodium acetate present to neutralize liberated hydrogen chloride, a catalyst for the cyclization reaction (Perry et al., 1972).

**TABLE II**[a]

|          |               | Product composition (%) | |
| -------- | ------------- | --- | --- |
| Catalyst | Solvent       | 30  | 31  |
| PdO      | HOAc          | 28  | 58  |
| PdO      | THF           | 81  | 18  |
| PdCl$_2$ | THF(NaOAc)    | 75  | —   |

[a] Data of Perry et al. (1972). Used with permission.

Excellent yields of *all-cis*-3,4-dimethyl-2,5-bis(3,4-dimethoxyphenyl)-tetrahydrofuran (**32**) can be obtained by hydrogenation of **29** over 10% palladium-on-calcium carbonate. Further hydrogenation of **32** over palladium oxide gives a mixture of **30** and **31** in ratios that depend greatly on the solvent. The extremes range from 5% **30**–95% **31** in methanol to 79% **30**–21% **31** in tetrahydrofuran (Perry *et al.*, 1972).

29

32

(89%)

## REFERENCES

Adams, R., Miyano, S., and Fles, D., *J. Am. Chem. Soc.* **82**, 1466 (1960).
Adkins, H., and Coonradt, H. L., *J. Am. Chem. Soc.* **63**, 1563 (1941).
Aeberli, P., and Houlihan, W. J., *J. Org. Chem.* **32**, 3211 (1967).
Aeberli, P., Houlihan, W. J., and Takesue, E. I., *J. Med. Chem.* **12**, 51 (1969).
Aft, H., and Christensen, B. E., *J. Org. Chem.* **27**, 2170 (1962).
Andrews, L. H., and McElvain, S. M., *J. Am. Chem. Soc.* **51**, 887 (1929).
Atkinson, J. H., Grigg, R., and Johnson, A. W., *J. Chem. Soc.* p. 893 (1964).
Barringer, D. F., Jr., Berkelhammer, G., Carter, S. D., Goldman, L., and Lanzilotti, A. E.,
     *J. Org. Chem.* **38**, 1933 (1973).
Batt, R. D., Martin, J. K., Ploeser, J., McT., and Murray, J., *J. Am. Chem. Soc.* **76**, 3663 (1954).
Bauer, H., *J. Org. Chem.* **26**, 1649 (1961).
Behun, J. D., and Levine, R., *J. Org. Chem.* **26**, 3379 (1961).
Biel, J. H., and Aiman, C. E., U.S. Patent 3,310,567 Mar. 21, 1967.
Biel, J. H., and Blicke, F. F., U. S. Patent 3,051,715 Aug. 28, 1962.
Biel, J. H., Friedman, H. L., Leiser, H. A., and Sprengler, E. P., *J. Am. Chem. Soc.* **74**, 1485
     (1952).
Birkofer, L., *Ber. Dtsch. Chem. Ges.* **75**, 429 (1942).
Bissell, E. R., and Finger, M., *J. Org. Chem.* **24**, 1259 (1959).
Boekelheide, V., and Morrison, G. C., *J. Am. Chem. Soc.* **80**, 3905 (1958).
Bradsher, C. K., and Voigt, C. F., *J. Org. Chem.* **36**, 1603 (1971).
Britton, E. C., and Horsley, L. H., U.S. Patent 2,834,784 (1958).
Broadbent, H. S., Allred, E. L., Pendleton, L., and Whittle, C. W., *J. Am. Chem. Soc.* **82**, 189
     (1960).
Brown, D. J., and Evans, R. F., *J. Chem. Soc.* p. 527 (1962).
Brown, D. J., Hoerger, E., and Mason, S. F., *J. Chem. Soc.* p. 211 (1955).
Brust, D. P., Tarbell, D. S., Hecht, S. M., Hayward, E. C., and Colebrook, D. L., *J. Org. Chem.*
     **31**, 2192 (1966).

Büchi, G., and Wüest, H., *J. Org. Chem.* **36**, 609 (1971).
Butula, I., and Khun, R., *Angew. Chem., Int. Ed. Engl.* **7**, 208 (1968).
Cantor, P. A., and VanderWerf, C. A., *J. Am. Chem. Soc.* **80**, 970 (1958).
Carroll, F. I., Blackwell, J. T., Philip, A., and Twine, C. E., *J. Med. Chem.* **19**, 1111 (1976).
Cavagnol, J. C., and Wiselogle, F. Y., *J. Am. Chem. Soc.* **69**, 795 (1947).
Cavallito, C. J., and Haskell, T. H., *J. Am. Chem. Soc.* **66**, 1166 (1944).
Chouikine, N. I., and Belsky, I. F., *Actes Congr. Int. Catal. 2nd, Paris, 1960* **2**, 2625 (1961).
Cline, R. E., Fink, R. M., and Fink, K., *J. Am. Chem. Soc.* **81**, 2521 (1959).
Cohn, W. E., and Doherty, D. G., *J. Am. Chem. Soc.* **78**, 2863 (1956).
Craig, L. C., and Hixon, R. M., *J. Am. Chem. Soc.* **52**, 804 (1930).
Dann, O., and Dimmling, W., *Ber. Dtsch. Chem. Ges.* **86**, 1383 (1953).
Darling, S. D., and Wills, K. D., *J. Org. Chem.* **32**, 2794 (1967).
Dart, M. C., and Henbest, H. B., *Nature (London)* **183**, 817 (1959).
DeJong, M., and Wibaut, J. P., *Rec. Trav. Chim. Pays-Bas* **49**, 237 (1930).
Dolby, L. J., Nelson, S. J., and Senkovich, D., *J. Org. Chem.* **37**, 3691 (1972).
Duschinsky, R., and Dolan, L. A., *J. Am. Chem. Soc.* **67**, 2079 (1945).
Duschinsky, R., Dolan, L. A., Randall, L. O., and Lehmann, G., *J. Am. Chem. Soc.* **69**, 3150 (1947).
Eisner, U., *Chem. Commun.* p. 1348 (1969).
Evans, R. F., *J. Chem. Soc.* p. 2450 (1964).
Felder, E., Maffei, S., Pietra, S., and Pitré, D., *Helv. Chim. Acta* **43**, 888 (1960).
Fox, J. J., and Van Praag, D., *J. Am. Chem. Soc.* **82**, 486 (1960).
Freifelder, M., *J. Org. Chem.* **27**, 4046 (1962).
Freifelder, M., *Adv. Catal.* **14**, 203 (1963a).
Freifelder, M., *J. Org. Chem.* **28**, 1135 (1963b).
Freifelder, M., *J. Org. Chem.* **29**, 2895 (1964).
Freifelder, M., "Practical Catalytic Hydrogenation," p. 577. Wiley (Interscience), New York, 1971.
Freifelder, M., and Stone, G. R., *J. Org. Chem.* **26**, 3805 (1961).
Freifelder, M., Robinson, R. M., and Stong, G. R., *J. Org. Chem.* **27**, 284 (1962).
Freifelder, M., Mattoon, R. W., and Ng, Y. H., *J. Org. Chem.* **29**, 3730 (1964).
Fuhlhage, D. W., and VanderWerf, C. A., *J. Am. Chem. Soc.* **80**, 6249 (1958).
Ginos, J. Z., *J. Org. Chem.* **40**, 1191 (1975).
Graves, T. B., *J. Am. Chem. Soc.* **46**, 1460 (1924).
Grethe, G., Lee, H. L., Uskovic, M., and Brossi, A., *J. Org. Chem.* **33**, 494 (1968).
Grewe, R., Krueger, W., and Vangermain, E., *Ber. Dtsch. Chem. Ges.* **97**, (1), 119 (1964).
Grigor'eva, N. E., Oganes'yan, A. B., and Mysh, I. A., *Zh. Obshch. Khim.* **27**, 1565 (1957).
Hall, H. K., Jr., *J. Am. Chem. Soc.* **80**, 6412 (1958).
Hall, H. K., Jr., and Johnson, R. C., *J. Org. Chem.* **37**, 697 (1972).
Hamilton, T. S., and Adams, R., *J. Am. Chem. Soc.* **50**, 2260 (1928).
Hanze, A. R., *J. Am. Chem. Soc.* **89**, 6720 (1967).
Hartmann, M., and Panizzon, L., *Helv. Chim. Acta* **21**, 1692 (1938).
Hayes, F. N., King, L. C., and Peterson, D. E., *J. Am. Chem. Soc.* **78**, 2527 (1956).
Heitmeier, D. E., Hortenstine, J. T., Jr., and Gray, A. P., *J. Org. Chem.* **36**, 1449 (1971).
Henze, H. R., and Winthrop, S. O., *J. Am. Chem. Soc.* **79**, 2230 (1957).
Hoffman, K., "Imidazole and Its Derivatives," p. 16. Wiley (Interscience), New York, 1953.
Howton, D. R., and Golding, D. R. V., *J. Org. Chem.* **15**, 1 (1950).
Hückel, W., and Stepf, F., *Justus Liebigs Ann. Chem.* **453**, 163 (1927).
Huffman, J. W., and Browder, L. E., *J. Org. Chem.* **29**, 2598 (1964).
Inamura, Y., *Bull. Chem. Soc. Jpn.* **43**, 3271 (1970).

Janot, M., Keufer, J., and LeMen, J., *Bull. Soc. Chim. Fr.* p. 230 (1952).

Karrer, P., and Waser, P., *Helv. Chim. Acta* **32**, 409 (1949).

Kaufman, K. D., Worden, L. R., Lode, E. T., Strong, M. K., and Reitz, N. C., *J. Org. Chem.* **35**, 157 (1970).

Kipping, F. B., *J. Chem. Soc.* p. 2889 (1929).

Kipping, F. B., *J. Chem. Soc.* p. 1336 (1932).

Kny, H., and Witkop, B., *J. Am. Chem. Soc.* **81**, 6245 (1959).

Koelsch, C. F., and Carney, J. J., *J. Am. Chem. Soc.* **72**, 2285 (1950).

Kray, L. R., and Reinecke, M. G., *J. Org. Chem.* **32**, 225 (1967).

Lew, B. W., U.S. Patent 3,225,066, Dec. 21, 1965.

Liu, H. J., Valenta, Z., and Yu, T. T. J., *Chem. Commun.* p. 1116 (1970).

Londergan, T. E., Hause, N. L., and Schmitz, W. R., *J. Am. Chem. Soc.* **75**, 4456 (1953).

Lown, J. W., Akhtar, M. H., and McDaniel, R. S., *J. Org. Chem.* **39**, 1998 (1974).

Lyle, R. E., and Lyle, G. G., *J. Am. Chem. Soc.* **76**, 3536 (1954).

Lyle, R. E., and Mallett, S. E., *Ann. N.Y. Acad. Sci.* **145**, 83 (1967).

Lyle, R. E., Perlowski, E. F., Troscianiec, H. J., and Lyle, G. G., *J. Org. Chem.* **20**, 1761 (1955).

McCarthy, W. C., and Kahl, R. J., *J. Org. Chem.* **21**, 1118 (1956).

McElvain, S. M., and Adams, R., *J. Am. Chem. Soc.* **45**, 2738 (1923).

McKennis, H., Jr., and DuVigneaud, V., *J. Am. Chem. Soc.* **68**, 832 (1964).

Magers, H. I. X., and Berends, W., *Rec. Trav. Chim. Pays-Bas* **78**, 109 (1959).

Marion, L., and Cockburn, W. F., *J. Am. Chem. Soc.* **71**, 3402 (1949).

Mathison, I. W., and Morgan, P. H., *J. Org. Chem.* **39**, 3210 (1974).

Maxted, E. B., and Walker, A. G., *J. Chem. Soc.* p. 1093 (1948).

Meinwald, J., and Ottenheym, H. C. J., *Tetrahedron* **27**, 3307 (1971).

Mertes, M. P., and Patel, N. R., *J. Med. Chem.* **9**, 868 (1966).

Minor, W. F., Hoekstra, J. B., Fisher, D., and Sam, J., *J. Med. Chem.* **5**, 96 (1962).

Mistryukov, E. A., Ilkova, E. L., and Ryashentseva, M. A., *Tetrahedron Lett.* p. 1691 (1971).

Mitsui, S., and Saito, H., *Nippon Kagaku Zasshi* **81**, 289 (1960).

Mitsui, S., Ishikawa, Y., Takeuchi, Y., Saito, H., and Mamuro, H., *Nippon Kagaku Zasshi* **81**, 286 (1960).

Möhrle, H., and Weber, H., *Tetrahedron* **27**, 3241 (1971).

Moffatt, J. S., *J. Chem. Soc. C* p. 734 (1966).

Munk, M., and Schultz, H. P., *J. Am. Chem. Soc.* **74**, 3433 (1952).

Neumeyer, J. L., McCarthy, M., Weinhardt, K.K., and Levins, P. L., *J. Org. Chem.* **33**, 2890 (1968).

Nielsen, A. T., *J. Org. Chem.* **35**, 2498 (1970).

Nienburg, H., *Ber. Dtsch. Chem. Ges. B* **70**, 635 (1937).

Nishimura, S., *Bull. Chem. Soc. Jpn.* **33**, 566 (1960).

Nishimura, S., and Taguchi, H., *Bull. Chem. Soc. Jpn.* **36**, 353 (1963).

Ochiai, E., Tsuda, K., and Ikuma. S., *Ber. Dtsch. Chem. Ges.* **69**, 2238 (1936).

Ochiai, E., Kaneko, C., Shimada, I., Murata, Y., Kosuge, T., Miyashita, S., and Kawasaki, C., *Chem. Pharm. Bull.* **8**, 126 (1960).

Oldham, W., and Johns, I. B., *J. Am. Chem. Soc.* **61**, 3289 (1939).

Orthner, L., *Justus Liebigs Ann. Chem.* **456**, 225 (1927).

Overberger, C. G., Palmer, L. C., Marks, B. S., and Byrd, N. R., *J. Am. Chem. Soc.* **77**, 4100 (1955).

Overhoff, J., and Wibaut, J. P., *Rec. Trav. Chim. Pay-Bas* **50**, 957 (1931).

Paquette, L. A., Youssef. A. A., and Wise, M. L., *J. Am. Chem. Soc.* **89**, 5246 (1967).

Patterson, J. M., Brasch, J., and Drenchko, P., *J. Org. Chem.* **27**, 1652 (1962).

Paul, R., and Hilly, G., *C. R. Acad. Sci.* **208**, 259 (1939).

Perry, C. W., Kalnins, M. V., and Deitcher, K. H., *J. Org. Chem.* **37**, 4371 (1972).
Ponomarev, A. A., and Chegolya, A. S., *Dokl. Akad. Nauk SSSR* **145**, 812 (1962).
Ponomarev, A. A., Chegolya, A. S., Smirnova, N. S., and Dyukareva, V. N., *Khim. Geterotsikl. Soedin.* No. 2, p. 163 (1966).
Putokhin, N. J., *Zh. Russ. Fiz.-Khim. Ova.* **62**, 2216 (1930).
Raasch, M. S., *J. Org. Chem.* **27**, 1406 (1962).
Raines, S., Chai, S.Y., and Palopoli, F. P., *J. Org. Chem.* **36**, 3992 (1971).
Rapala, R. T., Lavagnino, E. R., Shepard, E. R., and Farkas, E., *J. Am. Chem. Soc.* **79**, 3770 (1957).
Rapoport, H., Christian, C. G., and Spencer, G., *J. Org. Chem.* **18**, 840 (1954).
Robinson, B., *Chem. Rev.* **69**, 785 (1969).
Robison, M. M., Butler, F. P., and Robison, B. L., *J. Am. Chem. Soc.* **79**, 2573 (1957).
Rogers, E. F., and Becker, H. J., U.S. Patent 3,281,423 Oct. 25. 1966.
Rylander, P. N., and Steele, D. R., *Engelhard Ind., Tech. Bull.* **3**, 19 (1962).
Rylander, P. N., and Steele, D. R., *Engelhard Ind., Tech. Bull.* **5**, 113 (1965).
Rylander, P. N., and Steele, D. R., *Engelhard Ind., Tech. Bull.* **7**, 153 (1967).
Schöpf, C., Herbert, G., Rausch, R., and Schröder, G., *Angew. Chem.* **69**, 391 (1957).
Schubert, H., Berg, W. V., and Andrae, H., *Wiss. Z. Martin-Luther-Univ., Halle-Wittenberg, Math.-Naturwiss. Reihe* **11**(5), 603 (1962).
Schweizer, E. E., and Light, K. K., *J. Org. Chem.* **31**, 870 (1966).
Shapiro, S. L., Weinberg, K., Bazga, T., and Freedman, L., *J. Am. Chem. Soc.* **81**, 5146 (1959).
Shapiro, S. L., Freedman, L., and Weinberg, K., U.S. Patent 3,056,797, Oct. 2, 1962.
Short, L. N., and Thompson, H. W., *J. Chem. Soc.* p. 168 (1952).
Shuikin, N. I., and Vasilevskaya, G. K., *Izv. Akad. Nauk SSSR, Ser. Khim.* p. 557 (1964).
Shuikin, N. I., Bel'skii, I. F., and Vasilevskaya, G. K., *Zh. Obshch. Khim.* **32**, 2911 (1962).
Skomoroski, R. M., and Schriesheim, A., *J. Phys. Chem.* **65**, 1340 (1961).
Smith, A., and Utley, J. H. P., *Chem. Commun.* p. 427 (1965).
Smith, H. A., *in* "Catalysis" (P.H. Emmett, ed.), Vol. 5, p. 175. Reinhold, New York, 1957.
Smith, H. A., and Fuzek, J. F., *J. Am. Chem. Soc.* **71**, 415 (1949).
Smith, V. H., and Christiensen, B. E., *J. Org. Chem.* **20**, 829 (1955).
Sorm. F., *Collect. Czech. Chem. Commun.* **13**, 57 (1948).
Späth, E., and Kuffner, F., *Ber. Dtsch. Chem. Ges.* **68**, 494 (1935).
Sprake, J. M., and Watson, K. D., *J. Chem. Soc., Perkin Trans. I* p. 5 (1976).
Starr, D., and Hixon, R. M., *J. Am. Chem. Soc.* **56**, 1595 (1934).
Steck, E. A., Fletcher, L. T., and Brundage, R. P., *J. Org. Chem.* **28**, 2233 (1963).
Sugimoto, N., and Kugita, H., *Chem. Pharm. Bull.* **6**, 432 (1958).
Sundberg, R. J., Russell, H. F., Ligon, W. V., Jr., and Lin, L. S., *J. Org. Chem.* **37**, 419 (1972).
Supniewski, J. V., and Serafinowna, M., *Arch. Chem. Farm.* **3**, 109 (1936).
Tarbell, D. S., and Weaver, C., *J. Am. Chem. Soc.* **63**, 2939 (1941).
Taylor, E. C., and McKillop, A., *J. Am. Chem. Soc.* **87**, 1984 (1965).
Toth, T., and Gerecs. A., *Acta Chim. Acad. Sci. Hung* **67**(2), 229 (1971); *Chem. Abstr.* **74**, 125322r (1971).
Turner, R. A., Huebner, C. F., and Scholz, C. R., *J. Am. Chem. Soc.* **71**, 2801 (1949).
van Bergen, T. J., and Kellogg, R. M., *J. Org. Chem.* **36**, 978 (1971).
Vierhapper, F. W., and Eliel, E. L., *J. Org. Chem.* **96**, 2256 (1974).
Vierhapper, F. W., and Eliel, E. L., *J. Org. Chem.* **40**, 2729 (1975).
von Braun, J., and Lemke, G., *Justus Liebigs Ann. Chem.* **478**, 176 (1930).
von Braun, J., Petzold, A., and Seeman, J., *Ber. Dtsch. Chem. Ges. B* **55**, 3779 (1922).
von Braun, J., Gmelin, W., and Schultheiss, A., *Ber. Dtsch. Chem. Ges. B* **56**, 1338 (1923).
von Braun, J., Gmelin, W., and Petzold, A., *Ber. Dtsch. Chem. Ges. B* **57**, 382 (1924).

Walker, G. N., *J. Org. Chem.* **27**, 1929 (1962).

Walker, G. N., Smith, R. T., and Weaver, B. N., *J. Med. Chem.* **8**, 626 (1965).

Walter, L. A., and Margolis, P., *J. Med. Chem.* **10**, 498 (1967).

Webb, I. D., and Borcherdt, G. T., *J. Am. Chem. Soc.* **73**, 752 (1951).

Weinstein, B., and Craig, A. R., *J. Org. Chem.* **41**, 875 (1976).

Wempen, I., Brown, G. B., Ueda, T., and Fox, J. J., *Biochemistry* **4**(1), 54 (1965).

Wenkert, E., and Wickberg, B., *J. Am. Chem. Soc.* **87**, 1580 (1965).

Wenkert, E., Dave, K. G., Haglid, F., Lewis, R. G., Oishi, T., Stevens, R. V., and Terashima, M., *J. Org. Chem.* **33**, 747 (1968).

Whitlock, B. J., Lipton, S. H., and Strong, F. M., *J. Org. Chem.* **30**, 115 (1963).

Wibaut, J. P., and Kloppenburg, C. C., *Rev. Trav. Chim. Pays-Bas* **65**, 100 (1946).

Wibaut, J. P., and Proost, W., *Rec. Trav. Chim. Pays-Bas* **52**, 333 (1933).

Wibaut, J. P., Kloppenburg, C. C., and Beets, M. G. J., *Rec. Trav. Chim. Pays-Bas* **63**, 134 (1944).

Wieland, H., Hettche, O., and Hoshina, T., *Ber. Dtsch. Chem. Ges.* **61**, 2371 (1928).

William, J. R., Cossey, J. J., and Adler, M., *J. Org. Chem.* **37**, 2963 (1927).

Witkop, B., *J. Am. Chem. Soc.* **70**, 2617 (1948).

Woodward, R. B., and Doering, W. E., *J. Am. Chem. Soc.* **67**, 860 (1945).

Yakhontov, L. N., Yatesenko, S. V., and Rubtsov, M. V., *Zh. Obshch. Khim.* **28**, 3115 (1958).

Yamakawa, K., and Moroe, M., *Tetrahedron* **24**, 3615 (1968).

Yamamoto, K., *Yakugaku Zasshi* **76**, 922 (1956).

Young, D. V., and Snyder, H. R., *J. Am. Chem. Soc.* **83**, 3160 (1961).

Zelenkova, V. V., and Totonomarev, A. A., *Usp. Khim.* **20**, 589 (1951); *Chem. Abstr.* **48**, 663 (1954).

*Chapter* 13

# Catalytic Dehydrohalogenation

Catalytic dehalogenation is a convenient way of removing halogen under mild conditions. The reaction has wide synthetic utility, for halogen is often introduced transiently in reaction sequences. Halogens also serve effectively as blocking groups.

## I. CATALYSTS

Catalysts differ widely in their effectiveness in dehalogenations. Palladium is very active and preferred by most investigators. Nickel has been used effectively, but usually massive amounts are required. Platinum and rhodium are relatively ineffective, and their use is indicated in reductions of compounds when dehalogenation is to be minimized.

## II. BASIC MEDIA

Catalytic dehalogenation is often inhibited by release of the halogen acid, and to circumvent this difficulty many workers add 1 mole or more of a base to the reaction system. It has been inferred that the inhibiting species is hydrogen ion rather than halide ion and that the function of added base is to neutralize the acid and maintain the catalytic metal in a low valency or nucleophilic state (Denton *et al.*, 1964).

$$n\text{Pd} + 2m\text{H}^+ \rightleftharpoons (n - 2m)\text{Pd} + 2m(\text{PdH})^+ \rightleftharpoons n\text{Pd}^{2+} + m\text{H}_2$$

A variety of halogen acceptors have been used successfully including sodium, potassium, calcium, and barium hydroxides, magnesium oxide, sodium acetate, amines, and ammonia. Bases may well have other functions besides neutralization of the liberated acid, for empirically it has been found that various bases are not always interchangeable, and the results sometimes depend on the base employed. Magnesium oxide has proved to be useful in limiting the depth of hydrogenation. For instance, dehalogenation of chloropyrimidines over palladium in the presence of sodium acetate (Whittaker, 1950) or barium hydroxide (Whittaker, 1953) afforded tetrahydropyrimidines, but in the presence of magnesium oxide the pyrimidine was obtained. Similarly, the presence of triethylamine was effective in preventing double-bond saturation during dehalogenation of a complex iodoolefin (Pfitzner and Moffatt, 1964).

Diamines may function differently than monoamines in dehalogenation. Reduction over nickel of either 6,6-dichlorobicyclo[3.1.0]hexane or 2-oxa-7,7-dichlorobicyclo[4.1.0]heptane in the presence of mono-, di-, or triethylamine affords the cis-chloropropane derivative, whereas, in the presence of alkali or ethylene-, propylene-, or hexamethylenediamine the trans isomer is the main product. The diamines are thought to function through chelate formation (Isogai et al., 1970). Reinecke (1964) found that potassium hydroxide and triethylamine produced different product compositions in the dehalogenation of 4-dichloromethyl-4-methylcyclohexanone over 10% palladium-on-carbon.

## III. EFFECT OF STRUCTURE

The rate of catalytic dehalogenation depends markedly on structure and on the halogen involved. The rate of hydrogenolysis decreases with increasing electronegativity of the halogen, so that fluoride is often removed only with difficulty. Dehalogenation is accelerated by neighboring electrophilic groups, as indicated in Table I, which gives selected data from Denton et al. (1964). The efficacy of added base is also indicated, the effect being largest with the more slowly reduced compounds.

The greater reactivity of the less electronegative halogens can be used to advantage. Attempted debenzylation and debromination of 1 over palladium-on-carbon gave mainly the debenzylated derivative 2. Successful dehalogenation (3) was achieved by replacement of bromine with iodine and reduction over 5% palladium-on-carbon in methanol containing triethylamine (Albano and Horton, 1969).

**TABLE I**

**Rate of Hydrogenation of Various Halides in Methanol**[a]

| | Rate ($mlH_2/min/mg$ Pd-on-C $\times 10^3$) | |
| --- | --- | --- |
| Compound | Without base | With 1 equivalent of KOAc |
| $n\text{-}C_7H_{15}Br$ | Nil | Nil |
| $\phi Cl$ | 10 | 75 |
| $\phi Br$ | 79 | 125 |
| $\phi CH_2CH_2Br$ | 6.5 | 25 |
| $\phi CH_2CH_2CH_2Br$ | 7.5 | 17 |
| $\phi COCH_2Br$ | 122 | 165 |
| $C_2H_5OCOCH_2Br$ | 11 | 68 |

[a] Data of Denton *et al.* (1964). Used with permission.

Kraus and Bazant (1973) examined the vapor-phase hydrogenolysis of halobenzenes and some derivatives over palladium-on-carbon and concluded that the displacement of halogen proceeds by attack of a hydridic species in an $S_{N^2}$ type of reaction. The suggestion is supported by a positive value of $\rho$, the reaction parameter in the Hammett relationship, and by the order of reactivities of halogen (Br > Cl > F), which is the same as that obtained in substitution reactions by nucleophilic agents. Substituents decrease the rate of hydrogenolysis of chlorobenzene in the order 3-$CF_3$ > 3-Cl > 3-F > H > 3-$CH_3$ > 4-$CH_3$. Other workers (Baltzly and Phillips, 1946; Ruzicka and Procházka, 1970) using polar solvents found somewhat different substituent effects, perhaps due to competition of solvent with substrate for catalyst sites.

Dehalogenation is greatly accelerated by an adjacent carbonyl function (Table I). Use was made of this enhanced reactivity in a synthesis of L-hydroproline that included a regioselective hydrogenolysis of the trichloro compound **4** to the dichloro compounds **5** and **6**. This reduction was readily achieved over 5% palladium-on-carbon in the presence of excess sodium acetate in acetic acid, but without much stereoselectivity (Equchi and Kakuta, 1974).

| **4** | **5** (30–40%) | **6** (40–45%) |
| --- | --- | --- |

The following scheme has been tentatively suggested to account for the activating influence of adjacent carbonyl groups (Denton *et al.*, 1964):

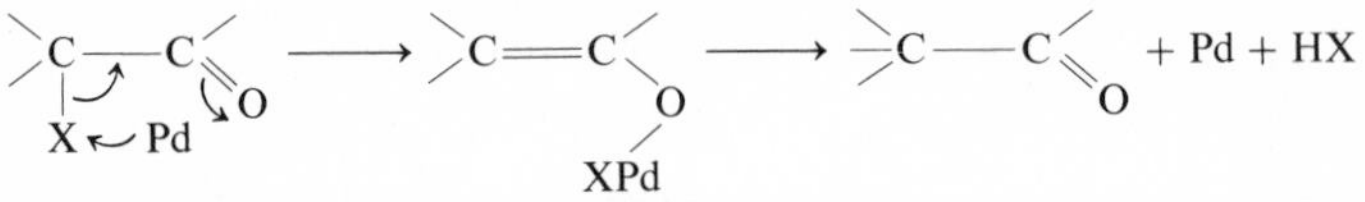

The causes of activation of halo ketones are complex, as was noted in the interesting differences in the rate of hydrogenolysis of 4α-chloro-4α-bromo- and 2α,4α-dibromocholestanones. Bromide in the former is lost slowly; in the latter it is lost so rapidly that no monobromo intermediate could be observed. The result suggests that reaction of the readily removed 2α-bromo group potentiates displacement of the 4α-bromo group (Denton *et al.*, 1964). Because of the enhanced reactivity of halo ketones, it is usually possible to carry out dehalogenations over palladium with little or no reduction of the ketonic function, even when it is adjacent to an aromatic ring (Sargent and Agar, 1958; Lyle and Troscianiec, 1959; Huffman, 1959; Barnes and Geber, 1961; Arkley *et al.*, 1963).

Another example of the activating effect of a bromo substituent on chlorine is shown by comparison of the reduction of **7** and **8**. In **7**, both halogens are

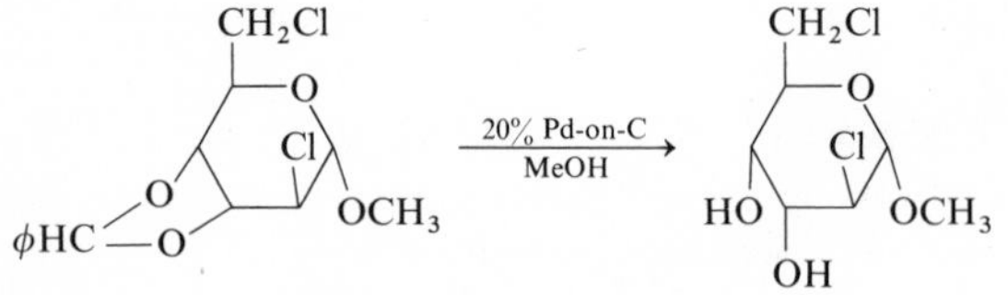

**7**

inert to hydrogenolysis, whereas in **8** both the bromine and the adjacent chlorine are removed (Hanessian and Plessas, 1969).

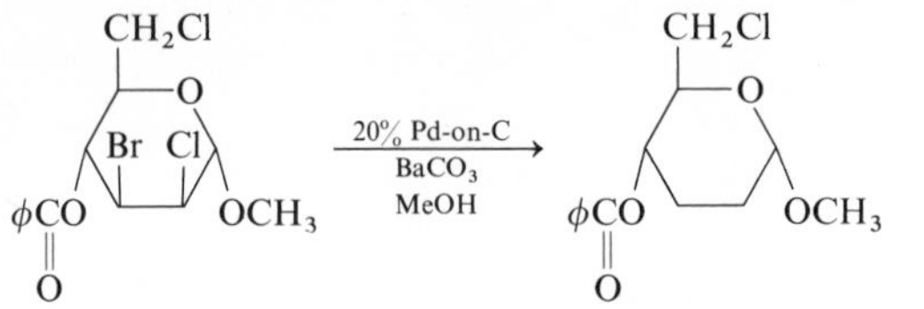

**8**

## IV. ELIMINATION REACTIONS

Catalytic dehalogenation may be accompanied by elimination of suitably disposed adjacent groups. Hydrogenolysis of **10** over palladium-on-carbon in methanol containing triethylamine proceeds smoothly, affording crystalline 3'-deoxytubercidin (**12**) in 62% yield. However, similar hydrogenolysis of the protected 3-bromonucleoside **9** followed by hydrolysis gave mainly **12** and 2',3'-dideoxytubercidin (**11**) in approximately equal amounts (Jain *et al.*, 1973).

The adjacent 2'-*O*-acetyl group is necessary for trans elimination, which does not occur with the 2'-hydroxyl group. The reaction has been pictured as involving an olefin intermediate (Russell *et al.*, 1973).

A similar elimination followed by reduction was observed during hydrogenolysis of a 2'-bromo-2'-deoxy-3'-*O*-methanesulfonyluridine derivative, but in that case the eliminated groups were cis to each other (Furukawa *et al.*, 1970). Olefin intermediates have been postulated to account for the loss of both hydroxyl and bromo functions during hydrogenolysis of a bromohydrin in the 16-acetyl-5β-androstane series (Taub *et al.*, 1961).

In compounds in which the olefin derived by elimination is resistant to hydrogenolysis, it may appear as the major product of the reaction. Cases in point are the conversion of the bromolactone **13** to dihydro-ψ-santonin (**14**) and of 12-bromooleanolic acid lactone (**15**) to oleanolic acid (**16**) (Denton *et al.*, 1964).

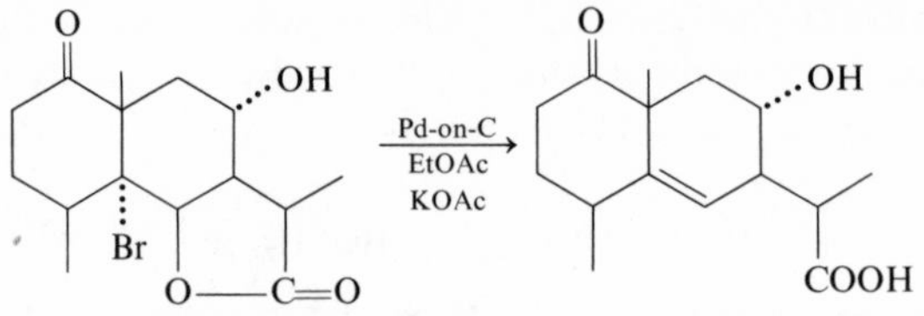

**13**

**14**
(100%)

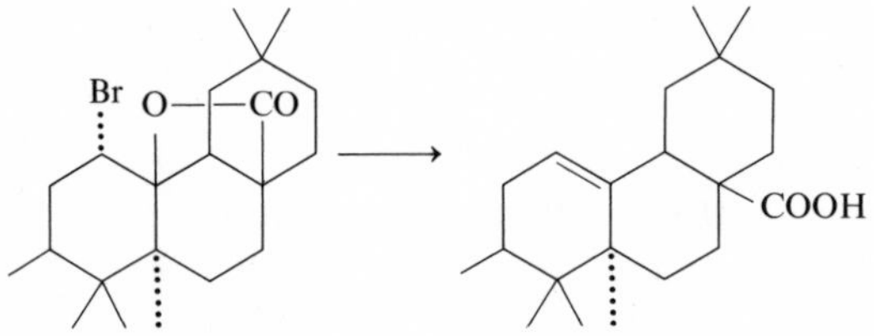

**15**                                                    **16**

## V.  HALOHYDRINS

A useful method for the hydroxylation of a double bond involves selective hydrogenation of an intermediate halohydrin. A preferred route to 14β-hydroxy steroids from 14-olefin precursors includes this sequence (Fritsch *et al.*, 1969; Stache *et al.*, 1969, 1971). As applied to bufalin (**18**), hydrogenolysis of either the bromohydrin or iodohydrin (**17**) proceeds easily over either Urushibara nickel A (Urushibara *et al.*, 1955) or Raney nickel. Comparable results were obtained in the conversion of 3β-acetoxy-14-dehydrobufalin to 3β-acetoxybufalin. However, low yields were obtained when the hydrogenolysis was carried out with the chlorohydrins (Kamano and Pettit, 1973). Saturation or hydrogenolysis of the unsaturated lactone was unimportant in these reactions.

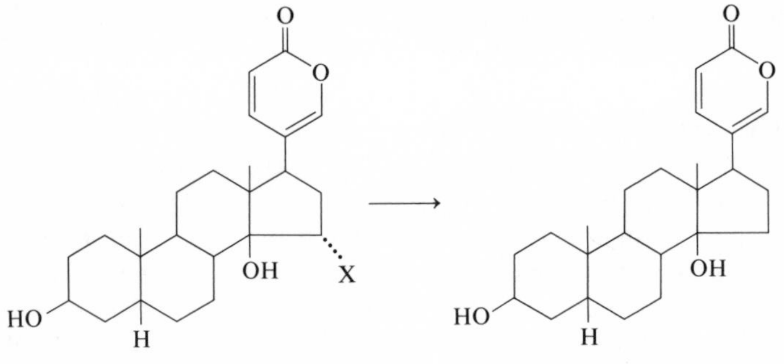

**17, X = Br, I**                                        **18**

Dehalogenation of **19** over palladium-on-calcium carbonate in ethanol proceeded with retention of configuration to afford *cis*-phenylcyclohexanol (**20**) (Berti *et al.*, 1968). Other reactions of halohydrins are discussed in Section IV.

**19**                    **20**

## VI.  HALOOLEFINS

Reduction of both functions in a haloolefin presents no problem. By employing a basic medium and palladium, the halogen can be removed even if its activation is lost by prior hydrogenation of the olefin. Dehalogenation without olefin saturation is more difficult; hydrogenation and hydrogenolysis often proceed concomitantly. The ratio of these reactions may be influenced by solvent (Ham and Coker, 1964). Palladium together with a hydrogen halide acceptor would seem to offer the best chance for a selective dehalogenation (Rosenmund and Zetzsche, 1918; Conroy, 1955). This technique was applied to the synthesis of a number of $\alpha,\beta$-unsaturated aldehydes derived from ketones by a Vilsmeier formylation (Traas *et al.*, 1977).

Bromoaldehydes give higher yields of unsaturated aldehydes than do the corresponding chloro compounds, reflecting the more facile hydrogenolysis of bromides. Palladium inhibited by quinoline and sulfur was also effective in this selective hydrogenation (Traas *et al.*, 1977).

## VII.  HALONITRO COMPOUNDS

Aromatic nitro functions are reduced very readily, and the reaction usually takes precedence over dehalogenation. Aliphatic nitro groups, on the other hand, are reduced with much more difficulty, and the reverse selectivity is apt to occur. An improved synthesis of secondary nitroparaffins involves this type of selectivity. The synthesis proceeds through chlorination of an oxime to a chloronitroso compound, oxidation with ozone of the nitroso group to a nitro group, and selective catalytic dehydrohalogenation (Barnes, 1976).

$$
\underset{\underset{\displaystyle NOH}{|}}{\overset{\displaystyle NO}{\underset{\|}{RCR'}}} \xrightarrow[CH_2Cl_2]{Cl_2} \underset{\underset{\displaystyle Cl}{|}}{\overset{\displaystyle NO}{\underset{|}{RCR'}}} \xrightarrow{O_3} \underset{\underset{\displaystyle Cl}{|}}{\overset{\displaystyle NO_2}{\underset{|}{RCR'}}} \xrightarrow[\substack{NaOH \\ H_2O,\ H_2}]{Pd\text{-on-}C} RCH(NO_2)R'
$$

Similarly, hydrogenolysis of the halogen in the chloronitro compound **23** occurs preferentially, and the reduction stops spontaneously after absorption of 1 mole of hydrogen, apparently due to poisoning of the catalyst by hydrogen chloride, to afford **24**. The dehalogenated amine **25** can be obtained directly from the nitro alcohol when the reduction is conducted in the presence of 1 mole of sodium acetate (Marquardt and Edwards, 1972). The amine **25** was not obtained by hydrogenolysis of the nitro alcohol **22** in ethanolic acetic acid over palladium-on-carbon because the hydroxyl function is resistant to catalytic cleavage due to stabilization of the benzylic function by a vicinal amino group (Zenitz *et al.*, 1948).

$$
\underset{\substack{|\\ CH_3}}{\overset{OH\ CH_3}{\underset{|\ \ |}{\phi CHCNH_2}}} \xleftarrow[H_2]{Pd\text{-on-}C} \underset{\substack{|\\ CH_3}}{\overset{OH\ CH_3}{\underset{|\ \ |}{\phi CHCNO_2}}} \longrightarrow \underset{\substack{|\\ CH_3}}{\overset{Cl\ CH_3}{\underset{|\ \ |}{\phi CHCNO_2}}} \xrightarrow[H_2]{Pd\text{-on-}C} \underset{\substack{|\\ CH_3}}{\overset{CH_3}{\underset{|}{\phi CH_2CNO_2}}}
$$

$$\qquad\quad \textbf{21} \qquad\qquad\qquad \textbf{22} \qquad\qquad\qquad \textbf{23} \qquad\qquad\qquad \textbf{24}$$

$$\Big\downarrow \substack{H_2 \\ NaOAc}$$

$$
\underset{\substack{|\\ CH_3}}{\overset{CH_3}{\underset{|}{\phi CH_2CNH_2}}}
$$

**25**

## VIII. POLYHALO COMPOUNDS

Selective dehydrohalogenation of polyhalo compounds is common. The major products often can be predicted. Sterically unhindered halogens are removed more readily than hindered halogens, and halogens activated by aryl (Wilson and Harris, 1951; Hayward *et al.*, 1968), benzyl (Overberger and Monagle, 1956), vinyl (Wiberg and Hess, 1966; Wilt and Vasiliauskas, 1972), allyl (McBee and Smith, 1955; Conroy, 1955), carbonyl (Cromwell and Cram, 1945; May and Mosettig, 1946), or other functions are more readily reduced than nonactivated halogen. Aromatization may also provide a driving force for dehalogenation (Gaertner, 1954). In compounds containing two different halogens the least electronegative is usually removed preferentially (Vavon and Mathieu, 1938; Britton and Keil, 1955; Florin et al., 1959; Johnson and Nasutavicus, 1963; Manatt et al., 1964; Lozeron et al., 1964).

An accumulation of halogen on a single carbon atom increases the ease of dehalogenation. Selective stepwise dehalogenation of this type of compound usually can be achieved readily, since each loss of halogen results in a compound of increased stability to hydrogenolysis (Batzly and Phillips, 1946; Wineman et al., 1958; Hofmann et al., 1959; Cohen et al., 1962; Fuqua et al., 1964). A case in point is the selective hydrogenation of 6-dichloromethyl-2,6-dimethylcyclohexa-2,4-dien-1-one (24), a product of an abnormal Reimer–Tieman reaction of 2,6-xylenol. In this case stepwise dehalogenation through 27 is the preferred route to the fully dehalogenated product 28 because, if complete dehalogenation of the dichloro compound 26 over palladium in the presence of potassium hydroxide is carried out in a single step, the yield of 28 is only 33% accompanied by 40% of the bicyclic ketone 29. Abnormal Reimer–Tieman products are obtained readily, and the sequence of reactions offers a convenient route to alicyclic ketones, bicyclic ketones, and chloromethylcyclohexanones (Brieger et al., 1969).

A useful procedure for the introduction of a methyl substituent involves dehydrohalogenation of a dibromomethylene derivative (Telschow and Reusch, 1975). This method was used earlier for the preparation of 6-methyl-$\Delta^4$-3-keto steroids (Liisberg et al., 1960).

Halogens attached to aromatic systems can be removed stepwise (Redman and Weimer, 1960) and, in unsymmetric structures, regioselectively. As a general guide, preferential hydrogenation will probably occur with the halogen that is most susceptible to nucleophilic displacement. In application of the guide due consideration should be given to the demanding steric requirements of catalytic hydrogenation.

Stepwise dehydrohalogenation of ethyl 2,6-dichloro-5-nitro-4-pyrimidine-carboxylate was achieved with palladium-on-carbon in dioxane in the presence of magnesium oxide, followed by reduction in ethanol containing triethylamine (Gallemaers *et al.*, 1976).

The detailed studies of Campbell *et al.* (1973) on the dehalogenation of 2,4-dichloro-5-(2,4-dichlorodiphenylmethyl)pyrimidine (**31**) illustrate the influence that catalyst, solvent, and additives can have on selectivity. All halogens are easily removed over palladium-on-carbon in ethanol containing potassium hydroxide. The use of a weaker base, sodium acetate, in ethanol resulted in the loss of three halogens with retention of an *o*-phenyl chlorine (**30**). A weak base, triethylamine, in a nonpolar solvent, ethyl acetate, gave the dichlorophenyl derivative **32**, in 75–90% yield over palladium-on-carbon. The yield of **32** rose to 90–100% when a less active catalyst, Raney nickel, was employed with triethylamine in a more polar solvent, methanol.

## IX. HALOGEN AS BLOCKING GROUP

Aromatically bound halogen is useful for blocking reactive positions on the aromatic nucleus. Huffman (1959) made use of this blocking group in the synthesis of 5-methoxy-1-tetralone through succinoylation of *p*-chloroanisole, cyclization, and dehydrohalogenation over palladium-on-carbon in ethanol containing triethylamine. Halogen blocking proved to be useful in the preparation of isomerically pure 1,3-disubstituted pyrazoles. The syntheses involved lithiation of **33** followed by interaction with a carbonyl compound to afford **34**. Without the 5-chloro substituent, metalation occurs exclusively at the 5 position. Dehalogenation to **35** without hydrogenolysis of the benzyl hydroxyl was achieved easily (Butler and Alexander, 1972).

## X.  TRAPPING REACTIONS

Controlled dehalogenation is a useful technique for trapping certain unstable compounds as their hydrochloride salts. Hydrogenolysis of a mixture of azidomethylpyrazine (**36**) and dichloromethylpyrazine (**31**) liberated hydrogen chloride at a proper rate to permit isolation of the aminomethylpyrazine hydrochloride **38** in good yields. Attempts to form the hydrochloride by substituting chloroform as the hydrogen chloride precursor or to use hydrogen chloride in the reaction mixture were unsuccessful (Abushanab *et al.*, 1973).

On the other hand, Secrist and Logue (1972) found chloroform to be quite effective as a source of hydrogen chloride. Hydrogenation of methyl-5-azido-5-deoxy-2,3-*O*-isopropylidene-*β*-D-ribofuranoside (**39**) which contains the acid-labile acetal and isopropylidene functions, cleanly afford the amine hydrochloride **40** without affecting either acid-sensitive group. In contrast, the use of methanolic hydrogen chloride resulted in removal of the isopropylidene group, whereas no reduction at all took place with hydrogen chloride in ether. The technique was shown to be applicable to nitro, oximino, and nitrile functions. Palladium-on-carbon and platinum oxide are both satisfactory catalysts.

N₃H₂C    O    OCH₃          Cl⁻ H₃ṄCH₂    O    OCH₃

⟶

            O   O                              O   O

            CH₃  CH₃                           CH₃  CH₃

              39                                  40

## XI. LABELING

Catalytic dehalogenation is an effective means of regiospecifically introducing deuterium or tritium. The usual catalysts are palladium-on-carbon, barium sulfate, calcium carbonate, or Raney nickel in conjugation with a base (Blackburn and Levinson, 1976). An unusual combination of catalysts was used in the synthesis of $\beta$-corticotrophin-(1-24)-tetracosapeptide labeled with tritium. Deiodination was carried out over a mixture of 5% palladium-on-carbon and 5% rhodium-on-calcium carbonate in dimethylformamide; neither catalyst was effective alone. Organic bases were unsatisfactory, and nothing was found that worked as well as the catalyst–base rhodium-on-calcium carbonate (Brundish and Wade, 1973).

A convenient, site-specific method for introducing deuterium into an aromatic nucleus involves treating an appropriate aryl halide with sodium borodeuteride in methanol-$d_1$ containing palladium chloride (Bosin et al., 1973).

## XII. DEHALOGENATION WITH HYDRAZINE

Hydrazine has been used as a source of hydrogen in dehalogenations, sometimes with excellent results (Mosby, 1959a,b; Chapman et al., 1960; Cremer and Tarbell, 1961; Parham et al., 1970). The subjects of dehydrohalogenation with hydrazine and catalytic hydrazine reductions in general have been reviewed comprehensively (Furst et al., 1965).

## XIII. ACID CHLORIDES—ROSENMUND REDUCTION

Selective dehalogenation of an acid chloride to an aldehyde is known as the Rosenmund reduction. It is usually carried out by bubbling hydrogen through a hot solution of acid chloride and suspended catalyst. This useful reaction was reviewed thoroughly by Mosettig and Mozingo (1948). The main difficulty encounted is overhydrogenation, since the aldehyde itself can be reduced to the alcohol, and in activated system the alcohol can be reduced to the hydrocarbon. The reactions represent yield losses per se, and the products, water and alcohol, decrease the yield still further by interaction with the acid chloride.

$$RCOCl + H_2 \longrightarrow RCHO + HCl$$
$$RCHO + H_2 \longrightarrow RCH_2OH$$
$$RCH_2OH + H_2 \longrightarrow RCH_3 + H_2O$$

## A.   Catalyst Regulators

Palladium is by for the preferred catalyst for Rosenmund reductions. It is often used with so-called catalyst regulators, such as quinoline–sulfur, thiourea, and thiophene, the function of which is to prevent overhydrogenation (Mosettig and Mozingo, 1948; Affrossman and Thomson, 1962). However, some workers (Foye and Lange, 1956) have preferred nonregulated 10% palladium-on-carbon to conventional inhibited catalysts. The problem of arriving at a preferred general procedure is complex because of catalyst variability and the possibility of accidental inhibition of the catalyst by impurities in the system, solvent, or substrate. It has been suggested that a poison always be added to ensure uniform conditions (Hershberg and Cason, 1955).

Detailed directions for carrying out a Rosenmund reduction of mesitoyl chloride over unregulated palladium-on-barium sulfate (Barnes, 1955) and β-naphthoyl chloride over quinoline–sulfur-regulated palladium-on-barium sulfate are given in *Organic Syntheses*.

## B.   Reduced Pressure

The Rosenmund reduction is usually carried out at reflux, frequently in xylene or toluene, to facilitate the removal of hydrogen chloride. Temperatures less than 100°C were preferred for maximal yields of o-hydroxybenzaldehyde from salicyloyl chlorides when these solvents and palladium-on-barium sulfate were used (Amakasu and Sato, 1967). Some workers have recommended that the temperature be kept at the lowest point at which hydrogen chloride will evolve to ensure maximal yield (Boehm and Schumann, 1933). Low-boiling solvents, such as acetone, can be used for this purpose (Peters and van Bekkum, 1971). Another way of reducing the refluxing temperature is to carry out the reduction at reduced pressures. In this way excellent yields of heat-sensitive dialdehydes have been obtained from diacid chlorides in refluxing benzene over unregulated 10% palladium-on-carbon at 30°–35°C (Johnson *et al.*, 1957). This useful technique is seldom used (Foye and Lange, 1956). It has given excellent results in several laboratories with compounds when other conventional procedures have failed. A study of the general applicability of reduced pressure in Rosenmund reductions seems worthwhile.

## C.  Added Bases

Rosenmund reductions have been carried out with good results in the presence of added bases such as dimethylaniline (Foye and Lange, 1956) or ethyldiisopropylamine (Peters and van Bekkum, 1971), which act as hydrogen acceptors. Anhydrous sodium acetate has been used as an acceptor for Rosenmund reductions carried out at elevated pressures (65 psig) and 35°C. The procedure is adaptable to large-scale preparations and is said to be superior to classic procedures in simplicity, ease of handling, and safety (Wagner *et al.*, 1970).

## REFERENCES

Abushanab, E., Bindra, A. P., Goodman, L., and Peterson, H., Jr., *J. Org. Chem.* **38**, 2049 (1973).
Affrossman, S., and Thomson, S. J., *J. Chem. Soc.* p. 2024 (1962).
Albano, E. L., and Horton, D., *J. Org. Chem.* **34**, 3519 (1969).
Amakasu, T., and Sato, K., *Bull. Chem. Soc. Jpn.* **40**, 1428 (1967).
Arkley, V., Gregory, G. I., and Walker, T., *J. Chem. Soc.* p. 1603 (1963).
Baltzly, R., and Phillips, A. P., *J. Am. Chem. Soc.* **68**, 261 (1946).
Barnes, M. W., *J. Org. Chem.* **41**, 733 (1976).
Barnes, R. A., and Gerber, N. N., *J. Org. Chem.* **26**, 4540 (1961).
Barnes, R. P., *Org. Synth., Collect. Vol.* 3, 551 (1955).
Berti, G., Macchia, B., Macchia, F., and Monti, L., *J. Org. Chem.* **33**, 4045 (1968).
Blackburn, D. W., and Levinson, S. H., *in* "Catalysis in Organic Syntheses, 1976" (P. N. Rylander and H. Greenfield, eds.), p. 75. Academic Press, New York, 1976.
Boehm, T., and Schumann, G., *Arch. Pharm.* (*Weinheim, Ger.*) **271**, 490 (1933).
Bosin, T. R., Raymond, M. G., and Buckpitt, A. R., *Tetrahedron Lett.* p. 4699 (1973).
Brieger, G., Hachey, D. L., and Ciaramitaro, D., *J. Org. Chem.* **34**, 220 (1969).
Britton, E. C., and Keil, T. R., U.S. Patent 2,725,402, Nov. 29, 1955.
Brundish, D. E., and Wade, R., *J. Chem. Soc., Perkin Trans. I* p. 2875 (1973).
Butler, D. E., and Alexander, S. M., *J. Org. Chem.* **37**, 215 (1972).
Campbell, J. B., Whitehead, C. W., Kress, T. J., and Moore, L. L., *Ann. N.Y. Acad. Sci.* **214**, 216 (1973).
Chapman, D. D., Cremer, S. E., Carman, R. M., Kunstmann, M., McNally, J. G., Jr., Rowsky, A., and Tarbell, D. S., *J. Am. Chem. Soc.* **82**, 1009 (1960).
Cohen, S., Thom, E., and Bendich, A., *J. Org. Chem.* **27**, 3545 (1962).
Conroy, H., *J. Am. Chem. Soc.* **77**, 5960 (1955).
Cremer, S. E., and Tarbell, D. S., *J. Org. Chem.* **26**, 3653 (1961).
Cromwell, N. H., and Cram, D. J., *J. Am. Chem. Soc.* **65**, 301 (1945).

Denton, D. A., McQuillin, F. J., and Simpson, P. L., *J. Chem. Soc.* p. 5535 (1964).

Eguchi, C., and Kakuta, A., *Bull. Chem. Soc. Jpn.* **47**, 1704 (1974).

Florin, R. E., Pummer, W. J., and Wall, L. A., *J. Res. Natl. Bur. Stand.* **62**, 119 (1959).

Foye, W. O., and Lange, W. E., *J. Am. Pharm. Assoc., Sci. Ed.* **45**, 742 (1956).

Fritsch, W., Stache, U., Haede, W., Radscheit, K., and Ruschig, H., *Justus Liebigs Ann. Chem.* **721**, 168 (1969).

Fuqua, S. A., Parkhurst, R. M., and Silverstein, R. M., *Tetrahedron* **20**, 1625 (1964).

Furst, A., Berlo, R. C., and Hooton, S., *Chem. Rev.* **65**, 51 (1965).

Furukawa, Y., Yoshioka, Y., Imai, K., and Honjo, M., *Chem. Pharm. Bull.* **18**, 554 (1970).

Gaertner, R., *J. Am. Chem. Soc.* **76**, 6150 (1954).

Gallemaers, J. P., Christophe, D., and Promel, R., *Tetrahedron Lett.* p. 693 (1976).

Ham, G. E., and Coker, W. P., *J. Org. Chem.* **29**, 194 (1964).

Hanessian, S., and Plessas, N. R., *J. Org. Chem.* **34**, 2163 (1969).

Hayward, E. D., Tarbell, D. S., and Colebrook, L. D., *J. Org. Chem.* **33**, 399 (1968).

Hershberg, E. B., and Cason, J., *Org. Synth., Collect. Vol.* 3, 627 (1955).

Hofmann, K., Orochena, S. F., Sax, S. M., and Jeffrey, G. A., *J. Am. Chem. Soc.* **81**, 992 (1959).

Huffman, J. W., *J. Org. Chem.* **24**, 1759 (1959).

Isogai, K., Kondo, S., Katsura, K., Sato, S., Yoshihara, N., Kawamura, Y., and Kazama, T., *Nippon Kagaku Zasshi* **91**, 561 (1970); *Chem. Abstr.* **74**, 3186d (1971).

Jain, T. C., Russell, A. F., and Moffatt, J. G., *J. Org. Chem.* **38**, 3179 (1973).

Johnson, F., and Nasutavicus, W. A., *J. Org. Chem.* **28**, 1877 (1963).

Johnson, W. S., Martin, D. G., Pappo, R., Darling, S. D., and Clement, R. A., *Proc. Chem. Soc.* p. 58 (1957).

Kamano, Y., and Pettit, G. R., *J. Org. Chem.* **38**, 2202 (1973).

Kraus, M., and Bazant, V., *in* "Catalysis" (J. W. Hightower, ed.), Vol. 2, p. 1073. Am. Elsevier, New York, 1973.

Liisberg, S., Godtfredsen, W. O., and Vangedal, S., *Tetrahedron* **9**, 149 (1960).

Lozeron, H. A., Gordon, M. P., Gabriel, T., Tautz, W., and Duschinsky, R., *Biochemistry* **3**, 1844 (1964).

Lyle, R. E., and Troscianiec, H. J., *J. Org. Chem.* **24**, 333 (1959).

McBee, E. T., and Smith, D. K., *J. Am. Chem. Soc.* **77**, 387 (1955).

Manatt, S. L., Vogel, M., Knutson, D., and Roberts, J. D., *J. Am. Chem. Soc.* **86**, 2645 (1964).

Marquardt, F. H., and Edwards, S., *J. Org. Chem.* **37**, 1861 (1972).

May, E. L., and Mosettig, E., *J. Org. Chem.* **11**, 429 (1946).

Mosby, W. L., *J. Org. Chem.* **24**, 421 (1959a).

Mosby, W. L., *Chem. Ind. (London)* p. 1348 (1959b).

Mosettig, E., and Mozingo, R., *Org. React.* **4**, 362 (1948).

Overberger, C. G., and Monagle, J. J., *J. Am. Chem. Soc.* **78**, 4470 (1956).

Parham, W. E., Davenport, R. W., and Biasotti, J. B., *J. Org. Chem.* **35**, 3775 (1970).

Peters, J. A., and van Bekkum, H., *Rec. Trav. Chim. Pays-Bas* **90**, 1323 (1971).

Pfitzner, K. E., and Moffatt, J. G., *J. Org. Chem.* **29**, 1508 (1964).

Redman, H. E., and Weimer, P. E., U.S. Patent 2,943,114, June 28, 1960.

Reinecke, M. G., *J. Org. Chem.* **29**, 299 (1964).

Rosenmund, K. W., and Zetzsche, F., *Chem. Ber.* **51**, 578 (1918).

Russell, A. F., Greenberg, S., and Moffatt, J., *J. Am. Chem. Soc.* **95**, 4025 (1973).

Ruzicka, V., and Procházka, J., *Collect. Czech. Chem. Commun.* **35**, 430 (1970); *Chem. Abstr.* **72**, 89634s (1970).

Sargent, L. J., and Agar, J. H., *J. Org. Chem.* **23**, 1938 (1958).

Secrist, J. A., III, and Logue, M. W., *J. Org. Chem.* **37**, 335 (1972).

Stache, U., Fritsch, W., Haede, W., and Radscheit, K., *Justus Liebigs Ann. Chem.* **726**, 136 (1969).

Stache, U., Radscheit, K., Fritsch, W., Haede, W., Kohl, H., and Ruschig, H., *Justus Liebigs Ann. Chem.* **750**, 149 (1971).

Taub, D., Hoffsommer, R. D., Slates, H. L., and Wendler, N. L., *J. Org. Chem.* **26**, 2852 (1961).

Telschow, J. E., and Reusch, W., *J. Org. Chem.* **40**, 862 (1975).

Traas, P. C., Takken, H. J., and Boelens, H., *Tetrahedron Lett.* p. 2027 (1977).

Urushibara, Y., Nishimura, S., and Uehara, H., *Bull. Chem. Soc. Jpn.* **28**, 447 (1955).

Vavon, G., and Mathieu, R., *C. R. Acad. Sci.* **206**, 1387 (1938).

Wagner, D. P., Gurien, H., and Rachlin, A. I., *Ann. N.Y. Acad. Sci.* **172**, 186 (1970).

Whittaker, N., *J. Chem. Soc.* p. 1565 (1950).

Whittaker, N., *J. Chem. Soc.* p. 1646 (1953).

Wiberg, K. B., and Hess, B. A., Jr., *J. Org. Chem.* **31**, 2250 (1966).

Wilson, A. N., and Harris, S. A., *J. Am. Chem. Soc.* **73**, 2388 (1951).

Wilt, J. W., and Vasiliauskas, E., *J. Org. Chem.* **37**, 1467 (1972).

Wineman, R. J., Hsu, E.-P. T., and Anagnostopoulos, C. E., *J. Am. Chem. Soc.* **80**, 6233 (1958).

Zenitz, B. L., Macks, E. B., and Moore, M. L., *J. Am. Chem. Soc.* **70**, 955 (1948).

# Chapter 14

# Hydrogenolysis of Small Rings

## I. CYCLOPROPANES

Carbon–carbon bonds rarely undergo hydrogenolysis under mild conditions, except those in cyclopropanes, which by virtue of ring strain are easily opened. Cyclopropanes are now readily available, and their hydrogenolysis has acquired new synthetic utility.

## A. Catalysts

The relative effectiveness of metals in the hydrogenolysis of cyclopropanes depends in part on the conditions under which the comparisons are made. Sinfelt (1973) found a decreasing order of reactivity for cyclopropane hydrogenolysis to be Rh > Pt > Pd > Ir > Os > Ru. In contrast, other workers (Irwin and McQuillin, 1968; Schultz, 1971) using substituted cyclopropanes found palladium to be very much more active than platinum.

As a working generality, it would appear that, in the hydrogenation of a cyclopropane carrying a reducible function, the function is best preserved and the ring opened over palladium, whereas the function is best reduced and the ring preserved over platinum or rhodium, if indeed the course of the reduction can be changed at all by the catalyst. Too few data exist to fit osmium, iridium, and ruthenium into this scheme. Illustrative of the effect of metal on selectivity is the hydrogenation of 1-phenylbicyclo[4.1.0]-heptane (Mitsui *et al.*, 1974) (Table I). Here all the metals formed only the configurationally retained product of C-1–C-7 fission, *trans*-1-phenyl-2-methylcyclohexane. The mode of attack at C-7 remains an open question (Kieboom *et al.*, 1974). The catalyst also affects to some extent which bond in the cyclopropane ring is broken. The degree of bond fission at C-1–C-6 is believed to reflect the atomic size of the metal and the effect that this has on the transition state of the π-benzyl complex (Mitsui *et al.*, 1974).

**TABLE I**[a]

|  | Product composition (%) | | |
|---|---|---|---|
| Catalyst | A | B | C |
| Raney Ni | 100 | — | — |
| Pd-on-C | 90 | — | 10 |
| Pd(OH)$_2$ | 95 | — | 4 |
| Rh-on-C | 60 | 30 | 10 |
| PtO$_2$ | 15 | 80 | 5 |

[a] Data of Mitsui et al. (1974). Reprinted with permission of Pergamon Press.

Acidic and bifunctional metal catalysts may exert profound influences on the direction of ring opening in gas-phase hydrogenolysis of cyclopropanes (Pines and Nogueira, 1972; Schlatter and Boudart, 1972).

## B. Direction of Ring Opening

Conjugative and inductive effects usually outweigh steric effects in both orienting and facilitating hydrogenolysis. In monosubstituted cyclopropanes, alkyl substituents tend to promote hydrogenolysis at the carbon–carbon bond opposite the substituent (Newham, 1963), whereas electron-attracting substituents favor ring opening at the carbon carrying the substituent (Irwin and McQuillin, 1968; Gray et al., 1968; Schultz, 1971; Roth, 1972; Kuehne and King, 1973). More highly substituted nonfunctional cyclopropanes tend to open at the carbon carrying the most hydrogens (Majerski and Schleyer, 1968; Hendrickson and Boeckman, 1971) or at the bond having the greatest strain (Freeman et al., 1973) or the least steric hindrance (Sauers and Shurpik, 1968; Cristol and Mayo, 1969); it is often difficult to judge a priori which effect is controlling. Cyclopropanes carrying only phenyl substituents tend to cleave exclusively adjacent to the phenyl-substituted carbon (Kazanskii et al., 1960; Schultz, 1971).

The above generalities are not without exception. Hydrogenolysis of the nitrile **1** cleaves the bonds indicated (Blanchard and Cairncross, 1966),

$$H_3C \underset{1}{-\underset{}{\bigtriangleup}} -CN \xrightarrow[\substack{5\% \ Pd\text{-}on\text{-}C \\ C_2H_5OH}]{2H_2} H_3C-\underset{CH_3}{\overset{H_3C}{\underset{|}{C}}}-CN + H_3C\underset{CH_3}{\overset{}{\underset{|}{\diagdown}}}-CN$$

**1**                                    79%                    2%

whereas **2** (Wilberg and Ciula, 1959) and **3** (Meinwald *et al.*, 1963) are cleaved at bonds opposite the electron-withdrawing substituent. These

**2**

**3**

differences may be due to steric effects; all three reductions occur at the less hindered site.

## C. Olefinic Cyclopropanes

Vinyl- and alkylidenecyclopropanes undergo hydrogenation readily with various degrees of simultaneous ring opening (deRopp *et al.*, 1958; Sarel and Breuer, 1959; Ullman, 1959; Hartzler, 1961; Laing and Sykes, 1968). The ratio of olefin saturation to ring hydrogenolysis depends partly on substrate structure but more importantly on the catalyst. Alkylated vinyl-cyclopropanes usually undergo almost complete ring hydrogenolysis over palladium regardless of the degree of double-bond substitution. Hydrogenolysis either precedes or occurs concomitantly with double-bond saturation (Kierstead *et al.*, 1952). In contrast, reductions over platinum or rhodium give mostly double-bond reduction (Cook *et al.*, 1970; Lambert *et al.*, 1972), with the percentage of hydrogenolysis gradually increasing as substitution at the double bond is increased (Poulter and Heathcock, 1968a). This marked difference in behavior between palladium, on the one hand, and platinum or rhodium, on the other, makes it possible to maintain a high degree of control over the products of reduction (Meinwald *et al.*, 1963; Cocker *et al.*, 1966).

Good yields of intermediate olefins can be obtained by arresting reductions over palladium after about 1 mole of hydrogen has been absorbed. Both thermodynamic and mechanistic selectivity over palladium are high, the former a consequence of the substrate being much more strongly adsorbed than the olefinic products (Poulter and Heathcock, 1968b).

4%                86%                10%

If this type of reaction is applied to the synthesis of an olefin, the products should be analyzed as the reaction progresses to obtain maximal yield. Due to isomerization of olefins after they are formed and to less than 100% mechanistic and thermodynamic selectivity, the maximal yield may not occur at exactly 1 mole of hydrogen absorbed. The major olefin is usually that obtained as if by 1,4 addition of hydrogen as, for example in the hydrogenation of thujopsene (Hochstetler, 1972).

Thujopsene                              95%

## D. Effect of Solvent

No systematic study of the effect of solvent on the hydrogenation of cyclopropanes has been made. Scattered reports suggest that both solvent and pH may have some effect on the reaction, although these variables apparently have less effect on cyclopropanes than they do on small heterocyclic rings (Mitsui et al., 1974). An illustration of the effect of solvent is given in Table II.

Parenthetically, in this reduction, scission of the four-membered ring is apparently much more rapid than that of the three-membered ring,

**TABLE II**

| Solvent | A (%) | B (%) |
|---------|-------|-------|
| β-Ethoxyethanol | 20 | 1 |
| Diglyme | 77 | 23 |

**TABLE III**[a]

| | A | B |
|---|---|---|
| Solvent | A (%) | B (%) |
| 50% aq. EtOH | 23 | 77 |
| Absolute EtOH | 43 | 57 |
| Tetrahydrofuran | 47 | 53 |
| Hexane | 68 | 32 |
| 85% aq. EtOH, 0.01 N NaOH | 84 | 16 |

[a] Data of Wuesthoff and Rickborn (1968). Used with permission.

underlining the high degree of angle strain incorporated into the tricyclo-[2.2.0.0$^{2,6}$] hexane molecule (Lemal and Shim, 1964).

Some pronounced solvent effects were reported in the hydrogenation of a vinylcyclopropane (Table III) (Wuesthoff and Rickborn, 1968). No trace of 4,4-dimethylcyclohexane is found in this reduction; spiro [2.5] octane itself is catalytically reduced to 1,1-dimethylcyclohexane (Shortridge *et al.*, 1948).

## E. Effect of Temperature

The temperature of reduction may have an effect on selectivity. Hydrogenation of (+)-α-thujene at room temperature afforded considerable amounts of a tetrahydro derivative with opening of the cyclopropane ring, even over platinum. However, hydrogenation at −20°C gave essentially pure dihydro derivative (Acharya *et al.*, 1969). Similarly, the hydrogenolysis product of (+)-car-3-ene, trimethylcycloheptane, increased in reduction over palladium from 64% at 0°C to 100% at 73°C (Cocker *et al.*, 1966). It would appear that in unsaturated cyclopropanes hydrogenation relative to hydrogenolysis is favored by lower temperatures.

## F.   Stereochemistry

Both stereospecific (Wissner and Meinwald, 1973) and nonstereospecific (Majerski and Schleyer, 1968) openings of cyclopropane rings have been recorded. The opening of a cyclopropane ring may occur by *suprafacial* hydrogen attack, in which the configuration at both carbon atoms is either retained or inverted, or by *antarafacial* hydrogen attack, in which the configuration of one carbon atom is retained and one is inverted. Testing the mode of attack requires a special situation, for stereoselective conversion to a single compound does not signify stereospecific attack at both carbons, nor does lack of stereoselectivity in the product necessarily point to non-specific attack (Kieboom *et al.*, 1974).

By hydrogenation of *trans*-1,2-dimethyl-1,2-diphenylcyclopropane, the relative contributions of suprafacial and antarafacial attack were quantified on the basis of the percentages of the resulting *dl*- and *meso*-2,4-diphenyl-pentanes. Over a range of solvent polarity, suprafacial hydrogen attack was 37–38% for platinum, 40–43% for rhodium, 45% for nickel, and 73–81% for palladium. Thus, platinum, rhodium, and nickel are similar, whereas palladium is distinctly different. However, by the addition of small amounts of strong acid to the system, predominantly suprafacial attack at palladium can be changed to largely (78–85%) antarafacial attack. The strong acid is believed to protonate the cyclopropane ring. Deuterium experiments established that the solvent participated in antarafacial attack, for there is a larger incorporation of deuterium in the meso form.

To account for the hydrogenolysis of lumitestosterone acetate (Augustine and Rearden, 1974) affording almost exclusively the 5α product rather than

the expected 5β compound, Augustine (1976) proposed, as an initial step in the hydrogenolysis, extraction of the C-6 hydrogen by the catalyst with concomitant double-bond formation between C-5 and C-6. The concept was used to account for a number of other stereochemical consequences

of hydrogenolysis of cyclopropanes (Prudhomme and Gault, 1966; Ghatak
et al., 1973).

Hydrogenolysis of 3-acetoxynortricyclene (4) over platinum in acetic
acid containing a small amount of perchloric acid at 80°C affords a mixture
of 32% 2-exo- (5) and 68% 7-acetoxynorbornanes (6) (Akhtar and Jackson,
1972). Deuteration experiments established that hydrogenolysis of the
rearranged 7-acetoxynortricyclene involves more than 90% cis hydrogen
addition from the outer edge of the cyclopropane ring.

It is of interest that hydrogenation of 7-tert-butyl-, 7-tert-butoxy-, and
7-acetoxynorbornadienes affords the corresponding nortricyclanes in yields
up to 30% (Baird and Surridge, 1972; Franzus et al., 1968). This type of
homoconjugative reduction has been ascribed to endocyclic catalyst–diene
complexation followed by hydrogen transfer through a π-homoallylic
metal–olefin complex.

## G.  Synthetic Applications of Cyclopropane Hydrogenolysis

The ease and specificity of the hydrogenolysis of substituted cyclopropanes
make it useful for the synthesis of compounds containing quaternary
carbons, gem-dialkyl, tert-butyl, and angular methyl substituents (Woodworth et al., 1968). Hydrogenolysis of 1-(1-adamantyl)-1-methylcyclopro-
pane (7) affords, in 96% yield, 1-tert-butyladamantane (8), a compound
difficult to prepare by conventional procedures.

Similarly, hydrogenolysis of tricyclo$[5.1.0.0^{4,8}]$octane gave bicyclo-[3.2.1]octane quantitatively (Schwarz *et al.*, 1965). No bicyclo[3.3.0]-octane, the other possible hydrogenolysis product, was obtained. The tendency for the most strained bond to break has been observed in other instances (Newham, 1963). Hydrogenolysis of tetracyclo$[4.3.0.0^{2,4}0^{3,7}]$-non-8-ene (9) gives brexane (10) in 89% yield (Schleyer and Wiskott, 1967), providing a convenient route to a compound difficult to obtain otherwise.

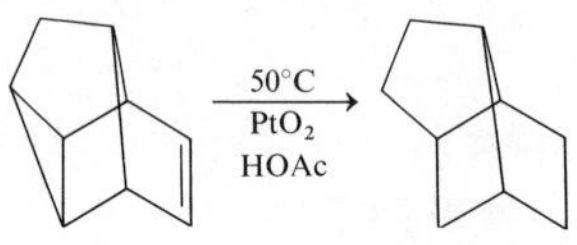

9                    10
(89%)

A facile synthesis of diamantane involves dimerization of norbornadiene to Binor-S (11), nearly quantitative conversion to a tetrahydro derivative (12), followed by aluminum bromide-catalyzed isomerization. The hydrogenated derivative isomerizes to diamantane (13) in an average yield of 65% (Gund *et al.*, 1970).

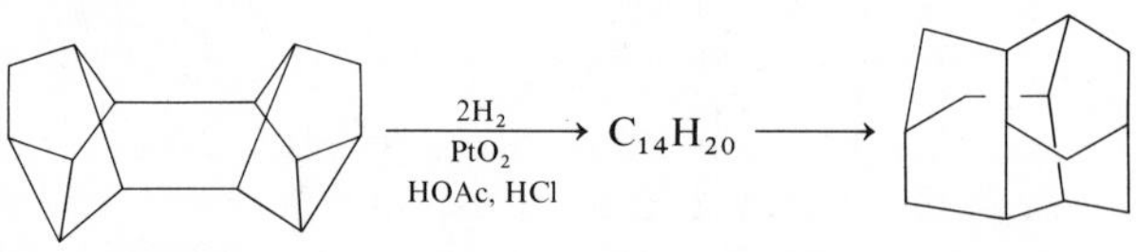

11                              12                              13

The synthesis of 1,1-dimethylcyclobutane is illustrative of the introduction of *gem*-dimethyl groups (Nametkin *et al.*, 1973). Hydrogenolysis is predominantly in the cyclopropyl ring at the bonds opposite the cyclobutane ring.

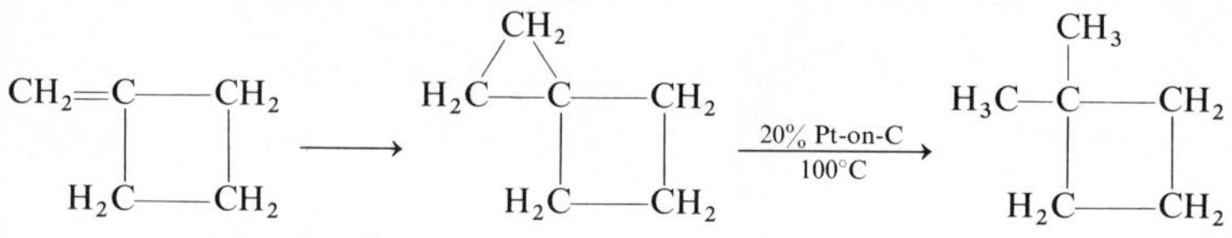

95%

Regiostereospecific hydrogenolysis of cyclopropyl ketones in the gibbene series proved to be a convenient means of obtaining 9α-gibbanes as the sole product. The reduction proceeds with inversion of configuration. In contrast to this clean reduction, hydrogenation of compounds in this series having double bonds at the ring juncture gives isomeric mixtures (Chakrabortty *et al.*, 1972; Ghatak *et al.*, 1973).

The work of Challand *et al.* (1969) provides an interesting example of the specificity of catalysts in the hydrogenolysis of cyclopropanes. Hydrogenation of **14** over platinum oxide (under the same conditions in which methylnorcarane is converted to 1,1-dimethylcyclohexane) failed to produce **15**. Instead, mixtures were produced containing eight-membered rings lacking the *tert*-hydroxyl. However, by the use of Nishimura's rhodium–platinum oxide (7:3) catalyst the cyclopropane ring was made to cleave in the desired direction, and, by incorporating sodium acetate in the acetic acid solvent, hydrogenolysis of the hydroxyl function was sharply curtailed. This catalyst (Nishimura, 1960) is frequently more effective than either platinum oxide or rhodium oxide alone (Rylander *et al.*, 1967a,b).

**14, R = H, CH$_3$**                **15**

A novel method for introducing an angular methyl group involves hydrogenolysis of a steroidal bicyclobutane derivative. The reaction, with the concomitant formation of a carbon–carbon double bond, has been termed a geminal hydrogenolysis. When deuterium is used both deuterium atoms appear in the 8α-methyl group (Galantay *et al.*, 1970).

## II.  OXIRANES

Most oxiranes readily undergo hydrogenolysis, affording as a principal product an alcohol or mixture of alcohols resulting from cleavage of a carbon–oxygen bond. Other products may arise by cleavage of the carbon–carbon bond in the oxirane ring (Gawron et al., 1963) or by loss of the oxygen function (Pigulevski and Sokolova, 1963; Mitsui et al., 1973). Epoxidation of an olefin followed by hydrogenolysis makes a useful synthetic sequence for converting an olefin to an alcohol.

The major problem connected with hydrogenolysis of epoxides is control of the direction of ring opening. Epoxides having a high degree of symmetry in the vicinity of the function, such as those derived from fatty acids or esters, tend to open randomly regardless of conditions (Fore and Bickford, 1959, 1961; Howton and Kaiser, 1964), but in unsymmetric epoxides one bond is usually cleaved preferentially to a degree that depends on substrate structure, catalyst, and conditions.

### A.  Multifunctional Compounds

Oxiranes usually undergo hydrogenolysis readily under mild conditions. The ring is sufficiently stable, however, so that easily reduced functions in the molecule often can be hydrogenated while the oxirane is kept intact (Poss and Rosenau, 1963). Failure to maintain the epoxide might be corrected by carrying out the reduction with a catalyst, such as rhodium-on-carbon, that has excellent hydrogenation activity for many functions and relatively little tendency toward hydrogenolysis of epoxides (Tarbell et al., 1961).

Nonconjugated olefinic epoxides are converted easily to the saturated epoxides over noble metals, but the reverse selectivity can also be effected if the double bond is protected against reduction by coordination with copper nitrate (Upendraro et al., 1972) or silver nitrate (Acharyya and Subbarao, 1970).

### B.  Direction of Ring Opening

The factors controlling the direction of epoxide ring opening are various, and it is not always easy to deduce, without an analogous reaction with which to compare, which bond will be broken. The ring may open at the weakest bond (Whitmore and Gebhart, 1942; Newman et al., 1949; Witkop and Foltz, 1957; Aizikovich et al., 1958; Aizikovich and Petrov, 1958; Adegoke et al., 1965; Herz et al., 1970) or at the carbon atom with the fewest substituents (Stavely, 1942) or at the carbon atom with the least steric hindrance (Ross *et al.*, 1956) or, in acidic solution, to afford the most stable carbonium ion (McQuillin and Ord, 1959). In addition, ring opening may be greatly influenced by the reaction environment and type of catalyst (Park and Fuchs, 1957).

## C.  Effect of Catalyst and Solvent

Catalysts can affect markedly the direction of ring opening, but the action of the catalyst should not be divorced from the medium in which it is used. Illustrative of the effects of catalysts and media are the results of the hydrogenation of benzalacetophenone oxide, a compound in which both carbon–oxygen bonds are activated (Table IV). The ratio of products changes markedly with the catalyst and with added sodium hydroxide, which is thought to decrease oxygen–metal bonding. Table IV shows the major products of hydrogenation after the absorption of 2 moles of hydrogen in neutral or alkaline ethanol (Sohma and Mitsui, 1970).

The hydrogenation of 1-phenyl-7-oxabicyclo[4.1.0]heptane provides another illustration of the effect of catalysts and media (Table V) (Mitsui *et al.*, 1974).

Both palladium and Raney nickel open the hindered benzyl–oxygen bond, palladium with inversion of configuration, and nickel in neutral media primarily with retention of configuration. Addition of alkali to nickel changes the stereochemistry of ring opening and curtails the extensive deoxygenation. Similarly, hydrogenation of *cis*-α,α$^1$-dimethylstilbene oxide over palladium proceeds with inversion, producing *threo*-2,3-diphenyl-butan-2-ol, whereas over nickel the configurationally retained erythro isomer is formed. The stereochemistry of the nickel-catalyzed reduction is changed by a small amount of alkali to produce predominantly the threo compound (Mitsui and Nagahisa, 1965).

Regiospecificity and stereospecificity are influenced by the presence or absence of alkali. Hydrogenation of 1,2-epoxydecane over Raney nickel without base present affords mainly 1-decanol; with base, it affords mainly 2-decanol (Newman *et al.*, 1949). Similarly, hydrogenation of 1,2-epoxy-butane over nickel affords 1-butanol, but, in contrast, over palladium the product is chiefly 2-butanol (Senechal and Cornet, 1971). Hydrogenation of styrene oxide over palladium-on-carbon (Mitsui *et al.*, 1973) or over palladium-on-barium sulfate in methanol gives exclusively 2-phenylethanol, but palladium-on-barium sulfate in buffered alkaline methanol gives 1-phenylethanol (Sokol$^1$skaya *et al.*, 1966). Nickel, on the other hand, in sodium carbonate solution at 40 atm and temperatures above 85°C gives

**TABLE IV**

**Hydrogenation of Benzalacetophenone Oxide**

| Amount of Catalyst (gm) | Catalyst | NaOH (mg) | $C_6H_5$-CHCH$_2$CH$_2$-$C_6H_5$ with OH (%) | CHCHCH$_2$ with OHOH — Erythro (%) | Threo (%) | CHCH$_2$CH with OH OH (%) |
|---|---|---|---|---|---|---|
| 0.2 | Pd-on-C | — | 5 | 62 | 25 | 4 |
| 0.4 | Pd-on-C | 40 | 3 | 77 | 13 | 1 |
| 1.0 | RaNi | — | 39 | 21 | 7 | 26 |
| 1.0 | RaNi | 80 | 36 | 27 | 17 | 26 |
| 1.0 | RaNi | 400 | 8 | 80 | 10 | 2 |
| 0.05 | PtO$_2$ | — | 27 | 32 | 10 | 31 |
| 0.05 | PtO$_2$ | 5 | 4 | 91 | 4 | 1 |

**TABLE V**

**Hydrogenation of 1-Phenyl-7-oxabicyclo[4.1.0]heptane$^a$**

| Catalyst | Amount of catalyst (gm) | NaOH (mmoles) | A Trans (%) | B Cis (%) | C (%) |
|---|---|---|---|---|---|
| RaNi | 1.0 | — | 30 | 11 | 59 |
| RaNi | 1.0 | 1.0 | 5 | 93 | 2 |
| Pd-on-C | 0.1 | — | 3 | 93 | 4 |
| Pd-on-C | 0.1 | 1.0 | — | 100 | Trace |

$^a$ Data of Mitsui *et al.* (1974). Reprinted with permission of Pergamon Press.

2-phenylethanol (Priese, 1972). A mixture of Raney nickel and palladium-on-carbon under mild conditions affords 2-phenylethanol in 96% yield, free of ethylbenzene (Wood, 1969).

## D.  Effect of Structure

Hydroxy ketones or diols are easily obtained from conjugated ketones by epoxidation followed by hydrogenolysis. Without other activating influence, the bond broken during hydrogenolysis is that adjacent to the carbonyl (McCurry and Singh, 1974). The sequence has proved to be useful in prostaglandin syntheses (Strike and Smith, 1970).

If other activating functions are present, the direction of ring opening may depend on subtle structural features, as suggested by comparison of the hydrogenation of the epoxy ketones **16** (Table VI) (Cromwell and Bambury, 1961) and **17** (Cromwell and Martin, 1968).

**TABLE VI**

| Catalyst | A (%) | B (%) | C (%) |
|----------|-------|-------|-------|
| Pd | 25 | 70 | — |
| Pt | 29 | — | > 50 |

In the palladium-catalyzed hydrogenolysis of the epoxide **16**, the benzyl–oxygen bond remains intact (unless the ketonic function is first reduced), whereas in **17** the benzyl–oxygen bond is cleaved, affording **18** in 80% yield.

The product composition in the reduction of **17** depends on substrate concentration; the percentage of diol (**19**) increases as the solution becomes more concentrated. Comparison of palladium and platinum in the reduction of **16** provides a useful working generality relating to ketonic epoxides: Platinum favors ketone reduction, and palladium favors epoxide hydrogenolysis (Temnikova and Kropachev, 1949; Lutz and Wood, 1938), if indeed the course of reduction can be changed at all by the catalyst (Herz, 1952). In contrast to **16**, the oxirane ring in **20** breaks cleanly at the benzyl–oxygen bond to afford **21** (Farkas et al., 1970). This cleavage is probably facilitated by the electron-releasing p-methoxy group (Kieboom et al., 1971).

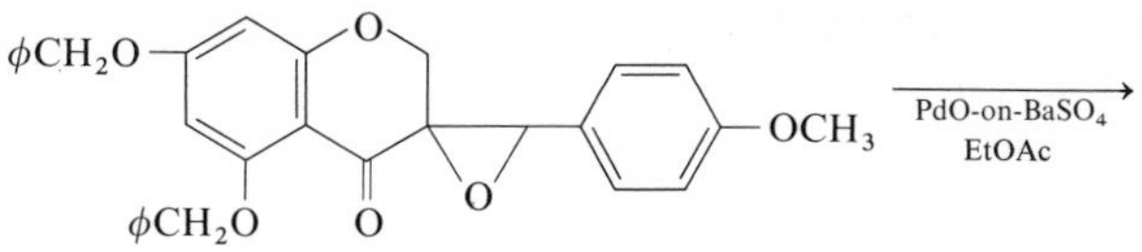

**20**

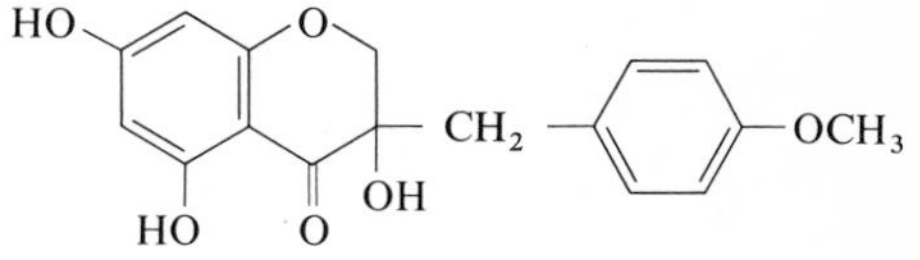

**21**

An unusual example of the effect of structure on the course of epoxide reduction is provided by a comparison of hydrogenation of 2-acetyl-1,4-naphthoquinone epoxide and the 3-methyl homologue over palladium-on-carbon. With the first compound, the acetyl group is retained, whereas in the latter it is lost as a ketone and the ring is aromatized.

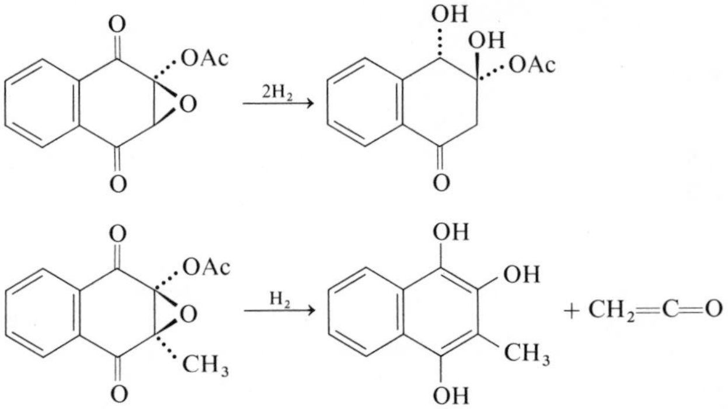

The authors (Read and Ruiz, 1973) advanced interesting arguments to account for these differences including the suggestion that the acetyl is lost through nucleophilic abstraction of a proton from the acetyl methyl group, leading to a ketene and a ketonic tautomer of the major product. On one occasion, conditions favored attack by the ketene on the newly formed C-1 hydroxyl, affording 4-acetoxy-2-methylnaphthalene-1,3-diol.

## E. Diols and Ethers

Hydrogenation of epoxides in water or alcohols may produce diols or hydroxy ethers through intervention of the solvent (Ushakov and Mikhailov,

1937). The unsaturated epoxide **22** in ethanol over Raney nickel produces
the ethoxy alcohol **23** (Pierson and Runquist, 1969).

**22**                    **23**

## F. Deoxygenation

Hydrogenolysis of oxiranes may result in deoxygenated products to an
extent that depends markedly on the catalyst and the acidity. The pronounced effect of these parameters is illustrated by the hydrogenation of
styrene oxide over nickel, palladium, and platinum (Table VII) (Mitsui *et
al.*, 1973).

**TABLE VII**[a]

| Catalyst | A (%) | B (%) |
|---|---|---|
| 5% Pd-on-C | 0 | 100 |
| RaNi, 0.4 gm | 35 | 65 |
| RaNi, 1.0 gm | 56 | 44 |
| PtO$_2$ | 8 | 92 |
| PtO$_2$ (trace HCl) | 70 | 30 |
| Pt black | 44 | 56 |

[a] Data of Mitsui *et al.* (1973). Reprinted with permission of Pergamon Press.

Deoxygenation over Raney nickel is sharply curtailed by elevated pressures
(7% at 100 atm) or by traces of sodium hydroxide (12% with 1 mmole). The
marked difference between platinum oxide and platinum black (obtained
by hydrogen reduction of platinum oxide) was attributed to differences in
alkalinity. Ethylbenzene was considered to arise via styrene by trans $\beta$ elimination and not by hydrogenolysis of intermediate 1-phenylethanol.

## III. AZIRIDINES

The aziridine ring is easily opened but can itself be prepared by saturation of an azirine ring (Morrow et al., 1965) although in some cases azirine ring hydrogenolysis may take precedence over ring saturation (Cram and Hatch, 1953). Palladium or platinum is usually used in the hydrogenolysis of aziridines.

### Direction of Ring Opening

Hydrogenolysis of the aziridine ring occurs primarily at the less hindered carbon–nitrogen bond, except when the more hindered carbon carries unsaturation or an aromatic substituent (Bottini et al., 1963; Campbell et al., 1946; Karabinos and Serijan, 1945; Kharasch and Priestley, 1939; Leonard and Jann, 1962; Leonard et al., 1963; Stevens and Pillai, 1972; Stevens et al., 1972; Sugi et al., 1975).

Aziridines, like oxiranes, may undergo ring opening with either retention or inversion of configuration in proportions that depend on the catalyst, solvent, and various additives (Mitsui and Sugi, 1969a,b; Sugi and Mitsui, 1969, 1970; Sugi et al., 1975). For instance, hydrogenolysis of 2-methyl-2-phenylaziridine in ethanol occurs mainly with inversion over palladium and mainly with retention over platinum, Raney nickel, and Raney cobalt (Sugi and Mitsui, 1969). However, in benzene, hydrogenolysis over palladium occurs with retention. Addition of sodium hydroxide to the reaction strongly promotes retention. Rationales for the complex interaction of various reaction parameters are given in the papers cited.

## REFERENCES

Acharya, S. P., Brown, H. C., Suzuki, A., Nozawa, S., and Itoh, M., J. Org. Chem. 34, 3015 (1969).
Acharyya, K. T., and Subbarao, R., Indian Patent 105,271, Feb. 14, 1970; Chem. Abstr. 76 46064w (1972).
Adegoke, E. A., Ojechi, P., and Taylor, D. A. H., J. Chem. Soc. p. 415 (1965).
Aizikovich, M. A., and Petrov, A. A., Zh. Obshch. Khim. 28, 3051 (1958).
Aizikovich, M. A., Maretina, I. A., and Petrov, A. A., Zh. Obshch. Khim. 28, 3046 (1958).
Akhtar, M. N., and Jackson, W. R., J. Chem. Soc. Chem. Commun. p. 813 (1972).
Augustine, R. L., in "Catalysis in Organic Syntheses, 1976" (P.N., Rylander, and H., Greenfield, eds.), pp. 325–341. Academic Press, New York, 1976.
Augustine, R. L., and Reardon, E. J., Jr., J. Org. Chem. 39, 1627 (1974).
Baird, W. C., Jr., and Surridge, J. H., J. Org. Chem. 37, 304 (1972).
Belorossov, E. L., Kryukov, S. I., and Farberov, M. I., Uch. Zap., Yarosl. Tekhnol. Inst. p. 64 (1971); Chem. Abstr. 78, 43157j (1973).
Blanchard, E. P., Jr., and Cairncross, A., J. Am. Chem. Soc. 88, 487 (1966).

Bottini, A. T., Dev, V., and Stewart, M., *J. Org. Chem.* **28**, 156 (1963).
Campbell, K. N., Sommers, A. H., and Campbell, B. K., *J. Am. Chem. Soc.* **68**, 140 (1964).
Chakrabortty, P. N., Dasgupta, R., Dasgupta, S. K., Ghosh, S. R., and Ghatak, U. R., *Tetrahedron* **28**, 4653 (1972).
Challand, B. D., Hikino, H., Kornis, G., Lange, G., and de Mayo, P., *J. Org. Chem.* **34**, 794 (1969).
Cocker, W., Shannon, P. V. R., and Staniland, P. A., *J. Chem. Soc.* C p. 41 (1966).
Cook, A. G., Herscher, S. G., Schultz, D. J., and Burke, J. A., *J. Org. Chem.* **35**, 1550 (1970).
Cram, D. L., and Hatch, M. J., *J. Am. Chem. Soc.* **75**, 33 (1953).
Cristol, S. J., and Mayo, G. O., *J. Org. Chem.* **34**, 2363 (1969).
Cromwell, N. H., and Bambury, R. E., *J. Org. Chem.* **26**, 997 (1961).
Cromwell, N. H., and Martin, J. L., *J. Org. Chem.* **33**, 1890 (1968).
deRopp, R. S., VanMeter, J. C., DeRenzo, E. C., McKerns, K. W., Pidacks, C., Bell, P. H., Ullman, E. F., Safir, S. R., Fanshawe, W. J., and Davis, S. B., *J. Am. Chem. Soc.* **80**, 1004 (1958).
Farkas, L., Gottsegen, A., and Nógrádi, M., *Tetrahedron* **26**, 2787 (1970).
Fore, S. P., and Bickford, W. G., *J. Org. Chem.* **24**, 620 (1959).
Fore, S. P., and Bickford, W. G., *J. Org. Chem.* **26**, 2104 (1961).
Franzus, B., Baird, W. C., Jr., and Surridge, J. H., *J. Org. Chem.* **33**, 1288 (1968).
Freeman, P. K., Ziebarth, T. D., and Rao, V. N. M., *J. Org. Chem.* **38**, 3823 (1973).
Galantay, E., Paolella, N., Barcza, S., and Coombs, R. V., *J. Am. Chem. Soc.* **92**, 5771 (1970).
Gawron, O., Fondy, T. P., and Parker, D. J., *J. Org. Chem.* **28**, 700 (1963).
Ghatak, U. R., Chakraborti, P. C., Ranu, B. C., and Sanyal, B., *J. Chem. Soc. Chem. Commun.* p. 548 (1973).
Gray, A. P., Kraus, H., Heitmeier, D. E., and Shiley, R. H., *J. Org. Chem.* **33**, 3007 (1968).
Gund, T. M., Williams, V. Z., Jr., Osawa, E., and Schleyer, P. von R., *Tetrahedron Lett.* p. p. 3877 (1970).
Hartzler, H. D., *J. Am. Chem. Soc.* **83**, 4990 (1961).
Henderickson, J. B., and Boeckman, R. K., Jr., *J. Org. Chem.* **36**, 2315 (1971).
Herz, W., *J. Am. Chem. Soc.* **74**, 2928 (1952).
Herz, W., Ligon, R. C., Kanno, H., Schuller, W. H., and Lawrence, R. V., *J. Org. Chem.* **35**, 3338 (1970).
Hochstetler, A. R., *J. Org. Chem.* **37**, 1883 (1972).
Howton, D. R., and Kaiser, R. W., Jr., *J. Org. Chem.* **29**, 2420 (1964).
Irwin, W. J., and McQuillin, F. J., *Tetrahedron Lett.* p. 2195 (1968).
Karabinos, J. V., and Serijan, K. T., *J. Am. Chem. Soc.* **67**, 1856 (1945).
Kazanskii, B. A., Lukina, M. Y., and Safonova, I. L., *Dokl. Akad. Nauk SSSR* **130**, 322 (1960).
Kharasch, M. S., and Priestly, H. M., *J. Am. Chem. Soc.* **61**, 3425 (1939).
Kieboom, A. P. G., de Kreuk, J. F., and van Bekkum, H., *J. Catal.* **20**, 58 (1971).
Kieboom, A. P. G., Breijer, A. J., and van Bekkum, H., *Rec. Trav. Chim. Pays-Bas* **93**, 186 (1974).
Kierstead, R. W., Linstead, R. P., and Weedon, B. C. L., *J. Chem. Soc.* p. 3610 (1952).
Kuehne, M. E., and King, J. C., *J. Org. Chem.* **38**, 304 (1973).
Laing, S. B., and Sykes, P. J., *J. Chem. Soc.* C p. 421 (1968).
Lambert, J. B., Koeng, F. R., and Jovanovich, A. P., *J. Org. Chem.* **37**, 374 (1972).
Lemal, D. M., and Shim, K. S., *Tetrahedron Lett.* p. 3231 (1964).
Leonard, N. J., and Jann, K., *J. Am. Chem. Soc.* **84**, 4806 (1962).
Leonard, N. J., Jann, K., Paukstelis, J. V., and Steinhardt, C. K., *J. Org. Chem.* **28**, 1499 (1963).
Lutz, R. E., and Wood, J. L., *J. Am. Chem. Soc.* **60**, 229 (1938).
McCurry, P. M., Jr., and Singh, R. K., *J. Org. Chem.* **39**, 2316 (1974).
McQuillin, F. J., and Ord, W. O., *J. Chem. Soc.* p. 3169 (1959).

Majerski, Z., and Schleyer, P. von R., *Tetrahedron Lett.* p. 6195 (1968).

Meinwald, J., Labana, S. S., and Chadha, M. S., *J. Am. Chem. Soc.* **85**, 582 (1963).

Mitsui, S., and Nagahisa, Y., *Chem. Ind.* (*London*) p. 1975 (1965).

Mitsui, S., and Sugi, Y., *Tetrahedron Lett.* p. 1287 (1969a).

Mitsui, S., and Sugi, Y., *Tetrahedron Lett.* p. 1291 (1969).

Mitsui, S., Imaizumi, S., Hisashige, M., and Sugi, Y., *Tetrahedron* **29**, 4093 (1973).

Mitsui, S., Sugi, Y., Fujimoto, M., and Yokoo, K., *Tetrahedron* **30**, 31 (1974).

Morrow, D. F., Butler, M. E., and Huang, H. C. Y., *J. Org. Chem.* **30**, 579 (1965).

Nametkin, N. S., Vdovin, V. M., Finkel'shtein, E. S., Popov, A. M., and Egorov, A. V., *Izv. Akad. Nauk SSSR, Ser. Khim.* p. 2806 (1973); *Chem. Abstr.* **80**, 82203g (1974).

Newham, J., *Chem. Rev.* **63**, 123 (1963).

Newman, M. S., Underwood, G., and Renoll, M., *J. Am. Chem. Soc.* **71**, 3362 (1949).

Nishimura, S., *Bull. Chem. Soc. Jpn.* **33**, 566 (1960).

Park, G. J., and Fuchs, R., *J. Org. Chem.* **22**, 93 (1957).

Pierson, G. O., and Runquist, O. A., *J. Org. Chem.* **34**, 3654 (1969).

Pigulevski, G. V., and Sokolova, A. E., *Zh. Prikl. Khim.* **36**, 455 (1963).

Pines, H., and Nogueira, L., *J. Catal.* **27**, 89 (1972).

Poos, G. I., and Rosenau, J. D., *J. Org. Chem.* **28**, 665 (1963).

Poulter, S. R., and Heathcock, C. H., *Tetrahedron Lett.* p. 5339 (1968a).

Poulter, S. R., and Heathcock, C. H., *Tetrahedron Lett.* p. 5343 (1968b).

Priese, O., Ger. (East) Patent 86,812, Jan. 5, 1972; *Chem. Abstr.* **78**, 97305z (1973).

Prudhomme, U. R., and Gault, F. G., *Bull. Soc. Chem. Fr.* p. 832 (1966).

Read, G., and Ruiz, V. M., *J. Chem. Soc., Perkin Trans. I* p. 368 (1973).

Ross, J. M., Tarbell, D. S., Lovett, W. E., and Cross, A. D., *J. Am. Chem. Soc.* **78**, 4675 (1956).

Roth, J. A., *J. Catal.* **26**, 97 (1972).

Rylander, P. N., Hasbrouck, L., Hindin, S. G., Iverson, R., Karpenko, I., and Pond, G., *Engelhard Ind., Tech. Bull.* **8**, 93 (1967a).

Rylander, P. N., Hasbrouck, L., Hindin, S. G., Karpenko, I., Pond, G., and Starrick, S., *Engelhard Ind., Tech. Bull.* **8**, 25 (1967b).

Sarel, S., and Breuer, E., *J. Am. Chem. Soc.* **81**, 6522 (1959).

Sauers, R. R., and Shurpik, A., *J. Org. Chem.* **33**, 799 (1968).

Schlatter, J. C., and Boudart, M., *J. Catal.* **25**, 93 (1972).

Schleyer, P. von R., and Wiskott, E., *Tetrahedron Lett.* p. 2845 (1967).

Schultz, A. L., *J. Org. Chem.* **36**, 383 (1971).

Schwarz, M., Besold, A., and Nelson, E. R., *J. Org. Chem.* **30**, 2425 (1965).

Senechal, G., and Cornet, D., *Bull. Soc. Chim. Fr.* p. 773 (1971).

Shortridge, R. W., Craig, R. A., Greenlee, K. W., Derfer, J. M., and Boord, C. E., *J. Am. Chem. Soc.* **70**, 946 (1948).

Sinfelt, J., *Adv. Catal.* **23**, 91 (1973).

Sohma, A., and Mitsui, S., *Bull. Chem. Soc. Jpn.* **43**, 448 (1970).

Sokol'skaya, A. M., Reshetnikov, S. M., Bakhanova, E. N., Kuzembaev, K. K., and Anchevskaya, M. N., *Khim. Khim. Tekhnol.* (*Alma-Ata*) p. 3 (1966); *Chem. Abstr.* **69**, 67027g (1968).

Stavely, H. E., *J. Am. Chem. Soc.* **64**, 2723 (1942).

Stevens, C. L., and Pillai, P. M., *J. Org. Chem.* **37**, 173 (1972).

Stevens, C. L., Cahoon, J. M., Potts, T. R., and Pillai, P. M., *J. Org. Chem.* **37**, 3130 (1972).

Strike, D. P., and Smith, H., *Tetrahedron Lett.* p. 4393 (1970).

Sugi, Y., and Mitsui, S., *Bull. Chem. Soc. Jpn.* **42**, 2984 (1969).

Sugi, Y., and Mitsui, S., *Bull. Chem. Soc. Jpn.* **43**, 1489 (1970).

Sugi, Y., Nagata, M., and Mitsui, S., *Bull. Chem. Soc. Jpn.* **48**, 1663 (1975).

Tarbell, D. S., Carman, R. M., Chapman, D. D., Cremer, S. E., Cross, A. D., Huffman, K. R., Kunstmann, M., McCorkindale, N. J., Ncnally, J. G., Jr., Rosowsky, A., Varino, F. H. L., and West, R. L., *J. Am. Chem. Soc.* **83**, 3096 (1961).
Temnikova, T. I., and Kropachev, V. A., *Zh. Obshch. Khim.* **19**, 2069 (1949).
Ullman, E. F., *J. Am. Chem. Soc.* **81**, 5386 (1959).
Upendraro, A., Chandrasekhararao, T., and Subbaro, R., *Fette, Seifen, Anstrichm.* **74**(2), 223 (1972); *Chem. Abstr.* **77**, 77037q (1972).
Ushakov, M. I., and Mikhailov, B. M., *Zh. Obshch. Khim.* **7**, 249 (1937).
Whitmore, W. F., and Gebhart, A. I., *J. Am. Chem. Soc.* **64**, 912 (1942).
Wiberg, K. B., and Ciula, R. P., *J. Am. Chem. Soc.* **81**, 5261 (1959).
Wissner, A., and Meinwald, J., *J. Org. Chem.* **38**, 1697 (1973).
Witkop, B., and Foltz, C. M., *J. Am. Chem. Soc.* **79**, 197 (1957).
Wood, T. F., Ger. Patent 1,918,852, Oct. 23, 1969; *Chem. Abstr.* **72**, 43111r (1970).
Woodworth, C. W., Buss, V., and Schleyer, P. von R., *Chem. Commun.* p. 569 (1968).
Wuesthoff, M. T., and Rickborn, B., *J. Org. Chem.* **33**, 1311 (1968).

*Chapter* **15**

---

# Miscellaneous Hydrogenolyses

This chapter discusses a number of reactions that either are not conveniently classified or do not warrant a separate chapter. All reactions discussed here are formally hydrogenolyses. The chapter is organized on the basis of the type of bond broken.

## I. BENZYL GROUPS ATTACHED TO OXYGEN

Hydrogenolysis of benzyl alcohols, ethers, esters, acetals, and phosphates can be achieved readily. Palladium, platinum, rhodium, nickel, cobalt, and copper have been used successfully in debenzylations, but palladium is by far the most favored catalyst. Palladium combines a high activity for debenzylation with a small tendency to saturate the aromatic ring (Rylander and Steele, 1965). On the other hand, ring saturation with minimal hydrogenolysis of the benzyl function apparently is best carried out over ruthenium (Rylander and Steele, 1965) and in some cases platinum (Richtmyer, 1934; Sowden and Kuenne, 1952; Hurd and Jenkins, 1966). Rhodium is useful in the hydrogenolysis of certain complex benzyl alcohols, especially when dehydrohalogenation is to be avoided (Fields *et al.*, 1961 ; McCormick and Jensen, 1962), but with simpler compounds ring saturation is apt to take precedence over hydrogenolysis (Stocker, 1962).

### A. Effect of Substrate Structure

Various workers have measured the effect of the leaving group on the rate of hydrogenolysis of benzyl–oxygen compounds. The rate of hydrogenolysis for acyl derivatives of atrolactic acid or its esters increases in the order $OH < OAc \ll OCOCF_3$ (Khan *et al.*, 1967). In the compound

$\phi CH_2OR$, the rate of hydrogenolysis increases in the order OH < O–alkyl ≪ O–aryl < $OH^+$–alkyl < $OH_2{}^+$ < OAc < $OCOCF_3$. The ease of displacement parallels the ability of the leaving group to bear a negative charge, suggesting a hydride displacement of the leaving group as its anion (Kieboom et al., 1971). The rate of hydrogenolysis has been correlated, by examination of the effect of ring substituents in benzyl alcohols (Kieboom et al., 1971) and the effect of ring size on rate in 1-phenylcycloalkanols (Khan et al., 1967), with the relative ease of $sp^3 \rightarrow sp^2$ change in coordination.

A general correlation exists between polar character of the leaving group and configuration of the product; less polar leaving groups favor retention and more polar groups favor inversion of configuration. This effect has been linked to hydrogen availability at the catalyst surface, the more rapid reactions being more hydrogen deficient (Khan et al., 1967). It has been noted in this regard that the steric outcome changes from retention to inversion with increasing palladium content (and lower hydrogen availability) of the catalyst (Mitsui and Imaizumi, 1963). Alternatively, it has been suggested that the chances of adsorption of the molecule on several sites become greater as the palladium concentration increases and the mechanism shifts from $S_{N1}$ at low metal concentrations to $S_{N2}$ at high concentrations (Senda and Mitsui, 1962).

## B.   Stereochemistry of Hydrogenolysis

The stereochemistry of hydrogenolysis of benzyl alcohol and its derivatives has been the subject of many investigations (Mitsui et al., 1963, 1969; Mitsui and Nagahisa, 1965; Garbisch et al., 1967). Hydrogenolysis usually proceeds with retention of configuration over Raney nickel (Bonner et al., 1952; Bonner and Zderic, 1956; Zderic et al., 1960; Garbisch, 1962; Mitsui et al., 1963, 1964), copper, and cobalt (Mitsui and Kudo, 1965) and with inversion over palladium and platinum catalysts. This generality is supported by many examples, but exceptions exist. The stereochemistry may be influenced by the type of catalyst, by its method of preparation, by additives, and by the substrate structure (Mitsui et al., 1962; Senda and Mitsui, 1962; Mitsui and Nagahisa, 1965). For instance, hydrogenolysis of 2-phenyl-2-methoxybutane or of ethyl 2-phenyl-2-methoxypropionate occurs with retention of configuration over Raney nickel, but predominantly with inversion if alkali is added. The result is attributed to the weakening of metal–oxygen bonding by alkali. Reduced nickel, in contrast to Raney nickel, also hydrogenates 2-phenyl-2-methoxybutane with inversion. Stereochemistry may also be influenced by some as yet undefined parameters as exemplified by conflicting results reported for the hydrogenolysis of 3-methyl-2-phenyl-2-butanol over palladium-on-carbon; inversion is reported by some workers, nonstereoselective attack by others (Kieboom, 1974).

Mitsui *et al.* (1969) relate the effect of metals on stereospecificity of hydrogenolysis of benzyl alcohols to the affinity of the metal for oxygen. Metals having a large affinity for oxygen, such as nickel, cobalt, and copper, tend to adsorb the hydroxyl function and promote retention of configuration, whereas metals with less affinity, palladium and platinum, tend not to adsorb the hydroxyl function and produce inversion.

The relative free-energy levels for the two transition states depend not only on the metal but also on steric and electronic factors of the substituents. The action of catalysts is thus not constant, but depends as well on the substrate structure. For example, ethyl atrolactate is stereospecifically hydrogenated with retention of configuration, whereas its acetate and propionate are inverted over both palladium-on-carbon and Raney nickel, a consequence of the bulkiness and electronegativity of the acyloxy groups (Imaizumi, 1960).

Differences in the behavior of metals can be illustrated by the hydrogenation of *exo*-2-phenyl-9-oxabicyclo[3.3.1]nonan-2-ol (Table I) (Cope *et al.*, 1970). Reduction over Raney nickel proceeds exclusively with inversion and over palladium with retention. No reduction over palladium in ethyl acetate occurred until a drop of perchloric acid was added, which is a common technique. The endo isomer was also formed in excellent yield by hydrogenation of 2-phenyl-9-oxabicyclo[3.3.1]non-2-ene over palladium-on-carbon.

TABLE I

Hydrogenation of *exo*-2-Phenyl-9-oxabicyclo[3.3.1]nonan-2-ol and 2-Phenyl-9-oxabicyclo[3.3.1]non-2-ene[a]

| RaNi | $\xrightarrow{\text{EtOH}}$ | 100% | 0 | |
| 10% Pd-on-C | $\xrightarrow[\text{HClO}_4]{\text{EtOH}}$ | 11% | 89% | |
| | | 5% | 95% | $\xleftarrow{\text{EtOAc}}$ 10% Pd-on-C |

[a] Data of Cope *et al.* (1970). Used with permission.

Similarly, catalytic hydrogenolysis of 1-benzyl-1-hydroxy-3-hydroxymethyltetralin over Raney nickel proceeds with retention of configuration and with inversion over palladium-on-carbon (Roberts *et al.*, 1973).

Different stereochemical results are not always obtained in reduction over Raney nickel and palladium. The lactone 1 with trans A/B ring junction opened with inversion very rapidly over palladium-on-carbon and slowly over Raney nickel to give only the cis acid, $1\alpha$-methyl-$1\beta$-carboxy-*cis*-1,2,3,4,9,10,11,12$\alpha$-octahydrophenanthrene (2) (Ghatak *et al.*, 1969). The preferred adsorption mode for nickel was probably barred by steric hindrance, as has been noted in other examples (Brewster and Braden, 1964).

## C.  Promoters

Hydrogenolysis of benzyl alcohols and their derivatives is promoted by small amounts of strong acids. Hydrochloric, sulfuric, and especially perchloric acids are often used for this purpose (Hartung and Simonoff, 1953). How effectively the acid will promote depends very much on the catalyst. Palladium-on-carbon catalysts frequently contain enough residual acid so that the addition of more strong acid produces only a two- or threefold increase in rate; with a neutral catalyst, addition of acid may increase the rate 20 or 30 times (Baltzly, 1976a). Divalent zinc, manganese, and cadmium salts at low concentration are also promoters, although they are not nearly as effective as acid (Baltzly, 1976b).

An effective use of sulfuric acid as a promoter was described by Marquardt (1973) in the hydrogenation of $\alpha$-(1-methyl-1-nitroethyl)benzyl alcohol (3) Reduction of this type of compound usually stops at the amino alcohol stage (4), but if the reduction is carried out at 70°C in acetic acid containing sulfuric acid over palladium black (Kindler *et al.*, 1948) the hydroxyl function is eliminated. The product is not the amine but rather the *N*-acetyl derivative 5, derived through an oxazoline intermediate. The temperature of reaction is critical, and 50°C was inadequate.

$$\underset{\textbf{3}}{\underset{\overset{|}{CH_3}}{\overset{\overset{OH\,NO_2}{|}}{\phi CHCCH_3}}} \longrightarrow \underset{\textbf{4}}{\underset{\overset{|}{CH_3}}{\overset{\overset{OH\,NH_2}{|}}{\phi CHCCH_3}}} \longrightarrow \left[ \underset{\overset{|}{CH_3}}{\overset{\overset{CH_3C=O}{\underset{O\ \ NH_2}{}}}{\phi CHCCH_3}} \longrightarrow \underset{\overset{|}{CH_3}}{\overset{\overset{CH_3}{\overset{O\quad N}{}}}{\phi CH-CCH_3}} \right] \longrightarrow$$

$$\underset{\textbf{5}}{\underset{\overset{|}{CH_3}}{\overset{\overset{NHCOCH_3}{|}}{\phi CH_2CCH_3}}}$$

Tertiary amines have proved to be effective promoters. Certain hindered benzyl esters may undero hydrogenolysis with difficulty under normal conditions, but very facile reductions can be achieved by the use of triethylamine as a promoter. The hydrogenolysis is viewed as a nucleophilic reaction in which an acylate anion is replaced by a hydride ion (Zymalskowski *et al.*, 1969).

$$\phi\text{-}\underset{\overset{|}{OCO\phi}}{\overset{\overset{CH_3}{|}}{CCOOC_2H_5}} \xrightarrow[\substack{C_2H_5OH \\ (C_2H_5)_3N}]{Pd\text{-}on\text{-}BaSO_4} \phi\text{-}\underset{\overset{|}{}}{\overset{\overset{CH_3}{|}}{CHCOOC_2H_5}} + \phi\underset{\overset{\|}{O}}{C}O^- \overset{+}{H}N(C_2H_5)_3$$

Water is sometimes an effective promoter. Debenzylation of **6** over palladium-on-carbon affords 4,5,6,7-tetrahydrobenzotriazole (**7**) in excellent yield with facile loss of morpholine and the establishment of a conjugated system. Reaction rates in acetic acid, dioxane, and methanol could be increased almost 15-fold by addition of 5–10% water (Oelschläger and Bremer, 1971).

## D.   Carbobenzyloxy Compounds

The carbobenzyloxy radical is widely used as a protecting group in organic syntheses. An advantage is that it can be removed easily by hydrogenolysis under the mildest conditions, usually without disruption of other portions of the molecule. It should be noted that the course of this hydrogenation cannot be followed by pressure decrease unless the reduction is carried out in the presence of alkali, inasmuch as 1 mole of carbon dioxide is liberated for each mole of hydrogen consumed. The extent of hydrogenolysis can be followed by measuring the carbon dioxide liberated. Palladium, supported or unsupported, is the best catalyst for this hydrogenolysis and is used by most investigators.

Hydrogenolysis of carbobenzyloxy compounds has been widely employed in the synthesis of peptide linkages (Bergmann and Zervas, 1932). Various derivatives of the carbobenzyloxy radical are often used to advantage. p-Bromobenzyl carbamates often have higher melting points and crystallize better than the corresponding benzyl carbamates (Channing et al., 1951). p-Nitrocarbobenzyloxy radicals are more labile to hydrogenolysis (Berse et al., 1957). It was thought that the lability of this function would permit its use as a sulfhydryl protecting agent, but apparently this is not so, for only the nitro group is reduced (Bachi and Ross-Petersen, 1972).

The N-carbobenzyloxy group of p-nitrophenyl esters of N-carbobenzyloxyamino acids and peptides can be removed by catalytic hydrogenation over palladium-on-carbon in the presence of 1 equivalent of hydrochloric acid without noticeable reduction of the nitro group (Kovacs and Robin, 1968). This method is useful for preparing oligo peptides by Goodman's "backing-off" procedure (Goodman and Steuben, 1959) and for preparing polyamino acids and sequential polypeptides when the acid-sensitive tert-butyl ester groups are present (Anderson and Callahan, 1960). For instance, hydrogenation of 5.18 mmoles N-carbobenzyloxy-α-tert-butyl-L-aspartic acid p-nitrophenyl ester in 60 ml methanol containing 5.22 mmoles hydrogen chloride and 100 mg 10% palladium-on-carbon produced after 5 min α-tert-butyl-L-aspartic acid p-nitrophenyl ester hydrochloride in 94% yield.

An important general procedure in the synthesis of peptides is the catalytic hydrogenolysis of N-benzyloxycarbonyl groups in peptides the side-chain functions of which have been protected by hydrogenation-resistant functions, tert-butyl esters, tert-butyl ethers, or tert-butyloxycarbonyl groups. This excellent procedure usually fails, however, when applied to cysteine- and methionine-containing peptides due to catalyst poisoning. Meienhofer and Kuromizu (1974) were able to circumvent this difficulty by using palladium black catalyst in dry refluxing (−33°C) liquid ammonia as solvent. Benzyl ester, benzyl ether, 2,6-dichlorobenzyl ether, N-benzyloxycarbonyl, N-2-bromobenzyloxycarbonyl, N-4-methoxybenzyloxycarbonyl, and the nitro group of nitroarginine were completely cleaved, whereas complete stability toward hydrogenolysis was found for tert-butyl ester, tert-butyl ether, N-tert-butyloxycarbonyl, N-p-toluenesulfonyl, and S-benzyl and S-acetamidomethyl groups. The method has been used in the synthesis of oxytocin

(Kuromizu and Meienhofer, 1974). In the hydrogenolysis of an intermediate tetrapeptide, the rate and yield were improved by the addition of either triethylamine (4 molar excess) or ammonium acetate (1.2 molar excess).

Catalytic hydrogenolysis has been applied successfully to solid-phase peptide synthesis (Merrifield, 1963) for removal of the peptide from the resin. The catalyst was prepared by dissolving palladium acetate in dimethylformamide, an excellent swelling solvent for resins, and the resin, with peptide chain attached, was added to the solution and allowed to equilibrate. Upon shaking with hydrogen at 40°C and 60 psig, hydrogenolysis of the benzyl binding function occurred readily (Jones, 1977; Schlatter and Mazur, 1977).

### E. Hydrogen Transfer Reactions

Hydrogenation through hydrogen donor compounds usually is only an alternative to the use of hydrogen, but sometimes the technique has advantages (Brieger and Nestrick, 1974). In the hydrogenolysis of **8** to **9** unsatisfactory results were obtained with platinum dioxide in ethyl acetate or palladium-on-carbon in ethanol. However, the desired product was achieved in 53% yield by refluxing **8** in tetrahydrofuran containing cyclohexene and palladium-on-carbon for 60 hr (Fahrenholtz, 1972).

### F. β Elimination

Hydrogenolysis of compounds containing both benzylic and homobenzylic oxygenated substituents may lead to the loss of either or both functions (Table II). The product distribution depends on the stereochemistry of the starting material, the substituents, and the metal. Loss of the homobenzylic substituent in 1-phenyl-1,2-diol derivatives was thought to be due to β elimination in suitably disposed adsorbed states (Table III). (Mitsui et al., 1969).

Hydrogenation of **10** over 10% palladium-on-carbon in acetic acid proceeds not at all unless perchloric acid is added, and then very slowly. The product **11**, obtained in quantitative yield, was derived by the unexpected loss of the C-4 equatorial hydroxyl (Craig et al., 1974).

**TABLE II**

**Hydrogenation of 1-Phenyl-1,2-Diacetoxycyclohexane**

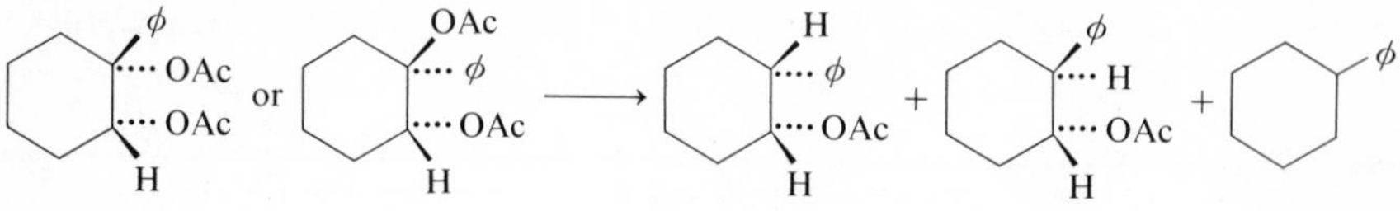

| | | | | |
|---|---|---|---|---|
| RaNi | | 29% | Trace | 71% |
| Pd | | 26% | 0 | 74% |
| | RaNi | 0 | 97% | 3% |
| | Pd | 0 | 98% | 2% |

**TABLE III**

**Hydrogenation of 1-Phenyl-2-Acetoxycyclohexanol[a]**

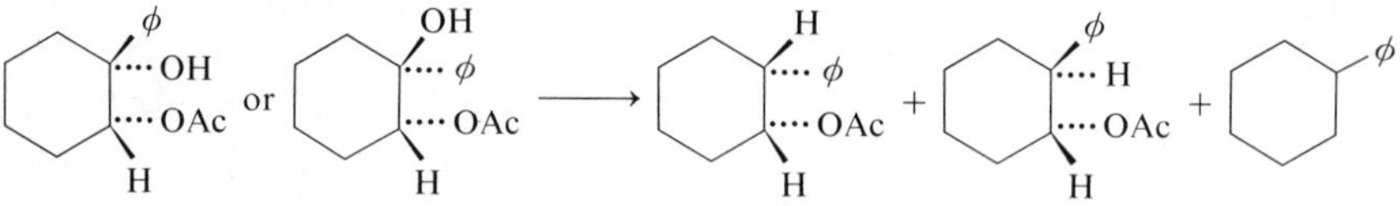

| | | | | |
|---|---|---|---|---|
| RaNi | | 0 | 93% | 7% |
| Pd | | 41% | 18% | 41% |
| | RaNi | 16% | 0 | 84% |
| | Pd | 15% | 85% | Trace |

[a] Data of Mitsui *et al.*, (1969). Used with permission.

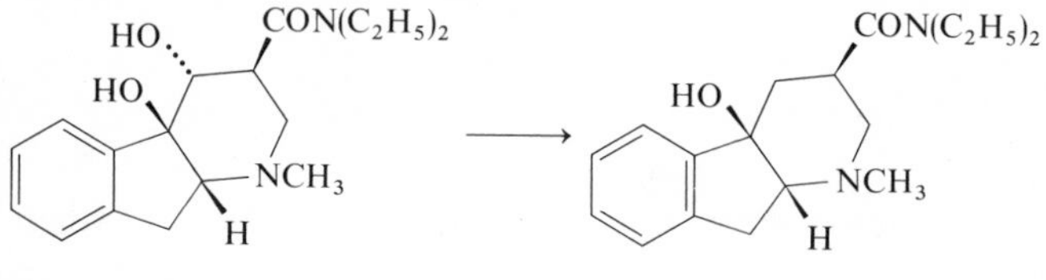

10            11

Loss of homobenzylic hydroxyl and acetoxyl functions is not uncommon (Zderic *et al.*, 1957; Greenlee and Bonner, 1959). Garbisch (1967) concluded from an examination of the hydrogenolysis of the homobenzylic oxygen functions in *cis-* and *trans*-4-*tert*-butyl-1-benzyl-1-cyclohexanol and the corresponding acetates that the reaction proceeds via $\beta$ elimination followed by hydrogenation of the resulting desorbed olefin.

## G.  Benzyloxy–Nitrogen Bonds

Both nitrogen–oxygen and benzyl–oxygen bonds undergo hydrogenolysis readily, but it appears that the benzyl–oxygen bond can be made to cleave selectively. Debenzylation of ethyl 1-hydroxy-2-pyridone-6-($\alpha$-oximino)-propionate proceeds rapidly over 10% palladium-on-carbon in methanol with little hydrogenation of either the oximino function or the pyridine oxide bond. The N-hydroxy group is useful as a protecting agent in the synthesis of pyridine derivatives (Greenwald and Zirkle, 1968).

A key step in the synthesis of $N^\varepsilon$-hydroxy-L-lysine (**14**) was the selective debenzylation of the intermediate **13** over 5% palladium-on-carbon in 0.5 $M$ hydrochloric acid. In contrast, hydrogenolysis in neutral medium afforded L-lysine (**12**) (Isowa and Ohmori, 1974). $N^\varepsilon$-Hydroxy-L-lysine is an important intermediate in the synthesis of mycobactines, sexadentate chelating agents for ferric iron, which occur naturally and stimulate the growth of mycobacteria.

## H.  Hydrogenolysis of Benzyl Phosphates

Benzyl esters of phosphoric acid have been employed with much success in the synthesis of phosphorylated amines and alcohols. Palladium catalysts are invariably used for the hydrogenolysis of the benzyl protecting group (Lies *et al.*, 1953; Ballou and Fischer, 1954; Friedman *et al.*, 1954; Friedman and Seligman, 1954; Tener and Khorana, 1958; Ukita *et al.*, 1958; Magerlein and Kagan, 1960; Westphal and Stadler, 1963; Dürckheimer and Cohen, 1964).

## II. BENZYL GROUPS ATTACHED TO NITROGEN

The benzyl–nitrogen bond is not as easily cleaved as the benzyl–oxygen bond, and competitive hydrogenations are apt to be more important. Attempted hydrogenolysis of ethyl 2-amino-2-phenylpropionate over Raney nickel at atmospheric pressure afforded only 4% ethyl phenylpropionate; the remaining material was derived from attack at the phenyl ring and/or ester function (Sugi and Mitsui, 1973). Nonetheless, a wide variety of benzylamines have been reduced successfully under mild conditions with avoidance of major side reactions (Hartung and Simonoff, 1953).

### A.  Catalysts

A variety of catalysts including palladium, platinum, nickel, and copper chromite have been used in the hydrogenolysis of benzylamines, but palladium is by far the preferred catalyst. The Pearlman (1967) catalyst (manufactured by Engelhard Ind., Newark, New Jersey), 20% palladium hydroxide-on-carbon, is particularly active for the hydrogenolysis of benzyl–nitrogen bonds and has proved to be successful even in cases in which other palladium-on-carbon catalysts have failed (Hiskey and Northrop, 1961; Cromwell and Rodebaugh, 1969; Anderson and Lok, 1972). Platinum catalysts might be indicated in reductions of benzyl compounds when dehydrohalogenation is to be avoided (Aeberli et al., 1975; Ning et al., 1970).

### B.  Effect of Structure

The effect of structure on the rate and course of catalytic hydrogenolysis of benzylamines has been examined by a number of investigators (Baltzly and Buck, 1943; Baltzly and Russell, 1950, 1953; Dahn et al., 1954). The ease of hydrogenolysis at room temperature and atmospheric pressure increases in the series primary < secondary < tertiary < quaternary ammonium salts (Dahn et al., 1954), but under more vigorous conditions (ca. 150°C and 60 atm) this order is reversed (Mitsui et al., 1954). This contrast suggests that hydrogenolysis proceeds by different pathways under different conditions (Murchú, 1969). The relative ease of cleavage of various benzylamines may also change with the solvent (Baltzly and Russell, 1954).

### C.  Stereochemistry

Benzylamines, unlike benzyl alcohols, tend to undergo hydrogenolysis with inversion over both Raney nickel and palladium catalysts (Murchú, 1969), an effect ascribed to the greater affinity of the nitrogen atom for nickel. The argument advanced in this regard is complex and probably would not have been made *a priori* (Sugi and Mitsui, 1973).

Anilino acids and esters may give different configurational products on hydrogenolysis. The ester, methyl 2-anilino-2-phenylpropionate, gave the inverted product over palladium and the retained product over nickel, whereas inversion occurred over both nickel and palladium when the free acid was used. The latter result is attributed to weakening of the nitrogen adsorption on nickel due to zwitterion formation (Mitsui and Sato, 1965; Sugi and Mitsui, 1973).

The anion associated with quaternary ammonium compounds may have marked effects on the stereochemical outcome of hydrogenolysis (Table IV) (Dahn *et al.*, 1970; Murchú, 1969; Sugi and Mitsui, 1973).

TABLE IV

$$\begin{array}{c} CH_3 \\ | \\ \phi CCOOCH_3 \\ | \\ N(CH_3)_3X \end{array} \longrightarrow \begin{array}{c} CH_3 \\ | \\ \phi CHCOOCH_3 + HN(CH_3)_3{}^+X^- \end{array}$$

| X | Catalyst | Configuration of product | | Reference |
|---|---|---|---|---|
|  |  | Retained (%) | Inverted (%) |  |
| $I^-$ | 10% Pd-on-C | 50 | 50 | Dahn *et al.* (1970) |
| $Br^-$ | 5% Pd-on-C | 30 | 70 | Sugi and Mitsui (1973) |
| $Br^-$ | Raney Nickel | 50 | 50 | Zymalkowski *et al.* (1969) |
| $OAc^-$ | 5% Pd-on-BaSO$_4$ | 17 | 83 | Zymalkowski *et al.* (1969) |

Racemization of the iodide over palladium and the bromide over nickel is attributed to the strong adsorption of these anions on the metal and its effect on the rate of racemization reactions relative to the rate of hydrogenolysis. Catalyst deactivators, such as sulfides, also tend to increase racemization (Murchú, 1969).

The facile catalytic cleavage of benzyl–nitrogen bonds has made the use of optically active benzylamines of interest as a means of inducing asymmetry. Optically active amino acids can be prepared by the Strecker reaction involving interaction of an aldehyde, hydrogen cyanide, and an optically active benzylamine. The resulting *N*-alkylaminonitriles are then hydrolyzed with 6 *M* hydrochloric acid and hydrogenolyzed over palladium hydroxide-on-carbon. Overall yields of amino acids range from 19 to 47%, with optical purities of 22–51% (Harada *et al.*, 1973). Marked solvent effects have been observed in hydrogenolyses of this type. The solvent is thought to control the configuration of the reacting species (Harada and Kataoka, 1978).

$$RCHO + \phi\overset{*}{C}HNH_2 + HCN \longrightarrow \underset{\substack{| \\ NH \\ | \\ \phi CHR'}}{RCHCN} \xrightarrow[2.\ H_2]{1.\ H^+} \underset{\substack{| \\ NH_2}}{R\overset{*}{C}HCOOH} + \phi CH_2R'$$

Optically active amino acids can also be prepared in high yield by stereo-selective decarboxylation of malonic acids. Optical purity ranges from 20 to 26% for low-temperature (20°C) decarboxylation and falls as the temperature is raised (Hayakawa and Shimizu, 1973).

$$\underset{\substack{| \\ CH_3}}{\phi\overset{*}{C}HNH_2} + \underset{\substack{| \\ COOH}}{\overset{COOH}{Br\overset{|}{C}CH(CH_3)_2}} \longrightarrow \underset{\substack{| \quad\quad | \\ CH_3 \quad COOH}}{\phi\overset{*}{C}HNH\overset{\overset{COOH}{|}}{C}CH(CH_3)_2} \xrightarrow[\substack{PdO\text{-}on\text{-}C \\ EtOH-HCl}]{1.\ -CO_2 \\ 2.\ H_2} \underset{\substack{| \\ NH_2}}{(CH_3)_2CH\overset{*}{C}HCOOH}$$

Hydrolysis and hydrogenolysis, over 5% palladium-on-carbon, of diketopiperazines prepared with optically active benzylamines can be used to prepare optically active alanine of 6–25% optical purity (Okawara and Harada, 1973). Other asymmetric syntheses of amino acids are discussed in Chapter 10.

A general synthesis of L-α-hydrazino acids from L-α-amino acids involves selective hydrogenolysis of an $N^\alpha$-benzyl bond without racemization. The hydrogenolysis is carried out in ethanol containing 1 equivalent of p-toluenesulfonic acid or hydrochloric acid. In the absence of acid, hydrogenolysis does not proceed (Achiwa and Yamada, 1975). The L-α-hydrazino acids thus obtained are useful for the asymmetric synthesis of peptides (Achiwa and Yamada, 1974) and modified peptides (Grupe et al., 1972).

$$\underset{\substack{| \\ NHCH_2\phi}}{RCHCOOR'} \xrightarrow[HCl]{NaNO_2} \underset{\substack{| \\ NCH_2\phi \\ | \\ NO}}{RCHCOOR'} \xrightarrow[Ac_2O]{Zn \\ HOAc} \underset{\substack{| \\ NCH_2\phi \\ | \\ NHCOCH_3}}{RCHCOOR'} \xrightarrow[H_2]{5\%\ Pd\text{-}on\text{-}C} \underset{\substack{| \\ NH \\ | \\ NHCOCH_3}}{RCHCOOR'}$$

$$85\%$$

It is of interest that the $N^\beta$-nitroso compound **16** could not be converted to the hydrazine **15** either by reduction with zinc dust in acetic acid or by hydrogenation with palladium-on-carbon catalysts. The sole product (**17**) was derived by hydrogenolysis of the nitrogen–nitrogen bond.

$$\underset{\substack{| \\ NCOCH_3 \\ | \\ NH_2}}{\phi CH_2CHCOOC_2H_5} \xleftarrow{\quad|\!\!\!\!\!/\quad} \underset{\substack{| \\ NCOCH_3 \\ | \\ NO}}{\phi CH_2CHCOOC_2H_5} \longrightarrow \underset{\substack{| \\ NHCOCH_3}}{\phi CH_2CHCOOC_2H_5}$$

$$\textbf{15} \qquad\qquad\qquad \textbf{16} \qquad\qquad\qquad \textbf{17}$$

## D.  Consecutive Reactions

Certain debenzylations are accompanied by further reactions that take place with or as a consequence of debenzylation. A case in point is the hydrogenolysis of 3-benzyl-1-benzyloxycarbonylazirin-2-one (**18**) over palladium black in ether to afford the intramolecular anhydride **19**. Fast intramolecular attack by the carboxyl group on the adjacent carbonyl carbon must take place before decarboxylation of the carbamic acid. Treatment of **18** with water leads instead to an intermolecular anhydride (Miyoshi, 1973).

**18**                                          **19**

Debenzylation of 1,4-dibenzyl-1,4-diazabicyclo[2.2.1]heptane diperchlorate over 10% palladium-on-carbon in water affords a mixture of 1-methylpiperazine and piperazine. The bridge opening requires a prior debenzylation, for there is no reaction when the *N,N'*-diethyl derivative is used (Pettit *et al.*, 1969).

The Diels–Alder addition product of furan and dibenzyl azodicarboxylate (**20**) is readily reduced over 10% palladium-on-carbon in ether to the dihydro derivative **21**, without appreciable hydrogenolysis of the allylic–oxygen

**20**                    **21**                    **22**                    **23**

bond. Further reduction affords, not the expected hydrazo compound (22), but rather a trimer (23) (Bandlish *et al.*, 1973).

Facile cyclization follows hydrogenolysis of 25, providing a convenient synthesis of a 1,4-benzoxazine-2,3-dione (26). Hydrogenolysis of 24 failed, apparently due to catalyst poisoning, but conversion of 24 to 25 with ethyl oxalyl chloride permitted ready hydrogenation (Smissman and Corbett, 1972).

24                                25                                26

## E. Benzylidene Compounds

Benzylidene compounds with the formula $C_6H_5CHX_2$, where $X = NRR$, OR, or OCOR, generally can be hydrogenated selectively over palladium catalysts to afford good yields of $C_6H_5CH_2X$. Ethanol and acetic acid are better solvents for this reaction than benzene, dioxane, or isopropyl ether. In mixed benzylidene compounds containing oxygen and nitrogen substituents,

TABLE V

A                                                          B

| Solvent | Product composition (%) | |
|---------|------|------|
|         | A    | B    |
| Ethanol | 36   | 64   |
| Butanol | 46   | 54   |
| Benzene | 96   | 4    |
| Dioxane | 97   | 3    |

the benzyl–oxygen bond is cleaved preferentially. *N*-(α-Aminobenzyl)amides lose the amine function preferentially over palladium under mild conditions,

but over nickel at elevated pressure (80 atm) and temperature (80°C) the direction of cleavage depends on the solvent (Table V) (Sakura *et al.*, 1972).

Hydrogenation of 9*b*-*p*-chlorophenyl-1,2,3,9*b*-tetrahydro-5*H*-imidazo-[2,1-*a*]isoindol-5-one (27; 14 gm) proceeds readily over platinum oxide in 150 ml acetic acid to afford **28** in 85% yield. Hydrogenolysis of the amino nitrogen occurred preferentially and without appreciable dehydrohalogenation, the latter a consequence of the use of platinum (Aeberli *et al.*, 1975).

27                                      28

### III.   ALLYLIC OXYGEN COMPOUNDS

Many compounds containing an allylic oxygen function have been hydrogenated. Both hydrogenation of the olefinic linkage and hydrogenolysis of the allylic carbon–oxygen bond occur readily, so that the outcome of the reduction is determined by the relative rates of these two competing reactions; the saturated oxygenated derivative is generally stable toward hydrogenolysis. This rate ratio is determined by substrate structure, catalyst, and environment in a partially predictable manner.

## A.  Solvents

The solvent may have an important influence on the products. In general, the ratio of hydrogenolysis to hydrogenation is expected to increase with increasing acidity and increasing polarity (Shoppee *et al.*, 1957; Cline *et al.*, 1959; Nishimura and Mori, 1959; Nishimura *et al.*, 1960; Taub *et al.*, 1963; Rylander and Himelstein, 1964).

## B.  Catalysts

Few comparisons of the effect of catalyst on the outcome of hydrogenation of allylic systems have been made, but it would appear that, if the course of reduction can be changed at all by the catalyst, hydrogenolysis should increase in the order Ru $\gtrsim$ Rh $\gg$ Pt $\approx$ Pd. Trace quantities of acid or base in the catalyst may drastically alter the results, which complicates comparisons of metals. Little or no hydrogenolysis occurred during hydrogenation of cinnamyl alcohol over 5% palladium-, platinum-, or rhodium-on-carbon in ethanol (Rylander and Himelstein, 1964), whereas over palladium chloride-on-carbon (Hartung, 1928) about equal amounts of hydrocinnamyl alcohol and propylbenzene were found (Baltzly and Buck, 1943). Reduction of cholest-4-en-3β-ol over platinum oxide gave markedly different product compositions depending on the alkali content of the catalyst (Dart and Henbest, 1960).

The following examples illustrate the effect that catalyst may some times have on the product composition. Palladium proved to be more efficient in the hydrogenolysis of some 6β-acetoxyl-$\Delta^4$ and 6β-hydroxy-$\Delta^4$ steroids than did platinum. Over 5% palladium-on-carbon a mixture of 6-deacetoxy-$\Delta^4$ and 6-deacetoxy-$\Delta^5$ steroids resulted, perhaps via a π-allyl complex, whereas over platinum in acetic acid the double bond was saturated with only partial hydrogenolysis of the acetoxy function (Annen *et al.*, 1973).

Hydrogenation of 3-*O*-acetyl-1,2:5,6-di-*O*-isopropylidene-α-D-erythrohex-3-enose (**29**) yields three different products depending on catalyst and reaction conditions. Hydrogenation in ether at room temperature with palladium-on-carbon as catalyst rapidly afforded **30** in 65% yield, in which double-bond isomerization converted the acetate function from a vinylic to an allylic one. A longer hydrogenation (2 hr) over palladium black affords the saturated derivative **31** in 75% yield, whereas over platinum the saturated deacetylated product **32** is formed in 67% yield (Slessor and Tracey, 1970).

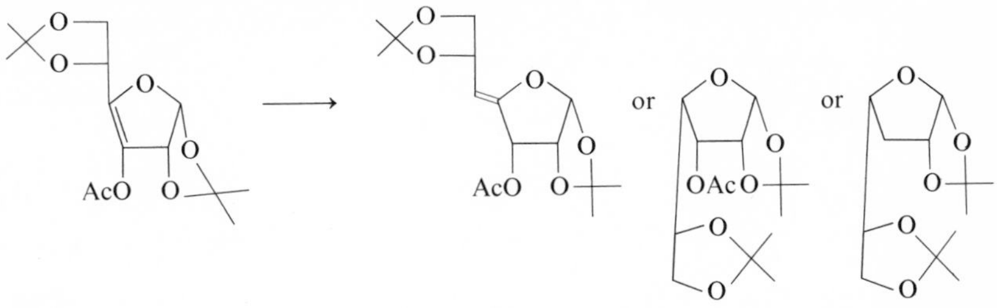

**29**                    **30**                    **31**                    **32**

Rhodium is useful in preventing hydrogenolysis of allylic hydroxyls that are easily cleaved over palladium. A case in point is the hydrogenation of 9-oxo-11,15-dihydroxyprosta-8(12),13-*trans*-dienoic acid (**33**) over 5% rhodium-on-alumina in methanol containing a small amount of acetic acid. Reduction under these conditions afforded the saturated diols **34** and **35**, whereas reduction over palladium-on-carbon resulted mainly in hydrogenolysis of the 11-hydroxy group before reduction of the $\Delta^{8(12)}$ double bond. Hydrogenation occurred predominantly from the side opposite the 11-hydroxyl group (Miyano and Dorn, 1972). The cyclopenten-3-on-5-ol system seems to be especially sensitive to hydrogenolysis (DePuy and Zaweski, 1959).

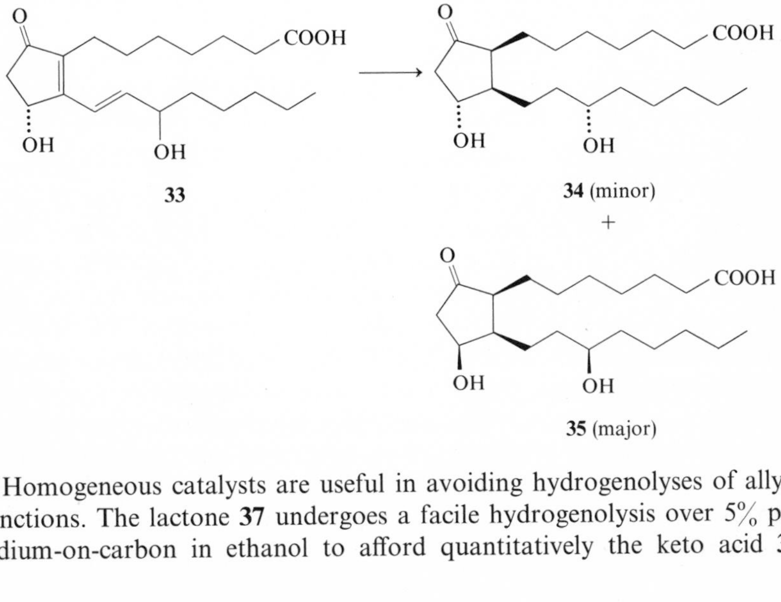

**33**                                            **34** (minor)

+

**35** (major)

Homogeneous catalysts are useful in avoiding hydrogenolyses of allylic functions. The lactone **37** undergoes a facile hydrogenolysis over 5% palladium-on-carbon in ethanol to afford quantitatively the keto acid **38**.

Hydrogenolysis was circumvented completely by the use of tris(triphenyl-phosphine) chlororhodium, which afforded the saturated lactone in 95% yield (Welch *et al.*, 1974).

**36**                              **37**                              **38**

Homogeneous catalysts have also been used to promote hydrogenolysis. Cholest-4-en-3$\beta$-ol undergoes hydrogenolysis and dehydration in the presence of hydrogen and dichlorobis(triphenylphosphine) platinum(11)–stannous chloride (Pt/Sn = 1:10) at room temperature and 30 atm hydrogen pressure to afford 5$\alpha$-cholest-3-ene (60%) and $\Delta^{2,4}$-cholestadiene (26%). The methanol solvent can also function as a source of hydrogen (Ichinohe *et al.*, 1969).

Tris(triphenylphosphine) rhodium chloride catalyzes the isomerization of allyl ethers to 1-propenyl ethers, hydrolysis of which at pH 2 rapidly gives an alcohol (Corey and Suggs, 1973).

$$ROCH_2CH{=}CH_2 \xrightarrow{Rh(I)} ROCH{=}CHCH_3 \xrightarrow{H^+} ROH + CH_3CH_2CHO$$

In anhydrous medium an olefin and aldehyde are formed (Wayaku *et al.*, 1975).

$$\phi CH{=}CHCH_2OH \xrightarrow{Rh(I)} [(\phi CH{=}CHCH_2)_2O] \longrightarrow \phi CH{=}CHCH_3 + \phi CH{=}CHCHO$$

## C.   Effect of Structure

Certain aspects of the effect of substrate structure on the course of the reduction of allylic compounds are clear. Compounds with bulky substituents in the neighborhood of the double bond (Wiberg and Hutton, 1954; Adams and Gianturco, 1956) or compounds with tetrasubstituted double bonds (Boekelheide and Chang, 1964; Godfrey *et al.*, 1955; Stedman *et al.*, 1964; Kupchan *et al.*, 1969) undergo, as would be expected, extensive hydrogenolysis inasmuch as the olefinic function is relatively inaccessible to approach by the catalyst. Conversely, structural features that tend to limit the access of the oxygen function or its adjacent carbon to the catalyst would be expected to diminish hydrogenolysis relative to hydrogenation. The effect of steric hindrance on oxygen adsorption and hydrogenolysis might change greatly with the catalyst (Callow and Thompson, 1964) depending on whether the catalyst typically produces hydrogenolysis with retention of configuration, like nickel, or with inversion of configuration, like platinum and palladium (Mitsui et al., 1963). Predicting the course of hydrogenation is difficult mainly with compounds in which neither the oxygen nor the olefin function has any obvious hindrance to adsorption on the catalyst. In compounds of this type small structural variations can almost completely change the type of products (Newman and VanderWerf, 1945; Meinwald et al., 1958; Noland et al., 1959; Meinwald and Frauenglass, 1960; Galantay et al., 1963).

Compounds containing allylic oxygen functions that are activated toward hydrogenolysis undergo loss of the function even after saturation of the double bond. A compound of this type (39), prepared from benzyne and 2,5-dimethylfuran, provides an improved route to cis-1,4-dimethyl-1,2,3,4-tetrahydronaphthalene (40) (Newman et al., 1975).

## D. Double-Bond Migration

Hydrogenation and hydrogenolysis of allylic and homoallylic systems may be accompanied by a double-bond migration (Dauben et al., 1960a,b). The isomerization can be masked easily due to subsequent saturation of the double bond (Dauben and Hance, 1955). For instance, formation of n-heptanol in the hydrogenation of hept-3-yne-1,7-diol over 5% palladium-on-barium sulfate is best accounted for by migration of an intermediate homoallylic double bond into an allylic position followed by hydrogenolysis and hydrogenation (Crombie and Jacklin, 1957).

$$HOCH_2CH_2C\equiv CCH_2CH_2CH_2OH \rightarrow HO(CH_2)_6CH_3 + HO(CH_2)_7OH$$

Allylic systems may form during hydrogenation even if the required migration involves isomerization of the double bond from a tetrasubstituted position. Both bemarivolide (41) and 6$\beta$-acetoxyconfertifolin (44) afford mixtures of the dihydro compound 42 and the hydrogenolysis product 43 on hydrogenation over platinum oxide. Formation of 43 from 44 is evidence that a prior migration of the tetrasubstituted double bond to 41 preceded hydrogenolysis. It can be argued that formation of 42 from 44 also requires

41                                42                            43

44

prior isomerization (Canonica *et al.*, 1969a,b). Isodrimenin (**45**) is substantially inert to hydrogenation, whereas confertifolin (**46**) gives *cis*-dihydro-confertifolin (**43**). Presumably, double-bond migration in isodrimenin is not possible because of the high energy needed to break the conjugated system (Appel *et al.*, 1960).

45                              46

Isomerization during the hydrogenation of allyl alcohols may result in aldehydes or ketones, which may or may not, depending on the catalyst, undergo further reduction. For instance, hydrogenation of cyclohexen-2-ol in ethanol over platinum affords only cyclohexanol, whereas over palladium, a catalyst with a high activity for double-bond isomerization and a low activity for carbonyl reduction, the product is 33% cyclohexanone and 67% cyclohexanol (Rylander and Himelstein, 1964). Allylic migration with sterically impeded double bonds is well documented (McQuillin, 1963), and, depending on the goal, the reaction can either be a nuisance or have synthetic utility (Barnes and MacMillan, 1967; Herz *et al.*, 1968). In general, if migration is desired, palladium is the preferred catalyst, operating optimally under hydrogen-poor conditions, i.e., high metal loadings, low pressure, and elevated temperatures (Rylander, 1973).

## IV.   HYDROGENOLYSIS OF THE CARBON–CARBON BOND

Hydrogenolysis of the carbon–carbon bond does not occur under mild conditions, unless the bond is weakened by some structural feature of the molecule. The most common example of carbon–carbon bond hydrogenolysis, that of cyclopropanes, is discussed separately in Chapter 14.

### A.   Cyclobutanes

Cyclobutanes undergo hydrogenolysis less readily than cyclopropanes, reflecting higher carbon–carbon bond energies (Newham, 1963). In both types of rings, hydrogenolysis is facilitated by adjacent unsaturation (Schreyer, 1963; Hasek *et al.*, 1964), by aromatic substituents (Cava and Pohlke, 1963; Jensen and Coleman, 1958), or by additional strain (Seebach, 1965; Vellturo and Griffin, 1965; Wiberg and Ciula, 1959; Lemal and Shim, 1964: Meinwald *et al.*, 1963).

The catalyst may influence the results of hydrogenolysis markedly. Hydrogenolysis of the homocubane **47** over palladium or platinum (catalysts commonly used for hydrogenolysis of strained rings) gave a complex mixture of many products, whereas over supported 5% rhodium only **48** and **49** are the major primary products (Table VI). Further hydrogenation causes **49** to be converted to **50**. The ratio of products also depends on the support (Toyne, 1976). Factors determining the sequence of bond cleavage in strained polycyclic compounds have been discussed by Musso (1975).

TABLE VI

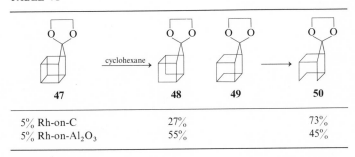

| 5% Rh-on-C | 27% | 73% |
| 5% Rh-on-Al$_2$O$_3$ | 55% | 45% |

### B.   Aromatization

Hydrogenolysis of a carbon–carbon bond occurs with particular ease when one of the resulting fragments is converted to an aromatic system in the process (Baird and Winstein, 1957). Hydrogenolysis of allyl or benzyl groups attached to a quaternary carbon in cross-conjugated or linearly conjugated cyclohexadienones is apt to be extensive. Hydrogenolysis relative to hydrogenation is increased greatly by an increase in the polarity of the solvent or by an increase in its hydrogen bonding ability, as illustrated by the reduction of **51** to **52** and **53** (Table VII). The results are consistent with a hydride ion-like species displacing a phenoxide ion from an allyl group (Miller and Lewis, 1974). The suggestion is a variation of that proposed by Anteunis and Verzele (1959) to account for similar behavior during the hydrogenolysis of lupulone (Carson, 1951). Benzyl groups have a greater tendency to cleave than do allyl groups. In the corresponding benzyl compound the yield of phenol is 100% even in hexane solvent (Miller and Lewis, 1974).

**TABLE VII**[a]

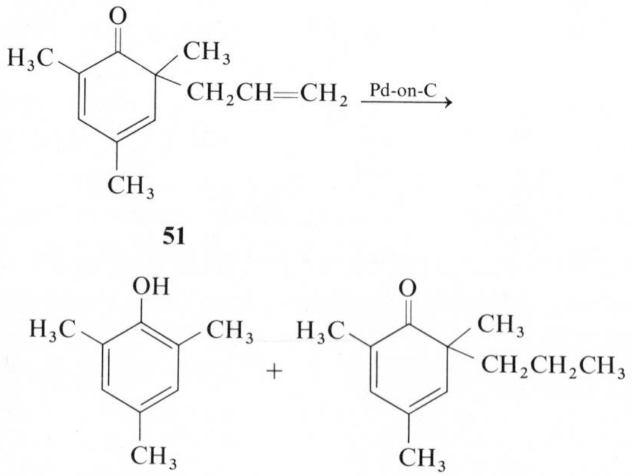

**51**

| Solvent | Yield of **52** (%) |
|---|---|
| Hexane | 13 |
| Methanol | 49 |
| Acetic acid–methanol (1:3) | 85 |

[a] Data of Miller (1974). Used with permission.

## V. HYDROGENOLYSIS OF THE OXYGEN–OXYGEN BOND

Peroxides, hydroperoxides, and ozonides are cleaved easily by hydrogen in the presence of hydrogenation catalysts. In synthetic work, platinum metals are often used without hydrogen in order to destroy excess hydrogen peroxide (Cope and Ciganek, 1963). Ruthenium is by far the most active metal for the decomposition of hydrogen peroxide, whereas supported platinum is more active for the decomposition of peroxy acids (Andersen and Romeo, 1964) and hydroperoxides (British Patent 1,009,939 Nov, 17, 1965).

Hydrogenolysis of ozonides of aldehydes and ketones is complex, and rearrangement products may form as well (Bailey, 1958; Thompson, 1962). Better yields can be obtained in reactive solvents such as methanol than in nonreactive solvents (Pryde *et al.*, 1960), and further improvement can be obtained by addition of pyridine (Pryde *et al.*, 1962). An unusual product of ozonide hydrogenation is the stable diol **55**, which is formed by reduction of **54** over 10% palladium-on-carbon (Marullo and Alford, 1968).

Hydrogenation of peroxides is carried out easily over palladium or platinum catalysts to afford alcohols or glycols (Cope *et al.*, 1957; McKay, *et al.*, 1964; Kende and Chu, 1970; Yamazaki *et al.*, 1975), often in excellent yield. When the resulting alcohol is itself sensitive to hydrogenolysis, its yield may be improved by the presence of an amine (Russell and Mayo, 1957). Palladium–lead-on-calcium carbonate (Lindlar catalyst) is useful for the hydrogenolysis of peroxides when olefin saturation is to be avoided (Agnello *et al.*, 1956; Laubach, 1957). Olefin saturation has also been controlled by solvent (Horinaka *et al.*, 1975).

Hydroperoxides are easily hydrogenated to the corresponding alcohols. Palladium or platinum catalysts are usually used (Lythgoe and Trippett, 1959; Fuson and Jackson, 1950; Fuchs, 1960; von Whittenau, 1964; Witkop and Partick, 1951; Stevens and Gasser, 1957). Unsaturated alcohols can be obtained from unsaturated hydroperoxides by the use of inhibited catalysts such as palladium–lead-on-calcium carbonate.

## VI.  HYDROGENOLYSIS OF THE NITROGEN–OXYGEN BOND

Catalytic hydrogenolysis of nitrogen–oxygen bonds is carried out very frequently, as exemplified by the many facile reductions of nitro, nitroso, hydroxylamine, and oxime functions. The reaction is one of very wide synthetic utility. The reduction of the more common functions has been discussed in other chapters.

### A.  N-Nitrosoamines

Compounds of this type are reduced readily to hydrazines, and, if the reduction is carried further, hydrogenolysis occurs; often, in fact, the major difficulty encountered is preventing overhydrogenation (Graefe, 1958).

The most important industrial hydrogenation of N-nitrosoamines is conversion of N-nitrosodimethylhydrazine to N,N-dimethylhydrazine, a component of rocket and missle fuels.

$$\begin{array}{c} H_3C \\ \phantom{H_3}\diagdown \\ \phantom{H_3C}\diagup \\ H_3C \end{array} NNO \longrightarrow \begin{array}{c} H_3C \\ \phantom{H_3}\diagdown \\ \phantom{H_3C}\diagup \\ H_3C \end{array} NNH_2$$

Palladium, platinum, rhodium, and nickel have been used in reductions of this type, but palladium-on-carbon is the preferred catalyst. In palladium-catalyzed reductions, dimethylamine by-product seems to be formed through decomposition of an intermediate tetramethyltetrazene (Paal and Yao, 1930); dimethylhydrazine is not easily cleaved over palladium, although it is over rhodium (Smith and Thatcher, 1962). Yields of the hydrazine may be improved through the use of added salts (Smith and Thatcher, 1962), exclusion of oxygen (Feldman and Frampton, 1964), and addition of iron compounds (Tuemmler and Winkler, 1961) or alkline earth hydroxides (Lima, 1964).

### B.  Amine Oxides

Amine oxides related to pyridine are more easily reduced than aliphatic tertiary amine oxides (Ochiai et al., 1943), but both types can be cleaved without difficulty. Palladium, platinum, rhodium, and ruthenium were used for the reduction of pyridine N-oxide; rhodium was the most active, but it was nonselective, reducing the ring as well (Rylander and Rakoncza, 1962). Rhodium is the preferred catalyst when a piperidine is the desired product. Palladium catalysts are used by most workers for N-oxide hydrogenolysis (Katritsky and Monro, 1958; Brown, 1957; Shaw, 1949; Lott and Shaw, 1949); platinum (Taylor and Driscoll, 1960; Lauer and Dyer, 1942) is also effective.

## C. Heterocyclic Compounds

Many heterocylic compounds containing an oxygen–nitrogen linkage undergo facile hydrogenolysis under mild conditions with ring opening. Palladium, platinum, and nickel are frequently used in these reductions (Mallory and Varimbi, 1963; Walker, 1962; Horner and Jürgens, 1957; Kochetkov and Sokolov, 1963; Lampe and Smolinska, 1958; Kochetkov et al., 1959).

### 1. Isoxazoles

Isoxazoles have several features that combine to make the grouping a useful one in organic syntheses. Isoxazoles are stable toward many reagents, yet undergo alkylation and hydrogenolysis readily. They may be considered masked $\beta$-diketones. Formation of isoxazoles from a $\beta$-diketone and hydroxylamine hydrochloride followed by hydrogenolysis and hydrolysis (D'Alcontres, 1950) makes a convenient sequence for protecting diketones.

The rate of hydrogenolysis of isoxazole rings is affected markedly by the pH of the medium. For instance, a 4-alkylated 3,5-dimethylisoxazole was reduced over palladium-on-carbon in 3 hr in 1 : 1 ethyl acetate–triethylamine, whereas it was essentially unchanged in 5 : 1 ethyl acetate–acetic acid in 20 hr (Stork et al., 1967). Palladium is usually used in the hydrogenolysis of isoxazoles, although platinum and nickel have been used successfully.

Alkylation of 3,5-dimethylisoxazole with sodium amide in liquid ammonia followed by hydrogenation to 2-amino-2-alken-4-ones and subsequent hydrolysis has been used to prepare a number of diketones (Kashima et al., 1973). For example,

95%

Di- and trialkylation at the same methyl can be achieved as easily using excess sodium amide and alkyl halides. Unsymmetric substitution is achieved by sequential use of alkyl halides. Diaminodiones and tetraones can be obtained from $\alpha,\omega$-dibromoalkanes as in the synthesis of 2,13-diaminotetradeca-2,12-diene-4,11-dione (**57**) and tetradecane-2,4,11,13-tetraone (**56**).

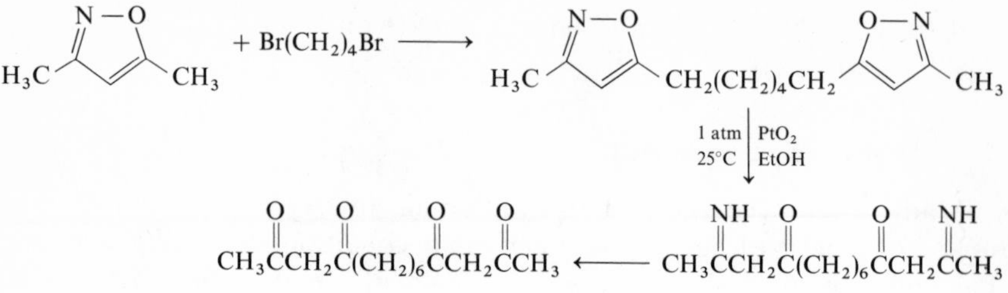

$$CH_3CCH_2C(CH_2)_6CCH_2CCH_3 \longleftarrow CH_3CCH_2C(CH_2)_6CCH_2CCH_3$$

**56**

**57**
(76%)

A synthesis of α,β-unsaturated diketones proceeds through methylisoxazoles. The procedure is illustrated here by the synthesis of 3,11-tetradecadiene-2,13-dione. Monoketones can be made as well (Kashima, 1975).

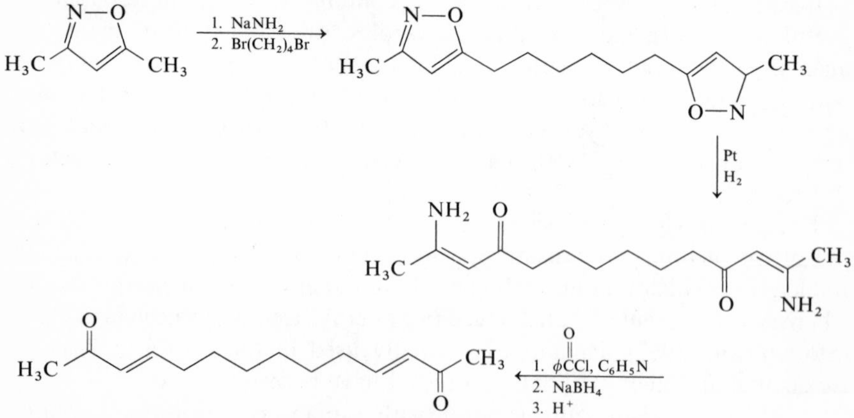

Hydrogenolysis of methylenediisoxazoles (**58**) to afford **59**, followed by cyclization in aqueous hydrochloric acid provides a convenient route to substituted resorcinols (**60**).

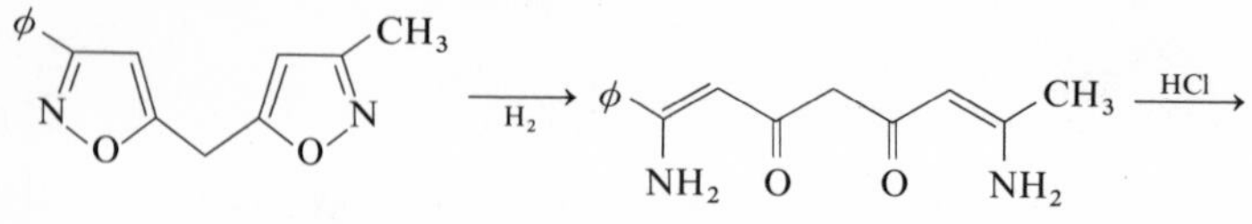

**58**

**59**

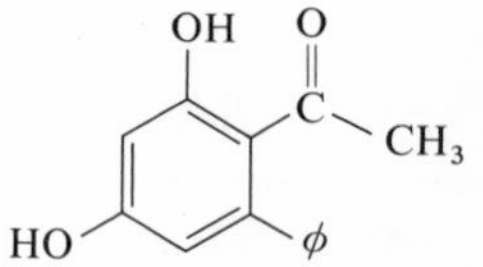

**60**
(80%)

Hydrogenolysis of the isomeric methylenediisoxazole **61** affords, on the other hand, an aminophenol (**62**) (Auricchio *et al.*, 1974).

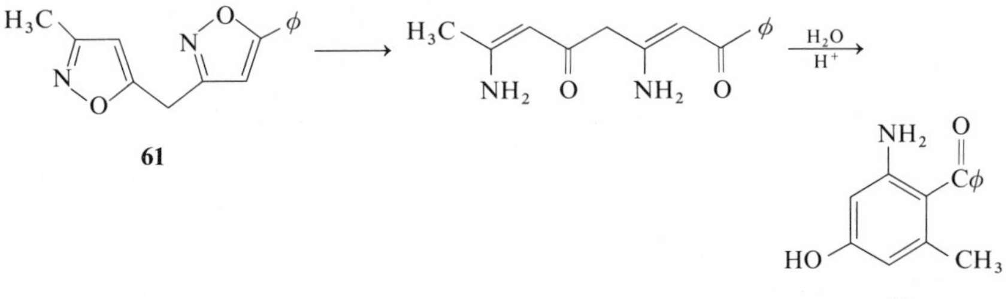

**61**

**62**

A general procedure for the synthesis of β-acylpyridines involves hydrogenolysis of 4-(3-oxoalkyl)isoxazoles. For instance, **63**, prepared from 4-chloromethyl-3,5-dimethylisoxazole and ethyl acetoacetate, gave the dihydropyridine **64** on hydrogenolysis over palladium-on-carbon in the presence of triethylamine. Oxidation of **64** with sodium nitrite and hydrochloric acid affords ethyl 5-acetyl-2,6-dimethyl-3-pyridinecarboxylate (**65**) (Ohashi *et al.*, 1967).

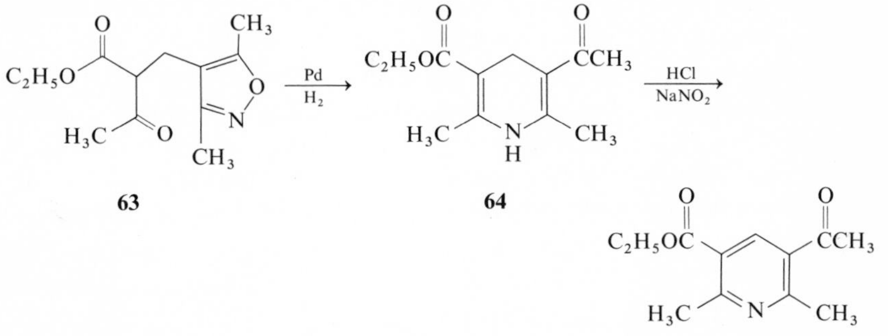

**63**                                                **64**

**65**

The isoxazole annelation reaction (Stork and McMurry, 1967a; Stork *et al.*, 1967),

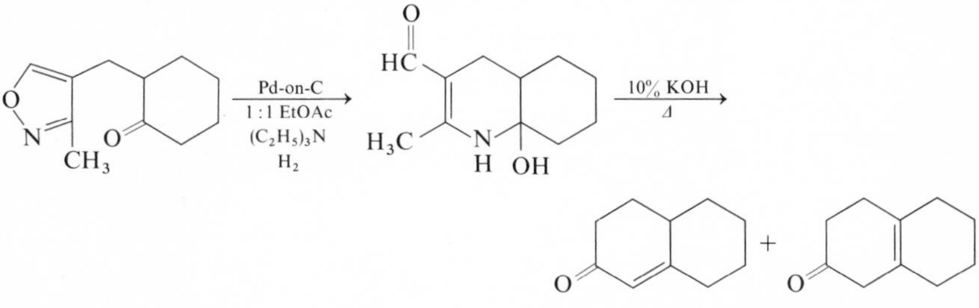

Equilibrium mixture

is well suited to the synthesis of steroids and other complex molecules (Stork and McMurry, 1967b; Ohashi *et al.*, 1967). An elaboration of ring A in the total synthesis of (±)-estr-4-ene-3,17-dione (69) and (±)-13β-ethylgon-4-ene-3,17-dione (70) involves selective hydrogenation of a $\Delta^9$ double bond and subsequent hydrogenolysis of an isoxazole ring.

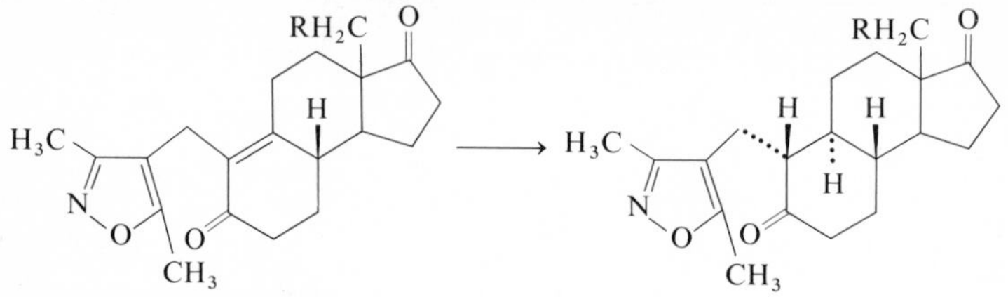

66, R = H, CH₃                                                67

Hydrogenation of 66 in 3:1 ethanol–triethylamine over 10% palladium-on-carbon under ambient conditions affords 67 by attack on the α face. This crude material was ketalized with ethylene glycol to give a bisketal, which was then reduced over palladium-on-carbon in 4% ethanolic potassium hydroxide to afford a vinylogous amine by hydrogenolysis of the nitrogen–oxygen bond. Without isolation the material was hydrolyzed by heating in base (68) then methanolic hydrochloric acid to effect deketalization and cyclization to 69 and 70. The overall yield for this five-step synthesis without purification of intermediates was 80–85% for 69 and 70% for 70 (Scott and Saucy, 1972; Scott *et al.*, 1972a). This synthesis was modified in order to prepare optically active 10-norsteroids (Scott *et al.*, 1972b). Ketalization before hydrogenolysis prevents yield loss by suppressing carbinolamine formation.

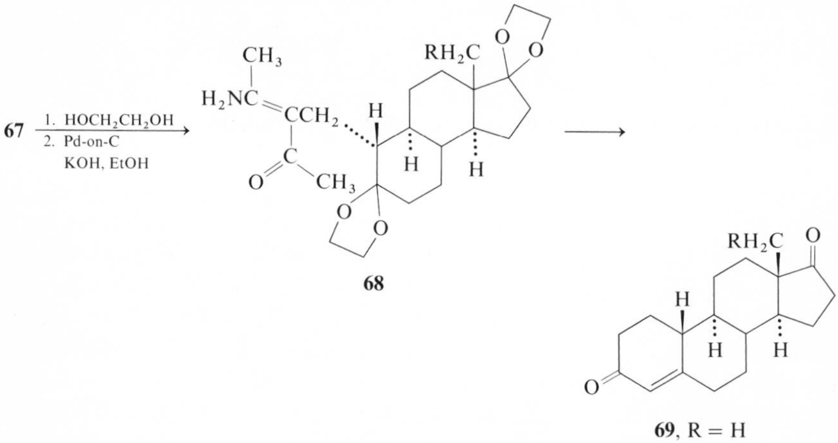

67 $\xrightarrow[\substack{2.\ \text{Pd-on-C} \\ \text{KOH, EtOH}}]{1.\ \text{HOCH}_2\text{CH}_2\text{OH}}$

68

69, R = H
70, R = CH₃

### 2. *Oxazolines*

*N*-Formylphenylalanine ethyl ester (**72**) (R = H) can be prepared in better than 95% yield by selective hydrogenolysis of 5-aryl-4-ethoxycarbonyl-$\Delta^2$-1,3-oxazolines (**71**) over palladium-on-carbon in ethanol. The formyl esters can be converted to phenylalanines by heating with 20% hydrochloric acid and to *N*-formylphenylalanines by treatment with potassium hydroxide in ethanol (Schöllkopf and Hoppe, 1970).

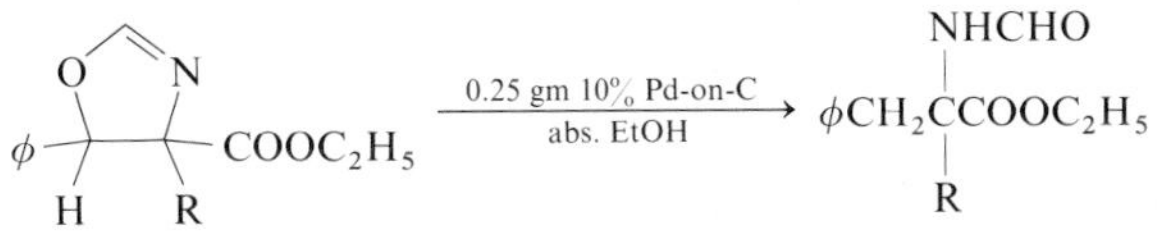
O
N
ϕ
COOC₂H₅
H
R
0.25 gm 10% Pd-on-C
abs. EtOH
NHCHO
ϕCH₂CCOOC₂H₅
R

**71** (5.0 gm) **72**

### 3. *Benzofurazans*

The course of reduction of benzofurazanes is sensitive to both substituents and catalysts (Cere *et al.*, 1972). However, reduction of 4- and 5-methoxy-benzofurazans over palladium affords high yields of 3- and 4-methoxy-1,2-phenylenediamine.

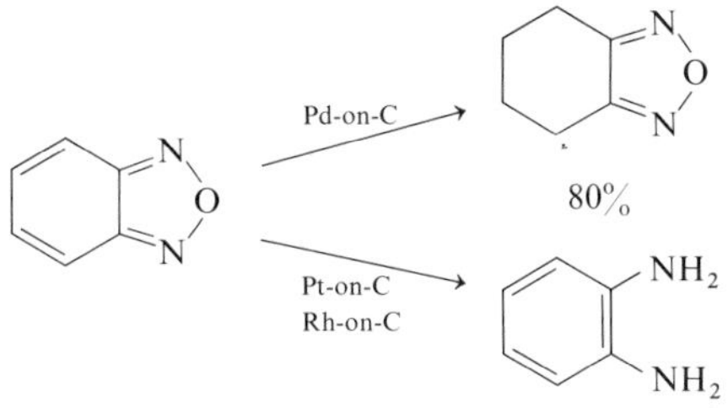
N
O
N
Pd-on-C
N
O
N
80%
Pt-on-C
Rh-on-C
NH₂
NH₂
ca. 100%

## VII. HYDROGENOLYSIS OF THE NITROGEN–NITROGEN BOND

Hydrazines (Cram and Bradshaw, 1963; Daves *et al.*, 1962; Overberger *et al*, 1955) and materials such as azines (Ugryumov, 1940; Panetta and Bunce, 1961; Berson *et al.*, 1962; Cohen *et al.*, 1950; Overberger and Gainer, 1958), hydrazones (Biel *et al.*, 1959; Overberger and Gainer, 1958; Stevens *et al.*, 1958), and azo compounds (Schmitz and Ohme, 1961; Moore and Marascia, 1959; Taylor *et al.*, 1958; Bodendorf and Wössner, 1959; Stevens and Freeman, 1964; Nagai *et al.*, 1973) that form hydrazines undergo hydrogenolysis of the nitrogen–nitrogen bond. However, the rate of hydrogenation to afford the hydrazines is usually sufficiently greater than the rate of hydrogenolysis so that hydrazines can be obtained from these materials in good yield (Heyman and Snyder, 1973). Palladium would seem to be the best catalyst for hydrogenolysis of aromatic azo compounds and compounds leading to it. Limited data suggest that rhodium may be more effective for cleavage of aliphatic hydrazines. Nickel is also useful for cleavage of nitrogen–nitrogen bonds (Panetta and Rahman, 1971; Panzica and Townsend, 1971).

Hydrazones of aromatic ketones can be converted directly to the hydrocarbons. The reaction is useful when carbonyl reagents have been used in the purification of the ketone (Burham and Eisenbraun, 1971).

Hydrogenolysis of optically active hydrazones was used as a means of preparing optically active amino acids (Kiyooka *et al.*, 1976). Hydrogenation of hydrazone failed completely over palladium-on-carbon or platinum oxide in ethanol but proceeded readily over platinum oxide in acetic acid.

### Azides

Organic azides have proved to be very useful as carriers of an incipient amine through a reaction sequence. Their use obviates the necessity of devising a suitable protecting group for the amine (Wagner *et al.*, 1972). The azido function is stable toward a variety of reagents and hydrolytic conditions, yet it can very easily be catalytically reduced to an amine. The reduction is often selective even though other easily reduced functional groups are present. Organic azide intermediates have also been useful when direct displacement of halogen by ammonia was not achieved (Loncrini and Walborsky, 1964; Holmes and Robins, 1965). Azido compounds are formed from inorganic azides in nucleophilic displacement reactions with inversion of configuration and are reduced with retention of configuration, resulting in amines of controlled stereochemistry (VanderWerf et al., 1954). Azido compounds have been useful in the preparation of amino sugars (Meyer zu Reckendorf, 1970; Koto et al., 1973; Kiely and Benzing-Nguyen, 1975).

Palladium, platinum, and nickel have been used successfully in the reduction of the azido function; the preferred catalyst depends on other features of the molecule. Hydrazine, as a source of hydrogen, is useful in limiting the depth of hydrogenation. The reduction of azides proceeds without pressure drop, and for this reason the reduction may be carried out longer than is probably necessary (McEwen et al., 1952).

## 1. Selectivity

The azido function often can be reduced selectively in the presence of other easily reduced functions, as illustrated by the following examples. A key step in the synthesis of 4-D-alanylamino-2-amino-2,4-dideoxy-D-galactopyranose (**76**), an analogue of the antifungal antibiotic Prumycin, involved reduction of an azido group in **73** without removal of a benzyloxycarbonyl and benzyl group. The reduction was accomplished at room temperature by the use of Raney nickel, which was added in several small portions.

Treatment of the amine **74** with 1 equivalent of N-benzyloxycarbonyl-D-alanine p-nitrophenyl ester gave **75**, which was then hydrogenated over 10% palladium-on-carbon, first in acetic acid–methanol and then in water, to give **76** (Kuzuhara et al., 1976).

An azide was a key intermediate in the synthesis of an imino derivative (78) of *Cecropia* juvenile hormones (79). Ring opening of the epoxide 79 with either azide ion or hydrazoic acid failed, but the azide could be obtained from the chloro ketone 77 in excellent yield. Reduction of the ketone with sodium borohydride followed by treatment with methanesulfonyl chloride gave the mesylate. The aziridine 78 was obtained by reduction of the azidomesylate with hydrazine hydrate in ethanol over Raney nickel. Reduction of olefinic linkages did not occur when this reducing system was used (Anderson *et al.*, 1972).

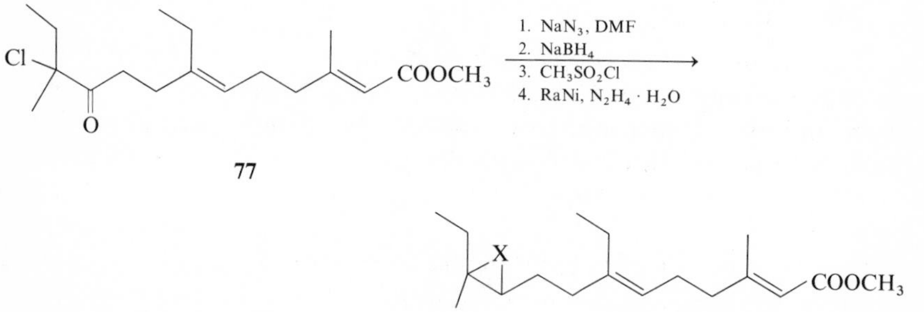

77

78, X = NH
79, X = O

An interesting difference in catalyst selectivity was found during a synthesis of 84 that involved displacement of a 5-bromo with azide followed by hydrogenation. The bromine atom of 80 was almost inert to displacement by various nucleophiles, but it could be readily replaced by conversion of 80 to

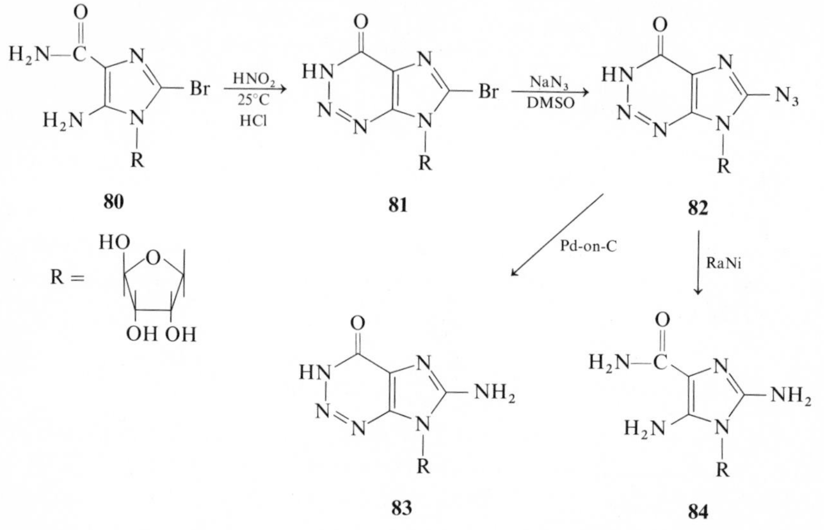

80

81

82

R =

Pd-on-C

RaNi

83

84

**81**. The 5-azido compound **82** was reduced only to the 5-amino derivative **83** over palladium-on-carbon, but over Raney nickel the original ring system was regenerated (**84**) (Ivanovics *et al.*, 1974).

Amino ketones can be prepared readily by hydrogenation of an azido ketone. Reduction of the aromatic ketone does not appear to be a troublesome side reaction (Spivak and Harris, 1972).

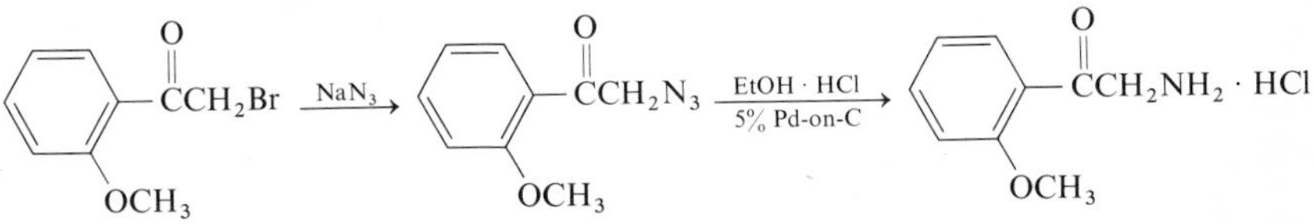

Although the azido function is easily reduced catalytically, it survives the attack of a number of other reducing agents such as sodium borohydride, lithium aluminum hydride, and, as illustrated below, sodium dithionite in the reduction of azidoquinones (**86**). Since azidoquinones are readily available, their easy conversion to hydroquinones (**87**) provides a convenient route to a large variety of highly substituted arylazides and amines (Pearce *et al.*, 1974). Reduction of the acetylated azidohydroquinone **88** is accompanied by an acetyl migration to afford **89**. The azidoquinone is reduced over platinum-on-carbon to the amino quinone **85**.

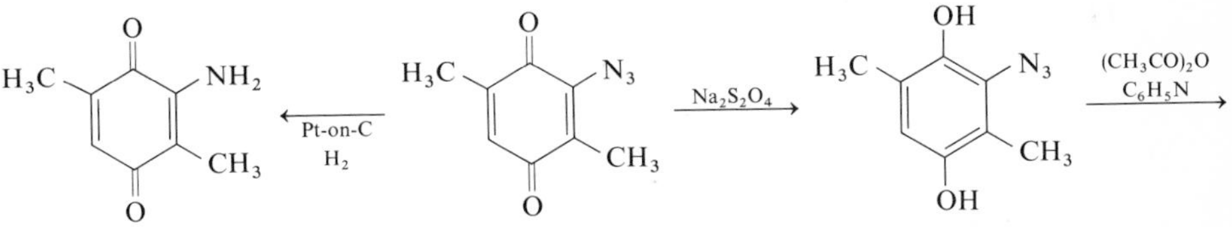

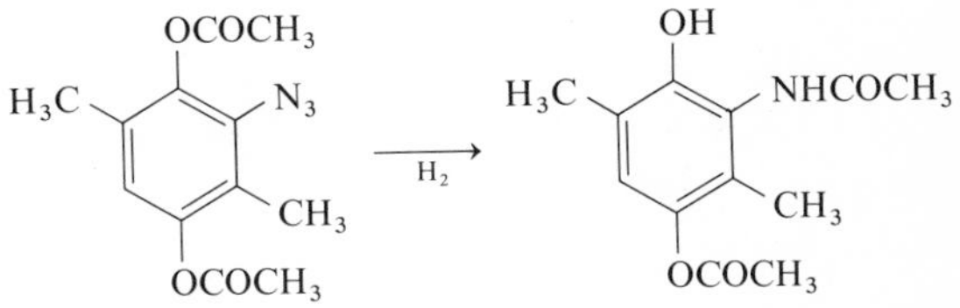

## 2. Stereochemistry

Azido compounds are formed in nucleophilic displacements with inversion of configuration and are reduced with retention of configuration. The use of azides in preparing amines of controlled stereochemistry is illustrated by the preparation in good yield of all four epimeric amino alcohols of estra-1,3,5(10)-triene-3-methyl ether with nitrogen at C-16 and oxygen at C-17. Hydrogenation of the azido alcohol was carried out in each case by the use of hydrazine hydrate and Raney nickel (Schönecker and Ponsold, 1975).

## 3.  Nucleotides

The use of an azido function has proved to be very effective in the synthesis of nucleotides and derivatives. The function survives phosphorylation reactions and acid and basic hydrolysis needed for the removal of protecting groups, and at the appropriate juncture it is converted easily to an amine by hydrogenolysis (Carrington *et al.*, 1965; Glinski *et al.*, 1970, 1973; Hobbs *et al.*, 1972; Torrence *et al.*, 1972). The synthesis of 2-amino-2′-deoxyuridine 3′-phosphate (**91**) illustrates the use of the azido function in this area of synthesis (Wagner *et al.*, 1972). The 5′-hydroxy group in 2′-azido-2′-deoxyuridine (**90**) is selectively protected by formation of the trityl ether, followed by phosphorylation with 2-cyanoethyl phosphate and dicyclohexylcarbodiimide in pyridine and hydrogenation over 5% palladium-on-barium sulfate in aqueous acid to afford **91**.

Chang and Coward (1976) employed an azido moiety as an incipient amine in the preparation of 5′-$N^{\gamma}$-adenoxyl-$\alpha,\gamma$-diaminobutyric acid.

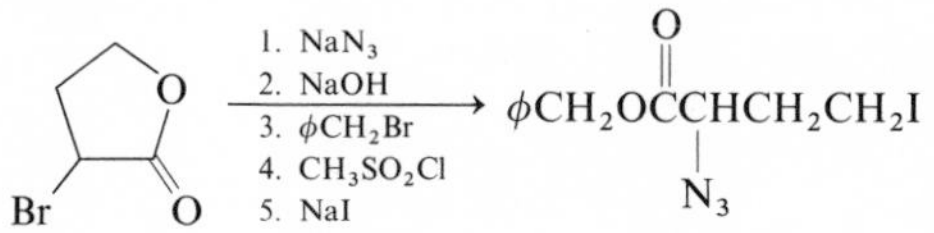

**92**

Alkylation of **93** with **92** gave $N$-benzyl-$N$-(2′,3′-$O$-isopropylideneadenos-5′-yl)-$\alpha$-azido-$\gamma$-aminobutyric acid benzyl ester (**94**). Removal of the benzyl groups, conversion of the azido function to an amine, and removal of the 2′,3′-$O$-isopropylidene group was accomplished in a single step by hydrogenation over palladium-on-carbon in aqueous formic acid, affording (**95**).

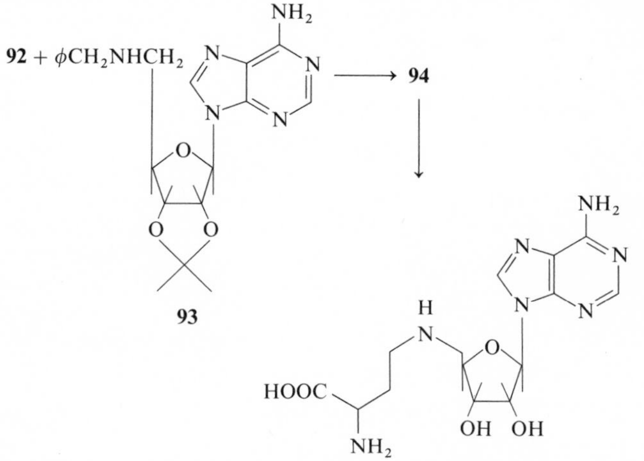

**95**

## 4.   Ring Enlargement

Sugar lactams having five-, six-, or seven-membered rings can be prepared by reductive cyclization of appropriate azidolactones or carboxylic acids.

The method is illustrated in the synthesis of D-ribo-3,4,5-trihydroxy-2-piperidone (98) from the azidolactone 96. The lactam 97 is formed in almost quantitative yield by reduction of 96 over platinum oxide or palladium-on-carbon in ethyl acetate or methanol. Hydrogenolysis of the benzylidene group over palladium-on-carbon in ethanol containing acetic acid affords 98 (Hanessian, 1969).

96                      97                      98

## 5.  Reductive Alkylation

Azido compounds containing a suitably disposed carbonyl undergo reductive alkylation on hydrogenation of the azido function. The usefulness of this reaction is illustrated by the preparation of the tranquilzer 7-chloro-1,3-dihydro-1-methyl-5-phenyl-2H-1,4-benzodiazepin-2-one. The reduction can be carried out with either hydrogen or hydrazine over palladium-on-carbon (Petersen, 1970).

**REFERENCES**

Achiwa, K., and Yamada, S., *Tetrahedron Lett.* p. 1799 (1974).
Achiwa, K., and Yamada, S., *Tetrahedron Lett.* p. 2701 (1975).
Adams, R., and Gianturco, M., *J. Am. Chem. Soc.* **78**, 1922 (1956).

Aeberli, P., Cooke, G., Houlihan, W. J., and Salmond, W. G., *J. Org. Chem.* **40**, 382 (1975).
Agnello, E. J., Pinson, R., Jr., and Laubach, G. D., *J. Am. Chem. Soc.* **78**, 4756 (1956).
Andersen, H. C., and Romeo, P. L., Sr., U.S. Patent 3,146,243, Aug. 25, 1964.
Anderson, A. G., Jr., and Lok, R., *J. Org. Chem.* **37**, 3953 (1972).
Anderson, G. W., and Callahan, F. M., *J. Am. Chem. Soc.* **82**, 3359 (1960).
Anderson, R. J., Henrick, C. A., and Siddall, J. B., *J. Org. Chem.* **37**, 1266 (1972).
Annen, K., Tschesche, R., and Welzel, P., *Ber. Dtsch. Chem. Ges.* **106**(2), 576 (1973).
Anteunis, M., and Verzele, M., *Bull. Soc. Chem. Belg.* **68**, 476 (1959).
Appel, H. H., Connolly, J. D., Overton, K. H., and Bond, R. P. M., *J. Chem. Soc.* p. 4685 (1960).
Auricchio, S., Morrocchi, S., and Ricca, A., *Tetrahedron Lett.* p. 2793 (1974).
Bachi, M. D., and Ross-Petersen, K. J., *J. Org. Chem.* **37**, 3550 (1972).
Bailey, P. S., *Chem. Rev.* **58**, 925 (1958).
Baird, R., and Winstein, S., *J. Am. Chem. Soc.* **79**, 4238 (1957).
Ballou, C. E., and Fischer, H. O. L., *J. Am. Chem. Soc.* **76**, 3188 (1954).
Baltzly, R., *J. Org. Chem.* **41**, 920 (1976a).
Baltzly, R., *J. Org. Chem.* **41**, 933 (1976b).
Baltzly, R., and Buck, J. S., *J. Am. Chem. Soc.* **65**, 1984 (1943).
Baltzly, R., and Russell, P. B., *J. Am. Chem. Soc.* **72**, 3410 (1950).
Baltzly, R., and Russell, P. B., *J. Am. Chem. Soc.* **75**, 5598 (1953).
Baltzly, R., and Russell, P. B., *J. Am. Chem. Soc.* **76**, 5776 (1954).
Bandlish, B. K., Brown, J. N., Timberlake, J. W., and Trefonas, L. M., *J. Org. Chem.* **38**, 1102 (1973).
Barnes, M. F., and MacMillan, J., *J. Chem. Soc. C.* p. 361 (1967).
Bergmann, M., and Zervas, L., *Ber. Dtsch. Chem. Ges. B* **65** 1192, 1201 (1932).
Berse, C., Boucher, R., and Piché, L., *J. Org. Chem.* **22**, 805 (1957).
Berson, J. A., Olsen, C. J., and Walia, J. S., *J. Am. Chem. Soc.* **84**, 3337 (1962).
Biel, J. H., Hoya, W. K., and Leiser, H. A., *J. Am. Chem. Soc.* **81**, 2527 (1959).
Bodendorf, K., and Wössner, W., *Justus Liebigs Ann. Chem.* **623**, 109 (1959).
Boekelheide, V., and Chang, M. Y., *J. Org. Chem.* **29**, 1303 (1964).
Bonner, W. A., and Zderic, A. A., *J. Am. Chem. Soc.* **78**, 3218 (1956).
Bonner, W. A., Zderic, A. A., and Casaletto, G. A., *J. Am. Chem. Soc.* **74**, 5086 (1952).
Brewster, J. H., and Braden, W. E., Jr., *Chem. Ind. (London)* p. 1759 (1964).
Brieger, G., and Nestrick, T. J., *Chem. Rev.* **74**, 567 (1974).
Brown, E. V., *J. Am. Chem. Soc.* **79**, 3565 (1957).
Burnham, J. W., and Eisenbraun, E. J., *J. Org. Chem.* **36**, 737 (1971).
Callow, R. K., and Thompson, G. A., *J. Chem. Soc.* p. 3106 (1964).
Canonica, L., Corbella, A., Gariboldi, P., Jommi, G., Krepinsky, J., Ferrari, G., and Casagrande, C., *Tetrahedron* **25**, 3895 (1969a).
Canonica, L., Corbella, A., Gariboldi, P., Jommi, G., Krepinsky, J., Ferrari, G., and Casagrande, C., *Tetrahedron* **25**, 3903 (1969b).
Carrington, R., Shaw, G., and Wilson, D. V., *J. Chem. Soc.* p. 6854 (1965).
Carson, J. F., *J. Am. Chem. Soc.* **73**, 1850 (1951).
Cava, M. P., and Pohlke, R., *J. Org. Chem.* **28**, 1012 (1963).
Ceré, V., Monte, D., and Sandri, E., *Tetrahedron* **28**, 3271 (1972).
Chang, C. D., and Coward, J. K., *J. Med. Chem.* **19**, 684 (1976).
Channing, D. M., Turner, P. B., and Young, G. T., *Nature (London)* **167**, 487 (1951).
Cline, R. E., Fink, R. M., and Fink, K., *J. Am. Chem. Soc.* **81**, 2521 (1959).
Cohen, S. G., Groszos, S. J., and Sparrow, D. B., *J. Am. Chem. Soc.* **72**, 3947 (1950).
Cope, A. C., and Ciganek, E., *Org. Synth., Collect. Vol.* **4**, 612 (1963).

Cope, A. C., Liss, T. A., and Wood, G. W., *J. Am. Chem. Soc.* **79**, 6287 (1957).

Cope, A. C., McKervey, M. A., Weinshenker, N. M., and Kinnel, R. B., *J. Org. Chem.* **35**, 2918 (1970).

Corey, E. J., and Suggs, J. W., *J. Org. Chem.* **38**, 3224 (1973).

Craig, J. C., Dinner, A., and Mulligan, P. J., *J. Org. Chem.* **39**, 1669 (1974).

Cram, D. J., and Bradshaw, J. S., *J. Am. Chem. Soc.* **85**, 1108 (1963).

Crombie, L., and Jacklin, A. G., *J. Chem. Soc.* p. 1622 (1957).

Cromwell, N. H., and Rodebaugh, R. M., *J. Heterocycl. Chem.* **6**, 435 (1969).

Dahn, H., Solms, U., and Zoller, P., *Helv. Chim. Acta* **37**, 565 (1954).

Dahn, H., Garbanino, J. A., and Murchú, C. O., *Helv. Chim. Acta* **53**, 1370 (1970).

D'Alcontres, S., *Gazz. Chim. Ital.* **80**, 441 (1950).

Dart, M. C., and Henbest, H. B., *J. Chem. Soc.* p. 3563 (1960).

Dauben, W. G., and Hance, P. D., *J. Am. Chem. Soc.* **77**, 2451 (1955).

Dauben, W. G., Hayes, W. K., Schwarz, J. S. P., and McFarland, J. W., *J. Am. Chem. Soc.* **82**, 2232 (1960a).

Dauben, W. G., Schwarz, J. S. P., Hayes, W. H., and Hance, P. D., *J. Am. Chem. Soc.* **82**, 2239 (1960b).

Daves, G. D., Jr., Robins, R. K., and Cheng, C. C., *J. Am. Chem. Soc.* **84**, 1724 (1962).

DePuy, C. H., and Zaweski, E. F., *J. Am. Chem. Soc.* **81**, 4920 (1959).

Dürckeimer, W., and Cohen, L. A., *Biochemistry* **3**, 1948 (1964).

Fahrenholtz, K. E., *J. Org. Chem.* **37**, 2204 (1972).

Feldman, J., and Frampton, O. D., U.S. Patent 3,129,263, Apr. 14, 1964.

Fields, T. L., Kende, A. S., and Boothe, J. H., *J. Am. Chem. Soc.* **83**, 4612 (1961).

Friedman, O. M., and Seligman, A. M., *J. Am. Chem. Soc.* **76**, 655 (1954).

Friedman, O. M., Klass, D. L., and Seligman, A. M., *J. Am. Chem. Soc.* **76**, 916 (1954).

Fuchs, J. J., U.S. Patent 3,068,275, Jan. 18, 1960.

Fuson, R. C., and Jackson, H. L., *J. Am. Chem. Soc.* **72**, 1637 (1950).

Galantay, E., Szabo, A., and Fried, J., *J. Org. Chem.* **28**, 98 (1963).

Garbisch, E. W., Jr., *J. Org. Chem.* **27**, 3363 (1962).

Garbisch, E. W., Jr., *Chem. Commun.* p. 806 (1967).

Garbisch, E. W., Jr., Schreader, L., and Frankel, J. J., *J. Am. Chem. Soc.* **89**, 4233 (1967)

Ghatak, U. R., Chatterjee, N. R., Banerjee, A. K., Chakravarty J., and Moore, R. E., *J. Org. Chem.* **34**, 3739 (1969).

Glinski, R. P., Khan, M. S., Kalamas, R. L., Stevens, C. L., and Sporn, M. B., *Chem. Commun.* p. 915 (1970).

Glinski, R. P., Khan, M. S., and Kalamas, R. L., *J. Org. Chem.* **38**, 4299 (1973).

Godfrey, J. C., Tarbell, D. S., and Boekelheide, V., *J. Am. Chem. Soc.* **77**, 3342 (1955).

Goodman, M., and Steuben, K. C., *J. Am. Chem. Soc.* **81**, 3980 (1959).

Graefe, A. F., *J. Org. Chem.* **23**, 1230 (1958).

Greenlee, T. W., and Bonner, W. A., *J. Am. Chem. Soc.* **81**, 4303 (1959).

Greenwald, R. B., and Zirkle, C. L., *J. Org. Chem.* **33**, 2118 (1968).

Grupe, R., Baeck, B., and Niedrich, H., *J. Prakt. Chem.* **314**, 751 (1972).

Hanessian, S., *J. Org. Chem.* **34**, 675 (1969).

Harada, K., and Kataoka, Y., *Tetrahedron Lett.* p. 2103 (1978).

Harada, K., Okáwara, T., and Matsumoto, K., *Bull. Chem. Soc. Jpn.* **46**, 1865 (1973).

Hartung, W. H., *J. Am. Chem. Soc.* **50**, 3370 (1928).

Hartung, W. H., and Simonoff, R., *Org. React.* **7**, 263 (1953).

Hasek, R. H., Gott, P. G., and Martin, J. C., *J. Org. Chem.* **29**, 2510 (1964).

Hayakawa, T., and Shimizu, K., *Bull. Chem. Soc. Jpn.* **46**, 1886 (1973).

Herz, W., Subramaniam, P. S., and Geissman, T. A., *J. Org. Chem.* **33**, 3743 (1968).

Heyman, M. L., and Snyder, J. P., *Tetrahedron Lett.* p. 2859 (1973).

Hiskey, R. G., and Northrop, R. C., *J. Am. Chem. Soc.* **83**, 4798 (1961).

Hobbs, J., Sternbach, H., and Eckstein, F., *Biochem. Biophys. Res. Commun.* **46**, 1509 (1972).

Holmes, R. E., and Robins, R. K., *J. Am. Chem. Soc.* **87**, 1772 (1965).

Horinaka, A., Nakashima, R., Yoshikawa, M., and Matsuura, T., *Bull. Chem. Soc. Jpn.* **48**, 2095 (1975).

Horner, L., and Jürgens, E., *Ber. Dtsch. Chem. Ges. B.* **90**, 2184 (1957).

Hurd, C. D., and Jenkins, H., *J. Org. Chem.* **31**, 2045 (1966).

Ichinohe, Y., Kameda, N., and Kujirai, M., *Bull. Chem. Soc. Jpn.* **42**, 3614 (1969).

Imaizumi, S., Nippon Kagaku Zasshi **81**, 631 (1960).

Isowa, Y., and Ohmori, M., *Bull. Chem. Soc. Jpn.* **47**, 2672 (1974).

Ivanovics, G. A., Rousseau, R. J., Kawana, M., Srivastava, P. C., and Robins, R. K., *J. Org. Chem.* **39**, 3651 (1974).

Jensen, F. R., and Coleman, W. E., *J. Am. Chem. Soc.* **80**, 6149 (1958).

Jones, D. A., Jr., *Tetrahedron Lett.* p. 2853 (1977).

Kashima, C., *J. Org. Chem.* **40**, 526 (1975).

Kashima, C., Tobe, S., Sugiyama, N., and Yamamoto, M., *Bull. Chem. Soc. Jpn.* **46**, 310 (1973).

Katritsky, A. R., and Monro, A. M., *J. Chem. Soc.* p. 1263 (1958).

Kende, A. S., and Chu, J. Y.-C., *Tetrahedron Lett.* p. 4837 (1970).

Khan, A. M., McQuillin, F. J., and Jardine, I., *J. Chem. Soc.* p. 136 (1967).

Kieboom, A. P. G., *React. Kinet. Catal. Lett.* **1**, 433 (1974).

Kieboom, A. P. G., de Kreuk, J. J., and van Bekkum, H., *J. Catal.* **20**, 58 (1971).

Kiely, D. E., and Benzing-Nguyen, L., *J. Org. Chem.* **40**, 2630 (1975).

Kindler, K., Hedemann, B., and Schärfe, E., *Justus Liebigs Ann. Chem.* **560**, 216 (1948).

Kiyooka, S., Takeshima, K., Yamamoto, H., and Suzuki, K., *Bull. Chem. Soc. Jpn.* **49**, 1897 (1976).

Kochetkov, N. K., and Sokolov, S. D., *Adv. Heterocycl. Chem.* **1**, 412 (1963).

Kochetkov, N. K., Khomatov, R. M., Budowsky, E. I., Karpeysky, M. J., and Severin, E. S., *Zh. Obshch. Khim.* **29**, 4069 (1959).

Koto, S., Kawakatsu, N., and Zen, S., *Bull. Chem. Soc. Jpn.* **46**, 876 (1973).

Kovacs, J., and Robin, R. L., *J. Org. Chem.* **33**, 2418 (1968).

Kupchan, S. M., Anderson, W. K., Bollinger, P., Doskotch, R. W., Smith, R. M., Renauld, J. A. S., Schnoes, H. K., Burlingame, A. L., and Smith, D. H., *J. Org. Chem.* **34**, 3858 (1969).

Kuromizu, K., and Meinhofer, J., *J. Am. Chem. Soc.* **96**, 4978 (1974).

Kuzuhara, H., Mori, O., and Emoto, S., *Tetrahedron Lett.* p. 379 (1976).

Lampe, W., and Smolinska, J., *Bull. Acad. Pol. Sci., Ser. Sci. Chim., Geol. Geograph.* **6**, 481 (1958).

Laubach, G. D., U.S. Patent 2,794,033, May 28, 1957.

Lauer, W. M., and Dyer, W. S., *J. Am. Chem. Soc.* **64**, 1453 (1942).

Lemal, D. M., and Shim, K. S., *J. Am. Chem. Soc.* **86**, 1550 (1964).

Lies, T., Plapinger, R. E., and Wagner-Jauregg, T., *J. Am. Chem. Soc.* **75**, 5755 (1953).

Lima, D. A., U.S. Patent 3,154,538, Oct. 27, 1964.

Loncrini, D. F., and Walborsky, H. M., *J. Med. Chem.* **7**, 369 (1964).

Lott, W. A., and Shaw, E., *J. Am. Chem. Soc.* **71**, 70 (1949).

Lythgoe, B., and Trippett, S., *J. Chem. Soc.* p. 471 (1959).

McCormick, J. R. D., and Jensen, E. R., U.S. Patent 3,019,260, Jan. 30, 1962.

McEwen, W. E., Conrad, W. E., and VanderWerf, C. A., *J. Am. Chem. Soc.* **74**, 1168 (1952).

McKay, A. F., Billy, J.-M., and Tarlton, E. J., *J. Org. Chem.* **29**, 291 (1964).

McQuillin, J., *Tech. Org. Chem.* **9**, 498 (1963).

Magerlein, B. J., and Kagan, F., *J. Am. Chem. Soc.* **82**, 593 (1960).
Mallory, F. B., and Varimbi, S. P., *J. Org. Chem.* **28**, 1656 (1963).
Marquardt, F.-H., *Ann. N.Y. Acad. Sci.* **214**, 110 (1973).
Marullo, N. P., and Alford, J. A., *J. Org. Chem.* **33**, 2368 (1968).
Meienhofer, J., and Kuromizu, K., *Tetrahedron Lett.* p. 3259 (1974).
Meinwald, J., and Frauenglass, E., *J. Am. Chem. Soc.* **82**, 5253 (1960).
Meinwald, J., Seidel, M. C., and Cadoff, B. C., *J. Am. Chem. Soc.* **80**, 6303 (1958).
Meinwald, J., Labana, S. S., and Chadha, M. S., *J. Am. Chem. Soc.* **85**, 582 (1963).
Merrifield, R. B., *J. Am. Chem. Soc.* **85**, 2149 (1963).
Meyer zu Reckendorf, W., *Tetrahedron Lett.* p. 287 (1970).
Miller, B., and Lewis, L., *J. Org. Chem.* **39**, 2605 (1974).
Mitsui, S., and Imaizumi, S., *Bull. Chem. Soc. Jpn.* **36**, 855 (1963).
Mitsui, S., and Kudo, Y., *Chem. Ind. (London)* p. 381 (1965).
Mitsui, S., and Nagahisa, Y., *Chem. Ind. (London)* p. 1975 (1965).
Mitsui, S., and Sato, E., *Nippon Kagaku Zasshi* **86**, 416 (1965).
Mitsui, S., Kasahara, A., and Endo, N., *Nippon Kagaku Zasshi* **75**, 234 (1954); *Chem. Abstr.* **49**, 10210d (1955).
Mitsui, S., Kamaishi, T., Imaizumi, S., and Takamura, I., *Nippon Kagaku Zasshi* **83**(10), 1115 (1962).
Mitsui, S., Senda, Y., and Konno, K., *Chem. Ind. (London)* p. 1354 (1963).
Mitsui, S., Imaizumi, S., Senda, Y., and Konno, K., *Chem. Ind. (London)* p. 233 (1964).
Mitsui, S., Fujimoto, M., Nagahisa, Y., and Sukegawa, T., *Chem. Ind. (London)* p. 241 (1969).
Miyano, M., and Dorn, C. R., *J. Org. Chem.* **37**, 1818 (1972).
Miyoshi, M., *Bull. Chem. Soc. Jpn.* **46**, 1489 (1973).
Moore, J. A., and Marascia, F. J., *J. Am. Chem. Soc.* **81**, 6049 (1959).
Murchú, C. Ó., *Tetrahedron Lett.* p. 3231 (1969).
Musso, H., *Ber. Dtsch. Chem. Ges. B.* **108**, 337 (1975).
Nagai, W., Kirk, K. L., and Cohen, L. A., *J. Org. Chem.* **38**, 1972 (1973).
Newham, J., *Chem. Rev.* **63**, 123 (1963).
Newman, M. S., and VanderWerf, C. A., *J. Am. Chem. Soc.* **67**, 233 (1945).
Newman, M. S., Dali, H. M., and Hung, W. M., *J. Org. Chem.* **40**, 262 (1975).
Ning, R. Y., Douvan, I., and Sternback, L. H., *J. Org. Chem.* **35**, 2243 (1970).
Nishimura, S., and Mori, K., *Bull. Chem. Soc. Jpn.* **32**, 102 (1959).
Nishimura, S., Onoda, S., and Nakamura, A., *Bull. Chem. Soc. Jpn.* **33**, 1356 (1960).
Noland, W. E., Cooley, J. H., and McVeigh, P. A., *J. Am. Chem. Soc.* **81**, 1209 (1959).
Ochiai, E., Ishikawa, M., and Katada, M., *Yakugaku Zasshi* **63**, 307 (1943).
Oelschläger, H., and Bremer, U., *Arch. Pharm. (Weinheim, Ger.)* **304**, 717 (1971).
Ohashi, M., Kamachi, H., Kakisawa, H., and Stork, G., *J. Am. Chem. Soc.* **89**, 5460 (1967).
Okawara, T., and Harada, K., *Bull. Chem. Soc. Jpn.* **46**, 1869 (1973).
Overberger, C. G., and Gainer, H., *J. Am. Chem. Soc.* **80**, 4556 (1958).
Overberger, C. G., Palmer, L. C., Marks, B. S., and Byrd, N. R., *J. Am. Chem. Soc.* **77**, 4100 (1955).
Paal, C., and Yao, W. N., *Ber. Dtsch. Chem. Ges. B* **63**, 57 (1930).
Panetta, C. A., and Bunce, S. C., *J. Org. Chem.* **26**, 4859 (1961).
Panetta, C. A., and Rahman, A-U., *J. Org. Chem.* **36**, 2250 (1971).
Panzica, R. P., and Townsend, L. B., *J. Org. Chem.* **36**, 1594 (1971).
Pearce, D. S., Lee. M-S., and Moore, H. W., *J. Org. Chem.* **39**, 1362 (1974).
Pearlman, W. M., *Tetrahedron Lett.* p. 1663 (1967).
Petersen, J. B., U.S. Patent 3,520,878, July 21, 1970.
Pettit, G. R., Fessler, D. C., and Settepani, J. A., *J. Org. Chem.* **34**, 2978 (1969).

Pryde, E. H., Anders, D. E., Teeter, H. M., and Cowan, J. C., *J. Org. Chem.* **25**, 618 (1960).

Pryde, E. H., Anders, D. E., Teeter, H. M., and Cowan, J. C., *J. Org. Chem.* **27**, 3055 (1962).

Richtmyer, N. K., *J. Am. Chem. Soc.* **56**, 1633 (1934).

Roberts, R. M., Bantel, K. H., and Low, C. E., *J. Org. Chem.* **38**, 1903 (1973).

Russell, G. A., and Mayo, F. R., U.S. Patent 2,794,055, May 28, 1957.

Rylander, P. N., "Organic Syntheses with Noble Metal Catalysts," p. 152. Academic Press, New York, 1973.

Rylander, P. N., and Himelstein, N., *Engelhard Ind., Tech. Bull.* **5**, 43 (1964).

Rylander, P. N., and Rakoncza, N., 1962; cited in Rylander, P. N., "Catalytic Hydrogenation over Platinum Metals," p. 485. Academic Press, New York, 1967.

Rylander, P. N., and Steele, D. R., *Engelhard Ind., Tech. Bull.* **6**, 41 (1965).

Sakura, N., Ito, K., and Sekiya, M., *Chem. Pharm. Bull.* **20**, 1156 (1972).

Schlatter, J. M., and Mazur, R. H., *Tetrahedron Lett.* p. 2851 (1977).

Schmitz, E., and Ohme, R., *Ber. Dtsch. Chem. Ges. B.* **94**, 2166 (1961).

Schöllkopf, U., and Hoppe, D., *Synthesis* p. 459 (1970).

Schönecker, B., and Ponsold, K., *Tetrahedron* **31**, 1113 (1975).

Schreyer, R. C., U. S. Patent 3,092,654, June 4, 1963.

Scott, J. W., and Saucy, G., *J. Org. Chem.* **37**, 1652 (1972).

Scott, J. W., Banner, B. L., and Saucy, G., *J. Org. Chem.* **37**, 1664 (1972a).

Scott, J. W., Borer, R., and Saucy, G., *J. Org. Chem.* **37**, 1659 (1972b).

Seebach, D., *Angew. Chem., Int. Ed. Engl.* **4**, 121 (1965).

Senda, Y., and Mitsui, S., *Nippon Kagaku Zasshi* **83**, 847 (1962).

Shaw, E., *J. Am. Chem. Soc.* **71**, 67 (1949).

Shoppee, C. W., Agashe, B. D., and Summers, G. H. R., *J. Chem. Soc.* p. 3107 (1957).

Slessor, K. N., and Tracey, A. S., *Can. J. Chem.* **48**, 2900 (1970).

Smissman, E. E., and Corbett, M. D., *J. Org. Chem.* **37**, 1704 (1972).

Smith, G. W., and Thatcher, D. N., *Ind. Eng Chem., Prod. Res. Dev.* **1**, 117 (1962).

Sowden, J. C., and Kuenne, D. J., *J. Am. Chem. Soc.* **74**, 686 (1952).

Spivak, C. E., and Harris, F. L., *J. Am. Chem. Soc.* **94**, 2494 (1972).

Stedman, R. J., Swered, K., and Hoover, J. R. E., *J. Med. Chem.* **7**, 117 (1964).

Stevens, C. L., and Gasser, R. J., *J. Am. Chem. Soc.* **79**, 6057 (1957).

Stevens, C. L., Gillis, B. T., French, J. C., and Haskell, T. H., *J. Am. Chem. Soc.* **80**, 6088 (1958).

Stevens, T. E., and Freeman, J. P., *J. Org. Chem.* **29**, 2279 (1964).

Stocker, J. H., *J. Org. Chem.* **27**, 2288 (1962).

Stork, G., and McMurry, J. E., *J. Am. Chem. Soc.* **89**, 5463 (1967a).

Stork, G., and McMurry, J. E., *J. Am. Chem. Soc.* **89**, 5464 (1967b).

Stork, G., Danishefsky, S., and Ohashi, M., *J. Am. Chem. Soc.* **89**, 5459 (1967).

Sugi, Y., and Mitsui, S., *Tetrahedron* **29**, 2041 (1973).

Taub, D., Kuo, C. H., Slates, H. L., and Wendler, N. J., *Tetrahedron* **19**, 1 (1963).

Taylor, E. C., and Driscoll, J. S., *J. Org. Chem.* **25**, 1716 (1960).

Taylor, E. C., Barton, J. W., and Osdene, T. S., *J. Am. Chem. Soc.* **80**, 421 (1958).

Tener, G. M., and Khorana, H. G., *J. Am. Chem. Soc.* **80**, 1999 (1958).

Thompson, Q. E., *J. Org. Chem.* **27**, 4498 (1962).

Torrence, P. F., Waters, J. A., and Witkop, B., *J. Am. Chem. Soc.* **94**, 3638 (1972).

Toyne, K. J., *J. Chem. Soc., Perkin Trans. I* p. 1346 (1976).

Tuemmler, W. B., and Winkler, H. J. S., U.S. Patent 2,979,505, Apr. 11, 1961.

Ugryumov, P. G., *J. Gen. Chem. USSR* **20**(22), 1985, 1995 (1940).

Ukita, T., Nagasawa, K., and Irie, M., *J. Am. Chem. Soc.* **80**, 1373 (1958).

VanderWerf, C. A., Heisler, R. Y., and McEwen, W. E., *J. Am. Chem. Soc.* **76**, 1231 (1954).

Vellturo, A. F., and Griffin, G. W., *J. Am. Chem. Soc.* **87**, 3021 (1965).

von Wittenau, M. S., *J. Org. Chem.* **29**, 2746 (1964).

Wagner, D., Verheyden, J. P. H., and Moffatt, J. G., *J. Org. Chem.* **37**, 1876 (1972).

Walker, G. N., *J. Org. Chem.* **27**, 1929 (1962).

Wayaku, M., Kaneda, K., Imanaka, T., and Teranishi, S., *Bull. Chem. Soc. Jpn.* **48**, 1957 (1975).

Welch, S. C., Hagan, C. P., White, D. H., and Fleming, W. P., *Synth. Commun.* **4**(6), 373 (1974).

Westphal, O., and Stadler, R., *Angew. Chem., Int. Ed. Eng.* **2**, 327 (1963).

Wiberg, K. B., and Ciula, R. P., *J. Am. Chem. Soc.* **81**, 5261 (1959).

Wiberg, K. B., and Hutton, T. W., *J. Am. Chem. Soc.* **76**, 5367 (1954).

Witkop, B., and Patrick, J. B., *J. Am. Chem. Soc.* **73**, 2188 (1951).

Yamazaki, M., Fujimoto, H., and Kawasaki, T., *Tetrahedron Lett.* p. 1241 (1975).

Zderic, J. A., Bonner, W. A., and Greenlee, T. W., *J. Am. Chem. Soc.* **79**, 1696 (1957).

Zderic, J. A., Rivera, M. E. C., and Limon, D. C., *J. Am. Chem. Soc.* **82**, 6373 (1960).

Zymalskowski, F., Schuster, T., and Scherer, H., *Arch. Pharm. (Weinheim, Ger.)* **302**, 272 (1969).

# Index